10/87

D0078807

Quantum mechanics and the particles of nature

Quantum mechanics and the particles of nature

AN OUTLINE FOR MATHEMATICIANS

ANTHONY SUDBERY

Department of Mathematics, University of York

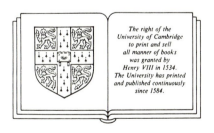

The right of the
University of Cambridge
to print and sell
all manner of books
was granted by
Henry VIII in 1534.
The University has printed
and published continuously
since 1584.

CAMBRIDGE UNIVERSITY PRESS

Cambridge

London   New York   New Rochelle

Melbourne   Sydney

Published by the Press Syndicate of the University of Cambridge
The Pitt Building, Trumpington Street, Cambridge CB2 1RP
32 East 57th Street, New York, NY 10022, USA
10 Stamford Road, Oakleigh, Melbourne 3166, Australia

First published 1986

Printed in Great Britain at the University Press, Cambridge

British Library cataloguing in publication data

Sudbery, Anthony
Quantum mechanics and the particles of nature.

1. Quantum theory
I. Title
530.1′2   QC174.12

Library of Congress cataloging-in-publication data

Sudbery, Anthony.
Quantum mechanics and the particles of nature.

Bibliography: p.
Includes index.
1. Quantum theory. 2. Particles (Nuclear physics)
I. Title.
QC174.12.S89 1986   530.1′2   85-29124

ISBN 0 521 25891 X hard covers
ISBN 0 521 27765 5 paperback

To my mother and father

Contents

Preface

This book is addressed to the reader who wants, as an educated person, to have an outline of the present state of knowledge of the constituents of the material world; who has a logical cast of mind and will follow a mathematical argument; but who may have little knowledge of physics and no intention of becoming deeply involved in the subject. In practice I have imagined this reader as a mathematics student taking a third-year undergraduate course in quantum mechanics such as is commonly offered as a part of the mathematics degree course in British universities. Only a minority of such students will be intending to pursue the subject further, and it seems more appropriate to aim for a wide survey of the interesting bits than to try to provide a sound basis for a training as a quantum mechanic.

The emphasis in the book, therefore, is on providing a coherent account of the basic theoretical concepts of quantum mechanics and particle physics. Experimental detail, mathematical rigour and calculational facility are all given lower priority than conceptual coherence. However, I hope that I have given sufficient experimental reason for every major statement of theory; that the mathematics is honest, with gaps acknowledged and without the inconsistencies which can puzzle and dishearten (or arouse the scorn of) mathematics students; and that there are enough problems at the end of chapters to enable readers to test their grasp of the concepts.

This approach to the subject has led me to omit several topics which would normally be included in a quantum mechanics course; for example, there is no scattering theory and little discussion of the Schrödinger equation as a differential equation. These topics may be indispensable to anyone who wants to work in the area, but they are not actually needed in explaining the results of the research in which they were tools. On the other hand, there are conceptual problems which can be (and often have been) ignored by the working physicist, but which seem much more important to the spectator who wants to understand more of the game. These metaphysical problems often arouse great interest in students, who find that it is poorly catered for in quantum mechanics textbooks (perhaps because the discussion is likely to be either inconclusive or unconvincing, and quite possibly both). I have devoted a chapter to such problems; it is indeed inconclusive, and may well be unconvincing.

The reading of mathematics and physics books is hopefully embarked upon more often than it is successfully completed. This fact of human nature is allowed for in the structure of this book, which has several points at which a reader can feel that they have reached the end of a journey. Thereafter the journey is started again, but in a different craft and at a different level. The first chapter is a general description of the structure of matter, leading to the introduction of quarks and leptons, and an account of the first ideas of quantum mechanics. This material will be familiar to many students, but not to all; it is included to make the book accessible to mathematics students who may have studied no physics, or have forgotten what they have studied, and to meet the complaint that books on particle physics always assume that you already know about particles. At the end of this chapter the reader will know what the particles of matter are, and what forces act between them.

The next two chapters contain the theoretical development of quantum mechanics, in the state-vector formalism with its standard interpretation. At the end of Chapter 3 the reader will know the basic assumptions and theoretical apparatus of quantum theory. Chapter 4 continues the study of quantum mechanics, but should perhaps be regarded as a prelude to the remaining chapters, being largely concerned with constructing more apparatus for later use (angular momentum theory, annihilation and creation operators), though it also contains the theory of the hydrogen atom as being of intrinsic interest.

The last three chapters provide three independent journeys, which can be taken in any order. Chapter 5 goes over the ground of Chapters 2 and 3 again, examining the concepts of quantum mechanics more critically. This journey ends in a muddy river delta, the mainstream having split into nine mouths. Chapter 6 goes over the ground of Chapter 1, the language of quantum mechanics now being available for a more detailed description of particles. Annihilation and creation operators are used to give a simplified treatment of the forces between particles – a kind of quantum field theory without space and time. Finally, the ideas of quantum field theory proper are described in Chapter 7. The first half of this chapter is a continuation of the formal development of quantum mechanics, and carries on directly from Chapter 4. The second half describes how quantum field theory is applied to particle physics in quantum chromodynamics and quantum flavourdynamics, and constitutes a third passage over the ground of particle physics.

Although the development is not formally axiomatic, the book does have a logical skeleton consisting of postulates (stated as such) and a chain of propositions (marked by the symbol ●) deduced from them. The bones are listed at the end of each chapter (except Chapters 1 and 6, which are cartilaginous). The mathematical arguments are not as rigorous as they might be, but the physical arguments are slightly more rigorous than they often are. The mathematical style is mainly algebraic, being based on commutation relations: the notion of a wave function is not needed in this logical skeleton. (The spectrum of the hydrogen atom is found by Pauli's original method, which predates the Schrödinger equation.) However, since it would be an impoverished idea of quantum mechanics that did not include wave functions, and since students are likely to have met them elsewhere, they are included from the beginning as an example of a type of state vector, and the usual assumptions about them (e.g. boundary conditions for the Schrödinger equation) are justified.

Teachers of quantum mechanics are divided, and no doubt always will be, about the suitability of the state-vector formalism for a first course in quantum mechanics. Those who, like myself, first learnt quantum mechanics by reading Dirac's immortal *Principles* have no doubt that the state-vector formalism is the best introduction to the subject. In deference to the other half of the world I have to say that this book *might* be found difficult by students who have not taken a first course on quantum mechanics based on wave functions. Formally, however, it requires no knowledge of any physics. Formally, also, the only mathematics required is vector algebra and vector calculus; but the reader with no knowledge of linear algebra will probably find it heavy going, and an acquaintance with the idea of a group and the elements of analytical mechanics will be helpful in places. Until Chapter 7 the only fact from special relativity that is used is the energy–momentum relation (1.5); for Chapter 7 the reader will need the 4-vector formalism and a knowledge of Maxwell's equations.

Bold type is used to indicate that a word or phrase is being defined, and the reader is not expected to know what it means. The symbol ■ denotes the end of a proof (or a proposition whose proof has already appeared). Complex conjugation is denoted by an overbar (not by an asterisk).

I would like to thank Mark Lawson, Chris Clarke, Richard Crossley, Peter Landshoff, Ian Drummond, Jeremy Rogers, Stephen McGahan, Alison Ramsay, Clifford Bishop, Denis Cronin, Roland Hall, Anne Thompson and Steve Roberts, all of whom read parts of the manuscript and made useful suggestions. I am grateful to the Scientific Information Service of CERN, Geneva, for supplying me with photographs and for their permission to use them. Finally, I would like to record my appreciation of the sensitive and patient editorship of Simon Capelin, and the care and forbearance of Sheila Shepherd and the other staff of Cambridge University Press.

Tony Sudbery
York, July 1985

1

Particles and forces

This book is primarily about the particles that make up the material universe, and the way that they interact with each other. To describe their behaviour, and even to name the particles themselves (i.e. to specify the properties which distinguish them from each other) requires the theoretical framework of quantum mechanics, which will be the concern of a large part of the book. Before embarking on the formal theory, however, we will give a general description of the constitution of matter and, in briefest outline, the reasons for believing that this description is true. This will introduce the particles which will be described in more detail in later chapters, and the forces between them; in the course of discussing the latter we will make a first qualitative encounter with the concepts of quantum mechanics.

A. THE ANALYSIS OF MATTER

1.1. **Molecules and atoms**
It is an old speculation, traceable in western thought alone to Greek thinkers who lived some centuries before Plato, that matter is made up of small, simple particles of which there are only a few distinct types, the variety of everyday substances being caused by the different ways in which these particles combine together. This idea remained an isolated speculation until the nineteenth century, when it became possible to relate it to various laws of physics and chemistry.

The laws of heat, and particularly the behaviour of gases, can be explained in terms of the laws of mechanics if it is assumed that any substance consists of a large number of particles called **molecules**, which in a gas are moving randomly. This explanation, which was demonstrated by means of the techniques of statistical mechanics developed by Maxwell, Boltzmann and Gibbs, is known as the **kinetic theory** of heat. It has the feature, usually regarded as a great advantage in a theory, that it reduces the number of primitive, unexplained concepts in physics: it enables heat and temperature to be identified with purely mechanical properties (heat being the total kinetic energy, and temperature the average kinetic energy, of the particles). However, this is a purely theoretical advantage, and was not universally regarded as a good reason to believe in the reality of the molecules. For this, independent evidence was required. It was provided by **Brownian motion**, in which a grain of pollen moving in a liquid is observed to make sudden and random changes of direction, as if it was being jostled by the molecules of the liquid. In 1905 Einstein showed that the observed quantitative features of this motion could be deduced from the hypothesis that the liquid consists of particles, using the same methods and assumptions of statistical mechanics as were used in the kinetic theory of heat.

From chemistry came the suggestion that molecules were themselves made up of smaller components. Every substance is a physical mixture of chemically pure substances, which in turn can be made by the combination of chemical elements. Dalton's laws of chemical combination (1903) can be understood if the molecules of a chemical compound are all alike, and are made up of smaller particles (called **atoms**) which are characteristic of the elements which combine to make the compound. This is the **atomic theory** of chemistry. Historically, it preceded the kinetic theory of heat, but it proceeds to a deeper level of analysis (atoms as opposed to molecules) and presents a simpler picture in that the number of different types of fundamental particle, instead of being equal to the enormous number of different chemical compounds, is replaced by the comparatively small number of chemical elements (the classical figure is 92, but this has been increased by the manufacture of artificial elements).

Nevertheless, 92 is rather a large figure for the number of basic constituents of matter. Moreover, the atoms are not simply featureless lumps of matter; there must be relations between them, as there are relations between the chemical behaviour of the elements. These are displayed in Mendeleev's **periodic table** (Fig. 1.1), in which the elements are laid out in a number of rows in order of increasing **atomic weight**, which is a measure of the mass of a single atom. In this table the elements in each column show similar chemical properties, with a slight but regular progression as one moves down the column; in each row there is a definite progression (e.g. a change in valency) as one moves along the row, and again the progression is regular. This pattern suggests that the atoms must have some internal structure in terms of which they can be compared, atoms in the same column having similar structures, while the structure changes in some regular way as one moves from left to right across the table.

Fig. 1.1.
The periodic table of the elements.

H																		He
Li	Be												B	C	N	O	F	Ne
Na	Mg												Al	Si	P	S	Cl	A
K	Ca	Sc		Ti	V	Cr	Mn	Fe	Co	Ni	Cu	Zn	Ga	Ge	As	Se	Br	Kr
Rb	Sr	Y		Zr	Nb	Mo	Tc	Ru	Rh	Pd	Ag	Cd	In	Sn	Sb	Te	I	Xe
Cs	Ba	La	...	Hf	Ta	W	Re	Os	Ir	Pt	Au	Hg	Tl	Pb	Bi	Po	At	
Fr	Ra	Ac		Th	Pa	U	Np	Pu	Am	Cm	Bk	Cf	...					

1.2. **Electrons, protons and neutrons**

The first subatomic particle to be discovered was the **electron** (symbol e^-); the discovery was announced by J. J. Thomson in 1897. Large numbers of this particle are given off by metals when they are heated or given a negative electric charge, or when light is shone on them. The particles are deflected by electric and magnetic fields; the direction of this deflection shows that they have a negative electric charge. By measuring the deflection of the particles in a magnetic field, their charge-to-mass ratio λ can be determined. Their charge is difficult to measure directly, but it is known from other sources (from Faraday's laws of electrolysis, and from Wilson's and Millikan's experiments on charged water and oil droplets suspended in an electric field) that electric charge always comes in integer multiples of a basic amount e, whose value in SI units is 1.6×10^{-19} coulomb. Assuming that this is the charge on the electron, its mass can be calculated as $m = e/\lambda = 9.1 \times 10^{-28}$ g. This is a tiny fraction (about 5.5×10^{-4}) of the mass of the lightest atom, the hydrogen atom.

Since electrons can be produced from many different kinds of matter, it must be assumed that they exist inside the atoms of all elements. But a normal atom is electrically neutral, so it must contain some positively charged material to balance the charge on the electrons, and since the electrons are so light this positive material must account for most of the mass of the atom. J. J. Thomson's 'plum-pudding' model of the atom pictured it as a cloud of positively charged material with the electrons orbiting inside it; but this picture was shown to be false by Rutherford's experiments on the scattering of α-particles. These are positively charged particles (with charge $2e$ and mass about equal to that of the helium atom) which are emitted by radium; Geiger and Marsden, under the direction of Rutherford, studied their motion when they were fired at thin sheets of gold foil. Since the electrons in the gold atoms are so much lighter than the α-particle, collisions with them will have little effect on the α-particle's motion; the main effect will be provided by the electrical repulsion of the massive positively charged material in the atom. If this is spread out throughout the atom, as in Thomson's model, most α-particles will encounter some of it and will be deflected by the encounter; but since the material is so diffuse the force on the α-particles will be small and so they will be deflected through small angles. The results of Geiger and Marsden were quite contrary to this: most of the α-particles went straight through the gold foil without being deflected at all, but of those that were deflected quite a high proportion turned through large angles, so that some bounced back in the direction they had come.

Rutherford's interpretation of this experiment was that rather than pushing their way through big, soft atoms, the α-particles were colliding against small, hard objects inside the atoms, which were otherwise empty. He showed that the distribution of the α-particles as a function of their angle of deflection (the **scattering angle**) agreed very well with the distribution calculated from the assumption that both the α-particles and the positive parts of the atoms were point particles. This led him to formulate his 'solar system' model of the atom, in which the positive charge is concentrated in a small **nucleus** around which the electrons circulate as the planets orbit round the sun, the electrostatic attraction of the nucleus for the negatively charged electrons replacing the gravitational attraction of the sun for the planets.

There are two difficulties with Rutherford's model. The first is that electric attraction is not quite like gravitational attraction, since it is associated with magnetism in a way which has no counterpart in gravity. According to Maxwell's theory of the electromagnetic field, an accelerating charged particle like the electron in Rutherford's atom has a changing electric field which causes a changing magnetic field which in turn causes a changing electric field, and this feedback results in oscillations in the form of electromagnetic radiation which carries energy away from the accelerating particle. Thus the electron ought to lose energy and fall into the nucleus.

The second difficulty is the nature of the radiation which is sometimes emitted by the atom. This happens when the electron, in its mysteriously stable orbit, receives energy from any source; it will lose it again by emitting radiation, but only at certain special frequencies which are characteristic of the atom. This set of frequencies is called the **spectrum** of the atom; the spectrum of hydrogen, for example, consists of the frequencies

$$ v_{mn} = R\left[\frac{1}{m^2} - \frac{1}{n^2}\right] \tag{1.1} $$

where R is a constant and m and n are integers. There is nothing in Rutherford's picture of an orbiting electron, which could have any frequency in its motion around the nucleus, to associate it with a discrete set of numbers like (1.1).

These difficulties were resolved by the adjustments to classical ideas of mechanics which were brought about by quantum mechanics. Then this structure, atom = nucleus + electrons, turned out to be sufficient for the explanation of the chemical relations between different elements, for the chemical behaviour of the atom could be explained purely in terms of the arrangement of its electrons (we shall see roughly how the explanation goes in Chapter 4). The model shows that an important characteristic of an atom will be the charge on its nucleus (in units of the electron charge e), which is equal to the number of electrons in the atom. This was identified as the number of the element, counting along the rows of the periodic table; it is called the **atomic number** of the element and usually denoted by Z.

But this could not be the end of the story; the nucleus, though small, must itself have an internal structure. If it did not, we would simply have exchanged 92 different kinds of atom for 92 different kinds of nucleus. As well as this theoretical preference, there was the empirical evidence of **radioactivity** to indicate that the nucleus was made of smaller parts.

Radioactive substances emit three different kinds of radiation, known as α-, β-, and γ-rays. α-rays consist of positively charged particles with charge $2e$, which we can now identify as the nuclei of helium atoms. β-rays consist of electrons, and γ-rays consist of electromagnetic radiation of extremely high frequency. Substances which emit α-rays or β-rays change their chemical identity. If an atom of the element with atomic number Z emits an α-particle, it loses electric charge $2e$ from its nucleus, whose charge becomes $(Z-2)e$; the atom then becomes an atom of the element with atomic number $Z-2$, with two extra orbiting electrons (which are likely to be removed from the atom soon after the radioactive emission). Similarly, if an atom emits an electron as part of β-radiation, its nucleus loses the charge $-e$ and the atom becomes an atom of the element with atomic number $Z+1$, with an overall positive charge of $+e$ because it has one electron too few. (Electrically charged atoms like these are called **ions**.) The source of radioactivity, then, is the atomic nucleus, which thus appears to contain α-particles and electrons inside it.

The process of emitting a particle is called the **decay** of the nucleus. Radioactive decay is inherently unpredictable; it is not possible to say when a particular nucleus will decay. However, it is possible to make a statement about the probability of a decay: for each type of radioactive nucleus there is a constant τ such that the probability that a nucleus of that type will decay in a small time dt is dt/τ (radioactive decay is what is known in probability theory as a **Poisson process**). It follows that of a large number N of similar radioactive nuclei, the number that will decay in time dt is $N\,dt/\tau$, and so the change in the number of nuclei of the original type is

$$dN = -\frac{N\,dt}{\tau}. \tag{1.2}$$

Hence the number of nuclei remaining undecayed at time t is

$$N = N_0 e^{-t/\tau} \tag{1.3}$$

where N_0 is the number of nuclei at time $t=0$. The constant τ is called the **lifetime** of the nucleus.

The mass of every nucleus is very close to an integer multiple A of the mass of the hydrogen nucleus, where A is always greater than the atomic number Z. This suggests that the hydrogen nucleus is a fundamental particle – it is called a **proton** (symbol p) for this reason – and that every nucleus is made up of A protons together with $A-Z$ electrons to bring the total electric charge down to Ze. The masses do not quite add up as one might expect from this picture – tha mass of the nucleus is not exactly $Am_p + (A-Z)m_e$, where m_p and m_e are the masses of the proton and the electron, but somewhat less – but this can be explained by special relativity. Relativity theory states that mass is equivalent to energy according to the famous formula $E = mc^2$. Now if the protons and electrons in the nucleus are stuck together, the nucleus must have less energy than its constituent parts would have when separated, since in order to separate them work must be done against the forces that stick them together. It follows that the mass of the nucleus must be somewhat less than the sum of the masses of its constituents. The difference is called the **binding energy** of the nucleus.

However, this simple picture cannot be quite right. As we will see in Chapter 4, the angular momentum of the nucleus has the wrong value for it to be made up of $2A + Z$ particles; rather it must be taken to be made up of just A particles. Thus we can still regard the electric charge of the nucleus as being contributed by Z protons, but the extra mass must come from $A - Z$ particles of a new type, whose mass is close to that of the proton and which has no electric charge.

This new particle is the **neutron** (symbol n). Its existence was conjectured by Rutherford in 1920 and experimentally demonstrated by Chadwick in 1932 (though they both thought at first that it was made up of a proton and an electron, and did not. recognise it as an independent fundamental particle).

Nuclei with the same number of protons but different numbers of neutrons are called **isotopes**. They constitute different forms of the same chemical element. The symbol for a nucleus is ^AX, where X is the symbol for the chemical element and A is the total number of protons and neutrons, as above. If it is desired to draw attention to the numbers of protons and neutrons separately, this can be expanded to $^A_ZX_{A-Z}$. Thus $^3_1H_2$ is an isotope of hydrogen which contains one proton and two neutrons (it is called tritium).

At this stage we have a very simple picture of the world in terms of just three elementary particles. Everything is made of molecules; molecules are made of atoms; atoms consist of electrons orbiting around a nucleus, which is made of protons and neutrons.

1.3. Neutrinos

There remains a problem associated with β-radiation in radioactivity: if the nucleus does not contain electrons, how can radioactive nuclei emit electrons as β-rays? Other puzzling questions about such radioactive decays arise from the following considerations of the velocity of the emitted electron.

Suppose a nucleus A decays into another nucleus B by emitting an electron:

$$A \to B + e^-. \tag{1.4}$$

We will apply the principles of conservation of energy and momentum to this process. The energy and momentum of each body involved are given by the relativistic formulae

$$E = \sqrt{(p^2 + m^2c^2)}, \tag{1.5}$$

$$p = \frac{Ev}{c^2} \tag{1.6}$$

where E, p and v are respectively the energy, momentum and velocity of the body, m is the mass (i.e. the rest-mass) of the particle, and c is the speed of light. If the original nucleus A was at rest, conservation of momentum requires that the emitted electron and the final nucleus B have equal and opposite momentum. Let the magnitude of this momentum be p; then the equation of conservation of energy becomes

$$m_A c^2 = \sqrt{(p^2 c^2 + m_B{}^2 c^4)} + \sqrt{(p^2 c^2 + m_e{}^2 c^4)}. \tag{1.7}$$

Thus p is uniquely determined in terms of the masses m_A, m_B and m_e of A, B and the electron. Hence all electrons emitted by stationary A nuclei should have the same energy, which is given by

$$E_e = \frac{(m_A{}^2 - m_B{}^2 + m_e{}^2)c^2}{2m_A}. \tag{1.8}$$

The experimental fact is that the electrons in a particular β-decay have varying energies, ranging from a minimum of $m_e c^2$ to the value which we have just calculated as a maximum (see Fig. 1.2). Thus energy appears not to be conserved in the decay. In 1930 Pauli suggested that the missing energy was carried away by another particle, which had not been observed since it was electrically neutral and had little interaction with matter. The existence of this particle would also resolve a discrepancy between the angular momentum of the original nucleus and that of the final nucleus and the emitted electron. Pauli called this particle a **neutrino** (symbol v), though today, for reasons which will emerge in the next section, it is known as an **antineutrino** (symbol $\bar{v}$). Thus the decay (1.4) should be written as

$$A \rightarrow B + e^- + \bar{v}. \tag{1.9}$$

The simplest example of such a process is the decay of the neutron, which, when outside the nucleus, is an unstable particle and decays into a proton with a lifetime of about 15 minutes:

$$n \rightarrow p + e^- + \bar{v}. \tag{1.10}$$

Since the energies of the electrons emitted in the decay (1.9) come arbitrarily close to the value (1.8) which they would have if no antineutrino was emitted, the energies of the antineutrinos must come arbitrarily close to zero. According to (1.5), this is only possible if the antineutrino's rest-mass m is zero.

Fig. 1.2.
The β-decay spectrum: $n(E)\,dE$ is the proportion of electrons emitted in the decay of a nucleus which have energy between E and $E + dE$.

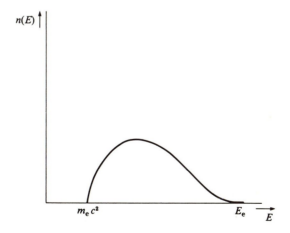

(1.5) and (1.6) together then give $v = c$; thus antineutrinos always travel at the speed of light.

Because the antineutrino has very little interaction with other forms of matter, its existence was not confirmed by independent evidence until 1956, when Reines and Cowan observed the rare processes which occur when antineutrinos collide with nuclei (see (1.15) in the next section).

1.4. Antiparticles: baryons and leptons

The motion of the particles we are interested in is to be described using special relativity, as we have just seen, and quantum mechanics, as we will be seeing at some length. For most of this book quantum mechanics will be discussed in a non-relativistic form, but in Chapter 7 we will see (rudimentarily) how it is combined with special relativity in quantum field theory. We will see that this requires that for every type of particle there should exist another type of particle with the same mechanical properties (viz. mass and spin), but with the opposite electric charge. This is called the **antiparticle** of the first particle.

The antiparticle of the electron is called the **positron** (symbol e^+). The theoretical necessity for its existence became apparent in the two years after the relativistic quantum equation describing the electron was discovered by Dirac† in 1928, and it was observed by Anderson in 1932. When an electron and a positron meet, they can annihilate each other, their mass being converted into the energy of electromagnetic radiation; conversely, in suitable circumstances an electron–positron pair can be created out of radiation. This pair creation leaves a characteristic signature in bubble chamber photographs; being oppositely charged, the electron and the positron will move with opposite curvatures in a magnetic field, thus making the ram's-horn shape that can be seen in Fig. 1.3.

The antiparticle of the proton, the **antiproton** (symbol $\bar{p}$) could only be produced in conjunction with a proton in a pair creation of the type just described (though the energy need not come from electromagnetic radiation, but could be in the form of the kinetic energy of a bombarding particle). Since the proton is so much more massive than the electron, it takes much more energy to create a proton–antiproton pair than an electron–positron pair, and it was not achieved experimentally until 1955.

Although they have zero electric charge, the neutron and the neutrino also have antiparticles $\bar{n}$ and $\bar{\nu}$. These are distinguished from the originals not by electric charge but by other properties which are best understood in terms of radioactive β-decay and various related processes, as follows.

Like the electron, the positron is emitted by some radioactive nuclei, and is accompanied by a particle with zero rest-mass; it is this particle which is now called the neutrino. For example, there is a radioactive isotope of oxygen (with a nucleus containing eight protons and six neutrons) which decays to nitrogen

---

† Dirac's original conception of the positron, which is still often presented as a valid description, was that it is a 'hole in a sea of negative-energy electrons'. There is no warrant for this in the present theory of antiparticles.

(seven protons, seven neutrons):

$$^{14}\text{O} \rightarrow {}^{14}\text{N} + \text{e}^+ + \nu. \tag{1.11}$$

This can be compared with the β-decay

$$^{14}\text{C} \rightarrow {}^{14}\text{N} + \text{e}^- + \bar{\nu} \tag{1.12}$$

(which is used in radiocarbon dating). The decay (1.12) can be regarded as being caused by the decay of one of the neutrons in the ^{14}C nucleus:

$$\text{n} \rightarrow \text{p} + \text{e}^- + \bar{\nu}, \tag{1.13}$$

which, as we have seen, is the process by which free neutrons decay. Similarly, the decay (1.11) can be regarded as being caused by the decay of one of the protons in the ^{14}O nucleus:

$$\text{p} \rightarrow \text{n} + \text{e}^+ + \nu. \tag{1.14}$$

Fig. 1.3.
Electron–positron pair creation (photo: CERN): the photons in an intense radiation field produce a large number of electron–positron pairs like the ones indicated.

Free protons do not decay this way, because the neutron has a greater mass than the proton and so energy would not be conserved if (1.14) occurred. However, inside the nucleus the energy of the neutron is reduced by the potential of the attractive nuclear force, so the process can occur. (This shows

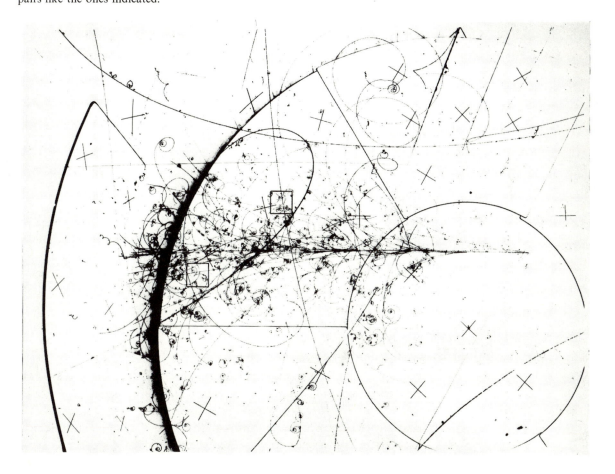

that the decay of the neutron as in (1.11) should not be taken as a reason for thinking that the neutron is a composite object containing the proton inside it; the proton and the neutron must be regarded as being on the same footing.)

Now consider the process by which Reines and Cowan first observed the antineutrino, in which an antineutrino from a decay like (1.12) collides with a proton, producing a neutron and a positron:

$$p + \bar{v} \rightarrow n + e^+. \tag{1.15}$$

A similar process is observed with neutrinos:

$$n + v \rightarrow p + e^-. \tag{1.16}$$

However, a process like (1.15) does not occur if the antineutrino is replaced by a neutrino (i.e. if particles produced by ^{14}O are used instead of ones produced by ^{14}C), and a process like (1.16) does not occur if the neutrino is replaced by an antineutrino. This shows that the neutrino and the antineutrino are definitely different particles, and it also suggests a way to describe the difference.

The processes (1.13)–(1.16) involve two distinct types of particle: the comparatively heavy particles n and p, and the light particles e^-, e^+, v and $\bar{v}$. These are called **baryons** and **leptons** respectively (from the Greek words for 'heavy' and 'light'). In each of the processes there is a baryon present both before and after, so the total number of baryons remains the same. This is not true of the leptons, but that is because antiparticles are involved; if we count the positron and the antineutrino as *negative* leptons, then the total number of leptons does remain the same in each case.

To put this less mysteriously, define a property called **lepton number** which has the value $+1$ for the electron and the neutrino, -1 for the positron and antineutrino, and 0 for all other particles. Calculate the lepton number of a set of particles by adding the lepton numbers of the individual particles, just as you calculate the total electric charge. Then there is a fundamental law of *conservation of lepton number* just like the law of conservation of electric charge.

Similar considerations apply to the baryons: if we define the **baryon number** to be $+1$ for the proton and the neutron, -1 for their antiparticles, and 0 for all other particles, then baryon number is conserved in all processes.

These three conserved quantities, the electric charge, the lepton number and the baryon number, are known as **additive quantum numbers**.

Summary: the first four particles

The properties of the particles that have been discussed so far are summarised in Table 1.1.

The unit of mass usually used for elementary particles is the **MeV**. This is actually a unit of energy – one electron volt (eV) being the energy acquired by an electron in being accelerated by a potential difference of one volt, and $1\,\text{MeV} = 10^6\,\text{eV}$ – but it can be used as a unit of mass because of the equivalence expressed in the relativistic equation $E = mc^2$. (Thus the unit of mass should be written MeV/c^2, but the c^2 is often dropped because of theoretical physicists' belief that $c = 1$ – which after all is right: $c = 1$ light-year per year.) The relation of the MeV to the standard unit of mass is that 1 MeV = 1.78×10^{-30} kg. However, the best way to appreciate its significance in this subject is probably to remember that the electron has a mass of about a half of an MeV, and the proton and the neutron have masses of about a thousand MeV.

Larger units of mass (or energy) are the **GeV** ($= 10^3$ MeV) and the **TeV** ($= 10^6$ MeV).

1.5. Quarks and leptons

The four particles listed in Table 1.1 seem to provide the elements of a description of matter which is both satisfyingly simple and sufficient to account for all forms of matter known before 1935. However, observations on cosmic rays and experiments with particle accelerators revealed the existence of other particles at the same level as these four in the analysis of matter. These new particles are not apparent in the make-up of ordinary matter, because they quickly decay (with lifetimes in the range from 10^{-6} to 10^{-10} sec) into protons, neutrons, electrons and neutrinos; nevertheless, they are just as fundamental as the first four.

The first of these new particles is the **muon** (symbol μ^-). This particle appears to be exactly like the electron in all respects but its mass (106 MeV) and the fact that it is unstable, decaying as follows:

$$\mu^- \to e^- + v + \bar{v}. \tag{1.17}$$

with a lifetime of 2×10^{-6} sec. This is not a very significant difference between the electron and the muon; it is simply a consequence of the muon's greater mass, which makes it possible for the decay to occur.

Like the electron, the muon has unit negative electric charge and has a positively charged antiparticle μ^+. It has lepton number $+1$, and it can be seen that in the decay (1.17) lepton number is conserved. In fact more than this is true, and (1.17) is an oversimplified representation of muon decay; for not only

Table 1.1. *Properties of the first four particles and their antiparticles*

Particle, antiparticle	Mass (MeV)	Charge	Lepton number	Baryon number	Date of discovery Theoretical	Date of discovery Experimental
Electron e^-	0.511	-1	$+1$	0	—	
Positron e^+		$+1$	-1	0	1928	1931
Proton p	938.3	$+1$	0	$+1$	—	1911
Antiproton $\bar{p}$		-1	0	-1	1928	1955
Neutron n	939.6	0	0	$+1$	1920	1932
Antineutron $\bar{n}$		0	0	-1	1928	1956
Neutrino v	0	0	$+1$	0	1930	1954
Antineutrino $\bar{v}$		0	-1	0	1930	1954

does the electron have a doppelganger in the muon, but also the neutrino is copied by another particle. This is called the **muon neutrino** (symbol v_μ), and the original neutrino of §1.4 is called the **electron neutrino** (symbol v_e) to distinguish it. The difference between the two lies solely in the processes they take part in: each of them will only undergo reactions involving its associated charged lepton, i.e. electron or muon. Thus we have

$$\left. \begin{array}{l} v_e + n \rightarrow p + e^- \\[2mm] v_\mu + n \rightarrow p + \mu^- \end{array} \right\}, \tag{1.18}$$

and

but

$$\left. \begin{array}{l} v_e + n \nrightarrow p + \mu^- \\[2mm] v_\mu + n \nrightarrow p + e^- \end{array} \right\}. \tag{1.19}$$

and

When the neutrino and antineutrino in muon decay (1.17) are identified as electron-type or muon-type, it is found that the decay is

$$\mu^- \rightarrow e^- + v_\mu + \bar{v}_e. \tag{1.20}$$

The facts represented by (1.18)–(1.20) can be organised by refining the concept of lepton number into two distinct quantum numbers, called **electron number** and **muon number**, with e^- and v_e having electron number $+1$ and muon number 0, μ^- and v_μ having electron number 0 and muon number $+1$, and the sign of both properties being reversed for antiparticles, as usual. Then in (1.18)–(1.20) both electron number and muon number are conserved separately.

In 1975 a third kind of lepton was discovered. It is known as the **tau** lepton (symbol τ^-, antiparticle τ^+); there is a third kind of neutrino v_τ associated with it, and a new independently conserved quantum number attached to both the tau and its neutrino.

New baryons were also discovered in the first cosmic ray investigations in the late 1940s. These carry another new quantum number called **strangeness** (the evidence for the existence of this quantum number is more complicated than for the lepton numbers, and will be discussed in the second part of this chapter); they also carry baryon number like the proton and neutron (unlike lepton number, baryon number is not further subdivided). There are many more of these new baryons than the new leptons, and they are related in more complicated ways; the relations between them are best seen by means of diagrams like Figs. 1.4 and 1.5, in which the particles are located on a plot of strangeness against electric charge. By using oblique axes we see that the particles fall into sets which have simple geometrical shapes (these diagrams have a mathematical significance which will emerge in Chapter 6). Figs. 1.4 and 1.5 show only the lightest of the particles; there are a number of other groups of particles, all forming similar patterns.

The same arguments that we used earlier to suggest that atoms must have an internal structure can now be applied again to these baryons. There are too many of them to be regarded as truly elementary, and the patterns of Figs. 1.4 and 1.5, like the periodic table, suggest that there must be some internal features which vary systematically from baryon to baryon. Also the electromagnetic properties of the proton and the neutron indicate that they contain finite distributions of charge and magnetisation extended over a region whose diameter is of the order of 10^{-13} cm (unlike the leptons, whose behaviour is that of point particles – at least on length scales down to 10^{-16} cm). A final analogy with investigations of the structure of the atom is provided by experiments in which very fast electrons are fired at protons and neutrons; like the α-particles in Rutherford's scattering experiment, an unexpectedly high proportion of the electrons are deflected through large

Fig. 1.4.
The octet of baryons: these particles have lifetimes of the order of 10^{-10} sec.

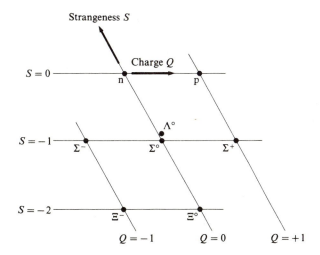

Fig. 1.5.
The decuplet of baryons: these particles have lifetimes of the order of 10^{-23} sec, except for the Ω^{-} (lifetime 10^{-10} sec).

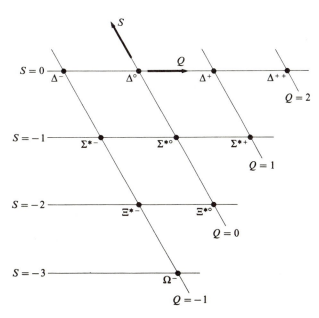

angles. Just as the Geiger–Marsden experiment demonstrated the existence of a small, hard nucleus inside the atom, these electron-scattering experiments indicate that there are smaller constituent particles inside baryons.

The positions of the baryons on the charge–strangeness plots of Figs. 1.4 and 1.5 can be understood if each baryon is made of three smaller particles, and if these smaller particles come in three varieties with values of charge and strangeness as plotted in Fig. 1.6. These particles are called **quarks** (the word was coined by James Joyce in *Finnegans Wake*, and is indefinite in meaning). Only two different kinds of quark are needed to make the proton and the neutron; they are called the **up** and **down** quarks (symbols u and d). Baryons with non-zero values of strangeness contain the third kind of quark, which is called the **strange** quark (symbol s). All these quarks have baryon number $\frac{1}{3}$, so that three of them make up an object with baryon number 1. The possible combinations of three quarks then give all the values of charge and strangeness seen in the plots of Fig. 1.5. The extra particles in Fig. 1.4 arise because the quarks have the further feature of spin: in the baryons of Fig. 1.5 all the quarks spin the same way, and Fig. 1.4 shows baryons in which two of the quarks spin in opposite directions.

The baryons at the corners of the triangle of Fig. 1.5 are made of three identical quarks, which are all spinning in the same direction and, as we will see in Chapter 6, all move in the same orbit inside the baryon. But there is a fundamental law, the Pauli exclusion principle (see §2.6) which states that identical particles cannot be in the same state of motion. It follows that the quarks in these baryons cannot in fact be identical; there must be some further feature of the quarks which distinguishes them from each other. This feature is called **colour**: each of the quarks u, d and s of Fig. 1.6 exists in three forms, as if there were a red variety, a blue one and a yellow one.

It is customary to warn the reader that the idea of the colour of a quark is not to be taken literally. I doubt if this caution is necessary. It is more interesting to point out that the colour of quarks shares with visual colour the property that any colour is a mixture of three primary colours, but the choice of these primary colours is arbitrary: any colour can be regarded as primary.

As well as combining with two other quarks to form a baryon, a quark can combine with an antiquark to form a particle with zero baryon number. Such

Fig. 1.6.
Quarks.

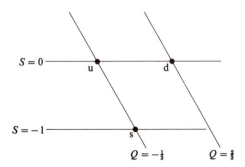

particles are called **mesons** (from the Greek for 'middle', their masses being intermediate between those of leptons and baryons). The mesons formed from the quarks u, d, s are shown in Fig. 6.7; it will be seen that they form the same hexagonal pattern as the octet of baryons, with one extra particle. Baryons and mesons are known collectively as **hadrons** (from the Greek for 'strong'; the reason for this will be explained in §1.7).

In 1974 the first particles in a new series of hadrons were discovered. These new particles have a new quantum number which is called **charm**; it is carried by a fourth quark, the charmed quark (symbol c), whose other properties are the same as the up quark. Three quarks of any kind can combine to form a baryon; thus we obtain new baryons with the property of charm (whose value for a baryon can be 0, 1, 2 or 3 according to how many charmed quarks it contains). Again, a quark and an antiquark of any kind can combine to form a meson.

The quantum numbers which distinguish the types of quark (like strangeness and charm) are called **flavours**. The four quarks u, d, c, s can be arranged in two pairs with identical properties except for their flavours: the pair (c, s) seems to be a copy of the pair (u, d), just as the pair of leptons (μ, v_μ) is a heavier copy of the pair (e, v_e). These repeated sets of apparently similar particles, differing only in mass and flavour, are known as **generations** or **families**.

Two further series of particles, discovered in 1979 and 1984, have shown the existence of fifth and sixth quarks b and t. Their flavours are sometimes called **truth** and **beauty**, but it is commoner, unfortunately, to use the clumsy nomenclature of **top** and **bottom** for these quarks. The full set of quarks and leptons is displayed in Table 1.2.

No quarks have ever been observed in isolation in the same way as leptons. It is thought that the forces which bind quarks in hadrons do not diminish with distance, so that quarks can never escape to become free particles.

Of the three pairs of quarks and leptons, one pair of each – the quarks u and d and the leptons e$^-$ and v_e – are necessary to make up the everyday world, and

Table 1.2. *Quarks and leptons*

| | Families | | | |
	1	2	3	Electric charge
Leptons	e$^-$	μ^-	τ^-	-1
	v_e	v_μ	v_τ	0
Quarks	u	c	t	2/3
	d	s	b	$-1/3$

a world which contained only these would seem to be quite possible. The existence of the other particles, and the relations between them, are mysteries.

1.6. Observation of particles

The most direct, and most informative, methods of observation of particles are those which make the particle leave a visible track. These include:

1. The cloud chamber. When an energetic charged particle passes through matter, it knocks electrons out of the atoms it meets, leaving a trail of positive ions behind it. In air which is supersaturated with water vapour, the condensation of the vapour is triggered by the ions, which thus become visible as a trail of water droplets.

2. The bubble chamber. This uses the same principle as the cloud chamber, but the supersaturated water vapour on the point of condensing is replaced by a superheated liquid on the point of boiling. Again, the ions in the track of a charged particle provide nuclei round which boiling occurs, and the track becomes visible as a trail of bubbles. The bubble chamber has now superseded the cloud chamber for most purposes.

3. The spark chamber. This uses the fact that electric discharges pass more easily through gas containing ions. Thus if a charged particle passes through the region between two plates and a voltage is applied between the plates, a spark will pass along the trail of ions left behind the particle, thus making its track visible. The gap between the plates must be fairly small, so spark chambers are usually used in large arrays.

4. Photographic emulsion. A charged particle has an effect on silver bromide similar to that of light, and so it leaves a record of its passage through a photographic emulsion just like the record of incident light which constitutes a photographic image. This has been particularly useful in recording **cosmic rays** (very high energy particles reaching the upper atmosphere from outer space).

Other kinds of particle detector are the **scintillation counter**, which uses the fact that some plastic materials emit a flash of light when a charged particle passes through them; the **Čerenkov counter**, which uses the characteristic electromagnetic radiation which is given off like a bow wave by a charged particle moving in a transparent medium at a velocity greater than that of light in the medium; and the **Geiger counter**, which is similar in principle to the spark chamber, but uses the current in the discharge rather than the visible spark. These three pieces of apparatus are called 'counters' because they simply register the passage of a particle through the apparatus and give no information about its path.

Most of these methods of observation will only give direct information about charged particles; neutral particles must be observed indirectly, by their effect on charged particles and nuclei. The properties of charged particles can be measured from the tracks in the first group of apparatus: the *charge* can be deduced from the curvature of the track in a magnetic field, the *momentum* from the density of ionisation in the track, and the *energy* from the distance travelled by the particle if it stops in the chamber.

B. THE ANALYSIS OF FORCE

1.7. Kinds of force   We now turn from the constitution of matter to its behaviour. The process of analysing matter into smaller components, which was discussed in the first half of this chapter, carries with it a process of explaining the behaviour of matter at one level of the analysis in terms of the behaviour of its components at the next level. We have already encountered two examples of this: the kinetic theory of heat explains the thermodynamic behaviour of large bodies of matter in terms of the mechanical behaviour of their molecules, and quantum theory explains the chemical properties of substances in terms of the behaviour of the electrons inside their molecules. Another example is the explanation of the shining of stars in terms of reactions in the nuclei of their atoms.

As different forms of matter are studied by different sciences, the effect of this analysis is apparently to reduce the number of sciences – in the above examples the study of heat is replaced by mechanics, chemistry by atomic physics and parts of astronomy by nuclear physics. This replacement is more apparent than real, because although the concepts of a higher science may be *analysed* in terms of a more fundamental science, they cannot be *eliminated* in favour of the latter without losing the understanding gained by the higher science. Nevertheless, in principle one can envisage a chain of analysis in which sociology is analysed into psychology, psychology into physiology, physiology into biology, biology into chemistry and chemistry into physics. (This view of science is called 'reductionism' by those who don't like it and 'the unity of science' by those who do.)

There is another dimension of analysis which applies to the behaviour of matter at a given level, aiming to represent any behaviour as a combination of certain basic kinds of behaviour. The aim is always to reduce the number of independent laws of nature, either by showing that some laws can be explained in terms of others, or by giving two laws a unified description, i.e. by showing that they are aspects of the same underlying process. The latter occurred when Faraday and Maxwell showed that electric and magnetic forces were intimately related; the former when Maxwell explained light in terms of this unified electromagnetic force.

At the level of the particles described in §1.4 (protons, neutrons, electrons and neutrinos), the result of this analysis is that there are the following four fundamental forces:

1. Gravitation. This acts on all particles, but is so weak compared with the other forces that it is only important when large numbers of particles are considered. Thus it is the dominant force in astronomy, it is significant in the behaviour of everyday macroscopic objects, and it is utterly negligible in considering individual elementary particles.

2. Electromagnetism. This acts on all charged particles, and also on the neutron because it has a magnetic moment. It holds the atom together, and is responsible for the configuration of the electrons in the atom, and hence for all chemical behaviour; it is also responsible for all forces between atoms and molecules. Thus all macroscopic forces can be reduced to gravitation and electromagnetism; between them these two account for all the behaviour of 'shoes and ships and sealing wax, of cabbages and kings'.

3. The weak force. This acts on all particles, and is responsible for the processes (1.11)–(1.16). Its main macroscopically significant effect is in radioactivity; it also plays a catalytic role in the chain of nuclear reactions which take place in stars.

4. The strong force. This does not act on leptons, but only on protons and neutrons (more generally, on baryons and mesons – this is the reason for the collective name 'hadrons'). It holds protons and neutrons together to form nuclei, and is insignificant at distances greater than 10^{-15} m. Its macroscopic manifestations are restricted to radioactivity and the release of nuclear energy.

The three forces which are relevant to elementary particles can be recognised in the three kinds of radioactivity: α-radiation is caused by the strong force, β-radiation by the weak force, and γ-radiation by the electromagnetic force.

These forces are summarised in Table 1.3. The figures given for the strength and range of the forces come from a comparison of the effects they produce on two protons. In some respects these resemble an ordinary Newtonian force between the protons, varying with the distance between them as if the force was derived from a potential function

$$V(r) = \frac{ke^{-r/R}}{r^n} \tag{1.21}$$

for some n. This is an inverse-power force which is diminished by an exponential factor at distances larger than a certain distance R, the **range** of the force. The **strength** of the force is measured by the constant k. Note that the weak force does not appear to be particularly weak on this reckoning; the reason for its apparent weakness is its very short range rather than its intrinsic strength.

Table 1.3. *The four fundamental forces*

Force	Particles affected	Range	Strength†
Gravitation	All	∞	10^{-39}
Electromagnetic	Electron Proton Neutron	∞	7×10^{-3}
Weak	All	10^{-17} m	4×10^{-3}
Strong	Proton Neutron	10^{-15} m	1

† The unit of strength is $hc/2\pi$ where h is Planck's constant and c is the speed of light.

On passing to the level of quarks and leptons a further simplification becomes apparent: the weak and electromagnetic forces are seen as aspects of a single force called the **electroweak** force. This unified theory is due to Weinberg and Salam. Before we can see how it works we will need to look more closely at the description of forces between elementary particles.

1.8. Particles of force

Two of the forces listed in the previous section, the forces of gravity and electromagnetism, are familiar from macroscopic physics. In that context they are described by **fields**. A field of force is a vector-valued function of position $\mathbf{F}(\mathbf{r})$ which gives the force that would be experienced by a particle at the point $\mathbf{r}$. That definition appears to give the field only a hypothetical existence, but the development of the theory of fields, and particularly Maxwell's theory of electromagnetism, gave reasons for thinking of fields as having their own independent reality. For one thing, Maxwell's theory shows that electromagnetic effects take a finite time to travel from one material body to another; as a result of this, the field must be regarded as having energy and momentum of its own, even at points where there is no matter.

A field is a continuous function of position, and so its energy and momentum are continuously distributed throughout space; they are like the energy and momentum of a fluid continuum and quite different from the energy and momentum of a system of discrete particles. But consideration of two separate problems, those of black-body radiation and the photo-electric effect, leads to the surprising and puzzling conclusion that although the energy in the field cannot be localised, nevertheless it only exists in discrete packets.

The problem of **black-body radiation** comprises a theoretical contradiction between electromagnetic theory and the statistical mechanics which is used in the kinetic theory of heat. Rayleigh and Jeans applied these theories to the problem of an insulated system of matter and radiation at a given temperature (for example, the inside of a closed oven which has settled down to a steady state, so that energy lost from the matter by emission of radiation is balanced by energy that it gains by absorbing radiation). They calculated the distribution of energy between the different frequencies of radiation. Their answer had the absurd feature that the total energy in the radiation is infinite – in other words, equilibrium between matter and radiation in an insulated enclosure is not possible at any temperature, and the matter will always cool to absolute zero by emitting all its energy in the form of radiation.

In 1900 Planck found a formula for the distribution in the radiation which fitted the experimental data, and showed that this formula would follow from electromagnetic theory and statistical mechanics if, instead of assuming that energy was a continuous quantity as Rayleigh and Jeans had, one assumed that it only took values which were integral multiples of a certain minimum quantity, called a **quantum** of energy. The size of this quantum varies with the frequency of the radiation; if the frequency is ν the quantum of energy is

$$E = h\nu,$$

(1.22)

where h is a universal constant which is now known as **Planck's constant**. This refers to the energy which is *exchanged* between radiation and matter. In 1905 Einstein, pursuing the theoretical implications of Planck's work, showed that it gave grounds for the stronger assumption that the energy in the radiation only *existed* in discrete quanta, and he supported this assumption† by showing how it explained the **photo-electric effect**. In the photo-electric effect electrons are emitted from the surface of a metal when electromagnetic radiation is incident on the metal. The effect is only observed if the frequency of the radiation is greater than a certain threshold value v_0 (which depends on the metal). If this condition is satisfied, radiation with a given frequency produces electrons with a range of velocities which depends only on the frequency (see Fig. 1.7). Varying the intensity of the radiation changes the number of electrons produced, but does not affect their velocities. Einstein explained these facts as follows: Suppose that to be liberated from a particular metal an electron needs an amount of energy W; this may be different for different electrons in the metal, but must be greater than a minimum value W_0 which is characteristic of the metal. Suppose also that radiation of frequency v consists of a collection of objects (called **photons**) each of which has energy hv. Then if $hv < W_0$, a photon cannot give any electron enough energy to escape from the metal. Thus the threshold is explained, and identified as $v_0 = W_0/h$. Now if $hv > W$, an electron which absorbs a photon acquires enough energy to leave the metal and has some energy left over, which appears in the form of kinetic energy. Thus its velocity is given by

$$\tfrac{1}{2}mv^2 = hv - W \leqslant hv - W_0 = h(v - v_0), \tag{1.23}$$

and so the maximum velocity is entirely determined by v. Finally, increasing

† Einstein's explanation of the photo-electric effect does not prove the existence of photons. In fact the photo-electric effect can be explained perfectly well on Planck's assumption that energy is only exchanged in multiples of hv. Einstein's paper (ter Haar 1967) was mainly concerned with the properties of radiation by itself. For a review of arguments for the existence of photons see Scully & Sargent 1972.

Fig. 1.7.
The photo-electric effect: the shaded region shows the possible values of the velocity of the emitted electron and the frequency of the incident radiation.

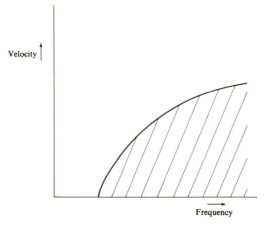

the intensity of the radiation increases the number of photons in it and therefore increases the number of electrons that are produced.

This is almost enough to show that the electromagnetic field is made up of material particles – almost, but not quite, since photons do not have definite positions like particles but retain the field-like characteristic of being spread through space. They also retain the property of a field that it can be cancelled by an opposite field; thus they exhibit the typically wavelike phenomena of **diffraction** and **interference**.

Diffraction is exhibited by waves when they meet an obstacle. Instead of being cut off sharply by the obstacle, they spread in all directions from every point on the edge of it. This has the result that the shadow cast by the obstacle is not clear-cut, but has a blurred border whose width is of the order of the wavelength of the waves casting the shadow.

Interference occurs in any form of wave motion when waves from a source can reach some point by two different routes. The phenomenon is clearest when the waves are **monochromatic**, i.e. when the waves repeat themselves regularly with a definite frequency and wavelength. Then at any instant the disturbance will be a function of distance from the source of the type shown in Fig. 1.8; the height of the crests may vary with distance from the source, but the distance between successive crests will always be the wavelength λ. At a given place the disturbance varies with time as a simple oscillation. For interference to occur the wave must be **coherent**, i.e. the pattern must be maintained over long distances and times. (An *exactly* monochromatic wave is automatically coherent; an incoherent wave is one which is almost monochromatic, but the regular alternation shown in Fig. 1.8 is occasionally disrupted, as in Fig. 1.9.)

Now if two waves arrive at the same point by two different routes the disturbance will be the algebraic sum of the disturbances in the individual waves. If they emanate from the same source and are part of the same coherent wave train, the result will depend on the difference between the distances that the two waves have travelled. If this is a whole number of wavelengths, the vibrations from the two waves will be in phase, i.e. they reach a crest or trough

Fig. 1.8.
A monochromatic wave.

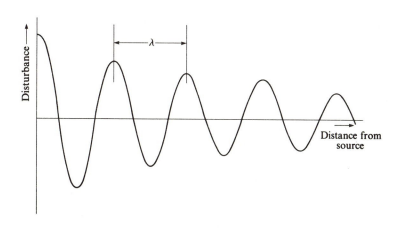

together and therefore reinforce each other. If, on the other hand, the path difference between the waves is half an odd number of wavelengths, then the vibration due to one wave reaches a crest when the other reaches a trough and therefore one wave reduces the disturbance due to the other, or may cancel it completely. This is illustrated in Fig. 1.10 for the **two-slit experiment**, in which the waves from the source at A can reach the line BC only by travelling through one of two holes H_1 or H_2 in a screen between the source and the line. At a point X on BC, the path difference is $AH_1X - AH_2X$; if this is a whole number of wavelengths the disturbance at X is large, while if it is half an odd number of wavelengths the disturbance at X is small. Fig. 1.10 shows the intensity of the wave at points on BC, which is related to the amplitude of the vibration at these points, as a function of position on BC.

Note that the two-slit experiment also involves diffraction, which occurs at

Fig. 1.9.
An incoherent wave: disruptions like that at X mean that the distance between successive crests sometimes differs from λ.

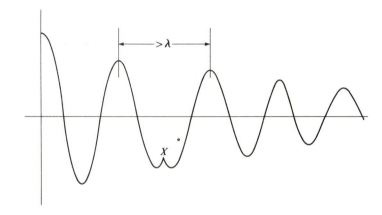

Fig. 1.10.
The two-slit experiment: at X the path difference is $AH_2 + H_2X - (AH_1 + H_1X)$ and is equal to one wavelength; at Y it is $AH_1 + H_1Y - (AH_2 + H_2Y)$ and is equal to half a wavelength.

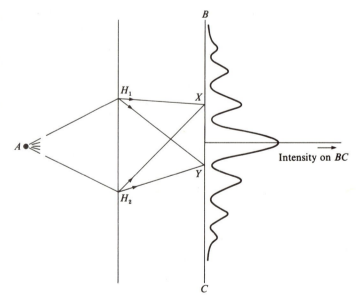

Intensity on BC

the holes H_1 and H_2. In general, whenever diffraction occurs it will be accompanied by interference: when a beam of waves spreads into the shadow of an obstacle, there is interference between waves coming from different parts of the edge of the obstacle. This causes a bright spot at the centre of the shadow of a small disc; in general, whatever the shape of the obstacle, there will be a pattern of alternating bands of high and low intensity at the edge of the shadow. This pattern can be enhanced if the obstacle has a regular structure, like the series of parallel scratches that make up a diffraction grating. Such a diffraction pattern is observed, for example, when X-rays pass through a crystal. The regularly spaced atoms in the crystal constitute the obstacle; the emerging X-rays are concentrated in some directions in which the electromagnetic waves reinforce each other, while in other directions they cancel.

These phenomena, which are characteristic of waves, are hard to reconcile with the idea that electromagnetic radiation consists of particles, but they ceased to give any reason for distinguishing between photons and the particles described in the first half of this chapter when it was discovered that the latter also displayed wavelike properties. In 1927 Davisson and Germer showed that a beam of electrons passed through a crystal will emerge in a diffraction pattern just as a beam of X-rays does. The electrons behave like a wave whose wavelength is determined by the momentum of the electrons according to a relationship previously proposed by de Broglie,

$$p = \frac{h}{\lambda}. \tag{1.24}$$

Diffraction by crystals shows electrons and photons both behaving in the same wavelike way; another indication that they are very similar in nature is provided by **Compton scattering**, which shows them both behaving in the same particle-like way. Compton investigated the response of free electrons to the incidence of a plane wave of monochromatic radiation, with frequency v and direction of propagation $\mathbf{k}$, say. He found that each electron moved as if it had collided with a particle with energy hv and momentum $hv\mathbf{k}/c$, energy and momentum being conserved in the collision. The momentum acquired by the electron may have a component perpendicular to $\mathbf{k}$ and is not predictable. It is related (via conservation of momentum) to the momentum of a photon which emerges, travelling in a definite direction, from the collision. (This is in sharp contrast with the predictions of the classical theory of the interaction between a charged particle and the electromagnetic field, according to which the particle should acquire momentum in the direction $\mathbf{k}$ and should emit radiation in the form of a spherical wave.)

Since the field was originally defined in terms of the force on a charged particle, saying that the field consists of photons amounts to saying that the force on a charged particle is caused by its absorption of or collision with a photon, as in the photo-electric effect or Compton scattering. Thus the electric repulsion between two electrons is understood as in Fig. 1.11. One electron emits a photon and recoils; the second electron absorbs the photon and acquires its momentum. (A picture like Fig. 1.11 is known as a **Feynman diagram**.)

We have now come full circle in our view of forces, and have returned to a pre-Newtonian view in which forces do not exist apart from matter, but consist of the action of particles of matter in contact. However, this view is only obtained at the cost of accepting apparently contradictory properties for matter, which behaves like particles in some circumstances and like a field in others.

Bohr and Heisenberg based their explanation of these contradictory properties on the principle that it is impossible to think of any physical system as having an independent reality, divorced from the observer; at the level of smallness we are considering, the physical processes used to observe the system will involve an inevitable interference with the system, which has an ineradicable minimum whose magnitude is of the order of Planck's constant h. This means that one must be chary of assuming that a property revealed by an experiment is simply a property of the system under study, as one would in classical physics; what an experiment reveals is a property of the system and apparatus together. Thus an experiment appropriate for waves may well show wavelike properties, while an experiment appropriate for particles shows particle-like properties. These are indeed contradictory in that they cannot be shown simultaneously, but they cannot be contradictory if regarded as properties of a collective (system + apparatus), since they refer to different collectives.

To illustrate this, consider the two-slit experiment (Fig. 1.10) with a beam of electrons, emanating from a source at A, passing through the holes H_1 and H_2, and impinging on a fluorescent screen at BC. This is an experiment appropriate to waves, and it elicits wavelike behaviour in the form of an interference pattern on BC. This can only be understood by saying that the wave passed through both holes. A particle must pass through one hole or the other; but this particle-like behaviour can only be demonstrated by changing the experiment and putting detectors next to the holes. The detector will register particles at either H_1 or H_2, never both simultaneously; but now of course the interference pattern is lost. When the interference pattern is present

Fig. 1.11.
Feynman diagram for the force between two electrons.

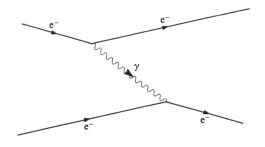

it contains an indication of particle-like behaviour, since it is not actually continuous but is made up like a television picture of dots caused by individual electrons, but it is not possible to understand the pattern on the assumption that each of these electrons must have passed through one slit or the other.

Heisenberg's **uncertainty principle** is a quantitative statement of these ideas. It asserts that it is impossible to set up an experiment which will prepare particles with precise values of both the position x and the momentum p in the x-direction. In general there will be uncertainty in both quantities, so that the position is only known to lie in a finite interval Δx, and the momentum similarly to lie in a finite interval Δp. The uncertainty principle states that these obey the inequality

$$\Delta x \cdot \Delta p \geqslant \frac{h}{4\pi}, \tag{1.25}$$

so that they cannot be reduced to zero simultaneously.

Heisenberg originally explained the uncertainty principle in terms of the uncontrollable change in momentum which is caused by determining the particle's position, as if the particle had definite values of both position and momentum but we were prevented from knowing them by the inevitable effect of our apparatus on the particle. The principles of quantum mechanics which have emerged since then deny that a particle can *have* definite values of position and momentum simultaneously: it can acquire a value for either one of them when an experiment is performed to measure it, but the value acquired will be unpredictable, and it is this forcing of a particular value on the particle that constitutes the uncontrollable effect of an observation on the system observed. As we will see in §2.4, insofar as (1.25) can be derived from the principles of quantum mechanics, the uncertainties Δx and Δp are statistical measures of the scatter in the precise values of x and p that the particle might take up on measurement.

The uncertainty principle can also be regarded as an expression of the conflict between wavelike and particle-like properties. If we use de Broglie's relation (1.24) to express momentum in terms of wavelength, the uncertainty principle (1.25) becomes

$$\Delta x \cdot \Delta k \geqslant \frac{1}{4\pi}, \tag{1.26}$$

where $k = \lambda^{-1}$ is the **wave number** (the number of waves per unit length). Eq. (1.26) can be proved to be automatically satisfied by any continuous function of x if Δx is a measure of the range of x for which the function takes values of an appreciable size, and Δk is a measure of the range of k which must be used if the function is represented as a superposition of waves with wave number k (i.e. functions $\sin 2\pi kx$). Thus (1.26) expresses the impossibility of describing simultaneously particle-like properties (with a function which is localised at a definite position x) and wavelike properties (with a function which has a definite wave number k).

It is intuitively reasonable that there should be some such relationship as this, since the function will need to have non-zero values over a range of values of x (at least a few wavelengths) in order to establish the existence of even an approximate repetition with that wavelength. Similar considerations should apply to a function of time, so that there is a corresponding inequality

$$\Delta t \cdot \Delta v \geqslant \frac{1}{4\pi}, \tag{1.27}$$

where v is the frequency of a phenomenon which varies in time. Using Planck's relation (1.22) to relate frequency to energy would give us

$$\Delta t \cdot \Delta E \geqslant \frac{h}{4\pi}. \tag{1.28}$$

However, it is hard to interpret (1.28) in quantum mechanics, since time is not a property of a particle as position is. The time–energy uncertainty relation does not have the character of a well-defined and reasonably simple statement which can be rigorously deduced from the basic postulates, as the position–momentum uncertainty relation (1.27) does; nevertheless, echoes of (1.28) can be heard throughout quantum mechanics.

The uncertainty relation makes it possible to understand how Fig. 1.11 can describe forces of attraction as well as repulsion. As it is drawn, Fig. 1.11 looks like a picture of repulsion, and the accompanying description seems to allow only for repulsion; both the recoil of the first electron and the impact of the second electron with the photon drive the electrons away from each other. But the attraction between an electron and a positron can now be described as follows: the electron emits a photon with momentum directed away from the positron, and recoils towards the positron. This statement entails a degree of definiteness in the momentum of the photon. By the uncertainty principle there is a corresponding uncertainty in its position: it might be on the other side of the positron, so that it can hit it and knock it towards the electron. This is pictured in Fig. 1.12, in which both wavy lines represent the same photon.

Fig. 1.12.
The force between an electron
and a positron.

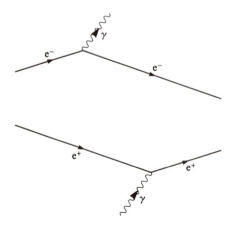

However, this process of attraction is usually represented by the same diagram (Fig. 1.11) as is used for repulsion.

Thus it is possible for a pair of particles to exchange a photon whose momentum can be directed in either direction along the line joining the particles. Which of the two possibilities actually takes place depends, as in classical mechanics, on the nature of the force; in the case of electromagnetism, on whether the particles have the same or opposite signs of electric charge.

In the processes depicted in Figs. 1.11 and 1.12 the photons have some of the potential energy classically associated with the force between the electrons, and so their energy is different from that of freely moving photons. For this reason they are called **virtual** (as opposed to 'real'); they cannot exist on their own, but must eventually be captured by a charged particle so as to form part of one of these processes.† This informal description of forces comes from the formalism of quantum field theory. It will not be possible in this book to show the full workings of this theory, but some of the mathematical features which are represented in Figs. 1.11 and 1.12 will be described in §6.5, and the way in which they are used in quantum field theory will be sketched in Chapter 7.

The basic events in electromagnetic interactions, according to this picture, are the emission and absorption of photons by charged particles. They are each represented by a single vertex in a Feynman diagram. The strength of the force between two particles is given by the intensity of the field, which is proportional to the number of photons; hence the greater the force exerted by (or on) a particle, the more photons it will emit (or absorb). It follows that the charge on a particle is proportional to the probability that it will emit or absorb a photon.

The two basic events are shown in Figs. 1.13(*a, b*). Figs. 1.13(*c, d*) show two further basic events involving electrons, positrons and photons: the annihilation of an electron and a positron to form a photon, and the creation of an electron–positron pair out of a photon. These events cannot occur with real particles (see problem 1.5); one of the particles involved must be a virtual particle which moves on to another basic event to complete the real process, as in Fig. 1.14 which shows a possible sequence of events by which Compton scattering can occur.

The relation between Figs. 1.13(*a*) and 1.13(*c*) is that the outgoing electron line in 13(*a*) is replaced by an incoming positron line in 13(*c*). Such a replacement is always possible at a vertex of a Feynman diagram. This leads us to adopt a new convention in drawing these diagrams. The arrows that we have put on the lines up to now are unnecessary if it is understood that time

---

† It is often stated that in these processes the principle of conservation of energy is violated by an amount ΔE for a time Δt satisfying (1.28). This is incorrect. The only nonconservation in these processes is of *kinetic* energy; as in classical mechanics, the total energy is the sum of the kinetic and potential energies and the principle of conservation of energy holds exactly at all times. It can appear to be violated because the energy does not always have a definite value.

always runs from left to right across the page. Instead, we will use arrows to identify electrons and positrons by putting an arrow on every electron line in the direction of time, and an arrow against the direction of time on every positron line. The basic vertex is then given by the single diagram of Fig. 1.15, but the lines can go in any direction on the page. A complete Feynman diagram will then contain solid lines of which some parts represent an electron and other parts a positron, but the arrows on the line follow each other from one end of the line to the other. Thus the Compton-scattering process of Fig. 1.14 is drawn as in Fig. 1.16(*a*). (It is amusing to think of this process as involving a single electron which at one stage travels backwards in time and appears as a positron). Another possible process for Compton scattering is shown in Fig. 1.16(*b*); both of these processes are included in Fig. 1.16(*c*). An indication of the mathematical justification for these Feynman diagrams will be given in Chapter 6.

A description of this type applies to each of the fundamental forces listed in §1.7; each of them is associated with a particle like the photon, a field quantum

Fig. 1.13.
Basic electromagnetic events.

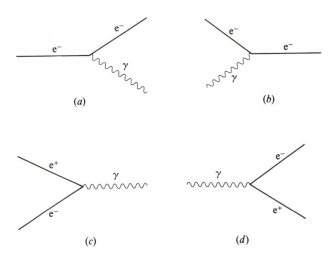

Fig. 1.14.
Compton scattering.

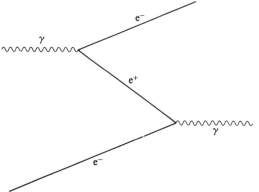

for the force. These particles belong to a class called **bosons**. The particles
introduced in Part A of this chapter, on the other hand, are called **fermions**.
The characteristics of these two classes of particles will be defined in §2.6.

The force of gravity, insofar as it is like the electromagnetic force, is expected
to be associated with a particle called the **graviton** which, like the photon,
travels at the speed of light. But there are two reservations to be made. First,
because of the extreme weakness of the gravitational force between elementary
particles, gravitons have not been and are not likely to be experimentally
observed. Secondly, the classical (i.e. non-quantum) theory of gravitation is *not*
like electromagnetic theory, but involves the subtleties of general relativity.
The problem of combining this with quantum theory has not been fully solved,
and so it is not absolutely certain that gravitons are required by theory (though
it does seem overwhelmingly likely).

The weak force is associated with three bosons, called W^+, W^- and Z^0. The
$W^\pm$ particles carry electric charge, as indicated by the superscript, and when

Fig. 1.15.
The basic electromagnetic
vertex.

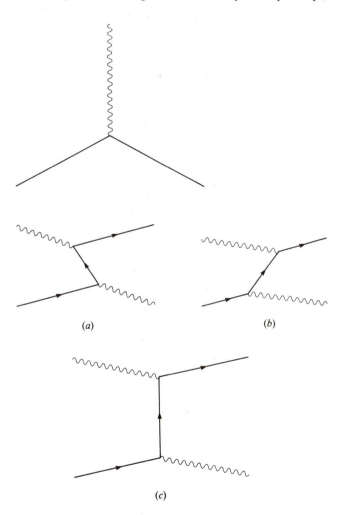

Fig. 1.16.
Compton scattering revisited.

(a)

(b)

(c)

they are emitted or absorbed by a particle they change its identity. The exchange of $W^{\pm}$ particles is responsible for processes like (1.16), as illustrated in Fig. 1.17.

A related description, also involving the W^- particle, can be given for neutron decay (1.13). The process is shown in Fig. 1.18; it involves the creation of an electron–antineutrino pair out of the W^- which is emitted by the neutron.

As we have seen, the strength of a force is proportional to the probability that the particle exerting the force will emit the associated boson. Thus the weakness of the weak force implies that the decay shown in Fig. 1.18 is a rare event, and therefore that the neutron has a long lifetime.

Fig. 1.18 incorporates the same convention as the Feynman diagrams for electromagnetic processes: a backward arrow on a fermion line denotes an antiparticle. It will be seen that Fig. 1.18 is obtained from Fig. 1.17 by rotating the neutrino line.

Since the neutron and the proton are not elementary particles, the processes shown in Figs. 1.17 and 1.18 are not fundamental processes but can be analysed in terms of processes involving quarks, the constituents of the proton and neutron. Now a neutron can be converted into a proton by changing a

Fig. 1.17.
$n + v \rightarrow p + e^-$.

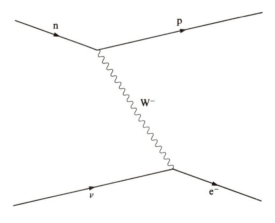

Fig. 1.18.
$n \rightarrow p + e^- + \bar{v}$.

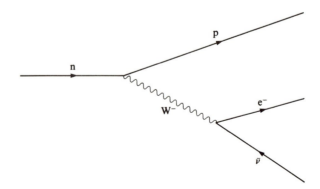

down quark into an up quark; thus the event at the vertex $n \rightarrow p + W^-$ of Figs. 1.17 and 1.18 can be understood as being due to an event $d \rightarrow u + W^-$ at the quark level, as shown in Fig. 1.19. This illustrates how emission or absorption of a $W^{\pm}$ by a quark changes the flavour of the quark.

The basic events of the weak force at the level of quarks and leptons are given by the vertices of Fig. 1.20. Like the basic electromagnetic vertices, they can occur in Feynman diagrams with the lines in any orientation. The vertices involving the Z^0, in which the fermion (quark or lepton) remains unchanged, give rise to forces between fermions in the same way as the similar events involving the photon give rise to electromagnetic forces; in fact there is a close relation between the Z^0 and the photon in the Weinberg–Salam theory of the **electroweak** force. A new feature of the Z^0 processes is that the force is experienced by the neutrino; the observation of this force on the neutrino provided the first experimental confirmation of the Weinberg–Salam theory in 1974. The $W^{\pm}$ and Z^0 particles were observed directly in 1983.

The vertices of Fig. 1.20 involve only the first family of quarks and leptons. Subsequent families are involved in very much the same way: for the second family, for example, the basic vertices can be obtained from those of Fig. 1.20 by substituting the corresponding particles (i.e. μ^- for e^-, ν_μ for ν_e, s for d and c for u). There are also vertices involving particles from different families, like that of Fig. 1.21(*a*). By means of processes in which these vertices occur, the particles of the higher families decay to those of the first family. Thus all hadrons containing a strange, charmed, beautiful or truthful quark eventually decay into nucleons.

In the unified Weinberg–Salam theory, the basic electromagnetic vertex of

Fig. 1.19.
$n \rightarrow p + W^-$ at the quark level.

Fig. 1.20.
The basic weak events: there are also events involving the muon and the τ in place of the electron.

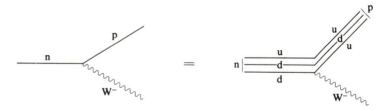

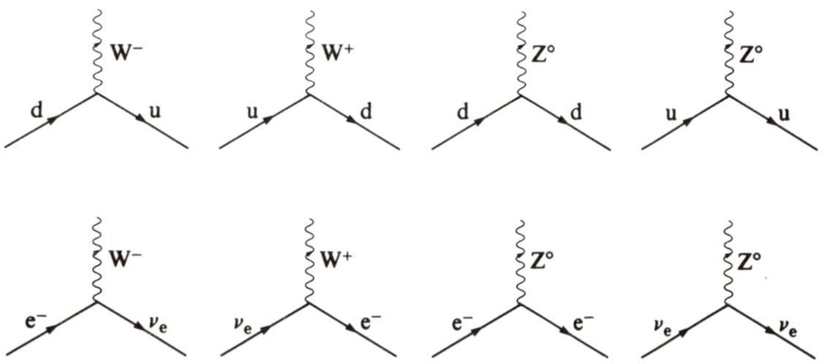

Fig. 1.15 goes together with the weak vertices of Fig. 1.20. Because the $W^{\pm}$ particles have electric charge, there are also vertices involving these particles and a photon, as shown in Fig. 1.21(*b*). There are also vertices in which the photon is replaced by a Z^0, and finally there are four-line vertices in which four bosons meet. The existence of these vertices which involve field quanta only is characteristic of the particular mathematical form of this theory, which is a *non-abelian gauge theory*. The meaning of this phrase will be explained in Chapter 7.

As we have seen, the *strength* of a force is proportional to the probability of the basic event which occurs in the Feynman diagram for that force. The other characteristic of the forces listed in Table 1.3, their *range*, can also be understood in terms of the exchange of field quanta, as in Feynman diagrams. The electromagnetic force, which has infinite range, is due to the exchange of photons, which travel at the speed of light and therefore have zero rest-mass. In 1935 Yukawa suggested that bosons with non-zero rest-mass μ, travelling slower than light, would not be able to move as far as a photon could from the particle that emitted them before being absorbed by another particle. The force they give rise to would therefore have a finite range R; it is given in terms of μ by

$$R = \frac{\hbar}{c\mu} \tag{1.29}$$

where $\hbar = h/2\pi$ (a combination which occurs in quantum mechanics more commonly than h itself). The inverse relation between distance and mass shown in this equation is characteristic of quantum mechanics; in general, short distances correspond to large mass, or equivalently high energy or momentum. The same relationship can be seen in de Broglie's equation (1.24) and the uncertainty relation (1.25).

Yukawa's proposal was made in the course of the search for a theory of the strong force. The range $R = 10^{-15}$ m quoted for this force in Table 1.3 would, according to (1.29), imply that the field quanta should have a mass of $\mu = 200$ MeV. Particles of about this mass, and with other properties appropriate to field quanta of the strong force, were discovered soon after Yukawa's suggestion; these are the **pions** (symbols π^-, π^0, π^+). They give rise to forces between baryons as in the Feynman diagrams of Fig. 1.22. However, at the deeper level of quark structure it appears that the pions are not elementary bosons but are composite objects each made up of a quark and an antiquark.

Fig. 1.21.
The remaining electroweak vertices.

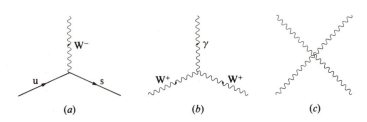

(*a*)   (*b*)   (*c*)

The processes of Fig. 1.22 can be analysed into more complicated processes, shown in Fig. 1.23, whose basic components are forces between quarks. The forces between baryons shown in Fig. 1.22 now appear as a residual effect of the forces between quarks which hold them together inside the baryon, in the same way as the forces between molecules which make ordinary matter cohere (the **van der Waals forces**) are a residual effect of the basic electromagnetic forces which hold electrons and nuclei together inside atoms.

The field quanta of the interquark force, represented by curly lines in Fig. 1.23, are called **gluons**. Emission or absorption of a gluon does not change the flavour of a quark, but it does affect its colour. There are six gluons corresponding to the six possible changes of primary colour; there are also two other gluons which do not change the colour of a quark. The mathematical description of the relation of these eight gluons to the colours of quarks will be given in Chapter 6; we will see that the gluons form the same pattern as the eight baryons shown in Fig. 1.4 (the gluons being arranged in a colour diagram, whereas Fig. 1.4 is a flavour diagram).

The basic vertex of this force as it affects quarks is shown in Fig. 1.24(*a*). The probability of this event, and therefore the strength of the force, is determined by the colour of the quark, in a generalisation of the way in which the strength

Fig. 1.22.
Nuclear forces.

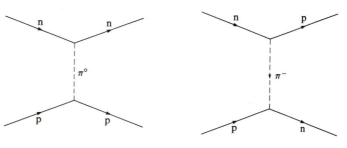

Fig. 1.23.
Nuclear forces at the quark level.

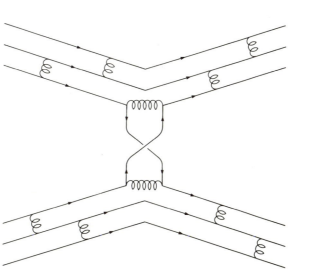

of the electromagnetic force on a particle is determined by its electric charge. Now gluons must also be regarded as coloured (since they carry colour from one quark to another), and so they themselves experience the same force as quarks. This means that there is a three-gluon vertex, Fig. 1.24(*b*). There is also a four-gluon vertex, Fig. 1.24(*c*), like the four-W vertex in the electroweak force: the theory of the interquark force is also a non-abelian gauge theory.

Like single quarks, single gluons have never been observed directly, and it is thought that they never will be because the force that binds them in hadrons does not fall off at large distances, so that they cannot escape. (However, there is speculation about the possible existence of particles consisting of a number of gluons and no quarks: these are called **glueballs**.) This feature of **confinement** is a complicated (and so far unproven) secondary effect in a force which is primarily an inverse-square force with infinite range, so that gluons are massless particles like photons.

The quantum theory of the electromagnetic force, which gave us our paradigm for the description of forces by means of Feynman diagrams, is called **quantum electrodynamics** (or **QED**). By analogy, the theory of the interquark force is called **quantum chromodynamics** or **QCD**. The unified theory of the electroweak force is sometimes called **quantum flavourdynamics**. These forces and their associated bosons (which are called **gauge bosons** because all the forces are described by gauge theories) are summarised in Table 1.4. Tables 1.2 and 1.4 include all particles which are currently regarded as elementary.

Table 1.4. *The fundamental forces and their bosons*

Force	Particles affected	Gauge bosons	Mass	Theoretical	Experimental
				Discovery	
Gravitational	All	Graviton	0	—	—
Electroweak	Leptons	Photon γ	0	1900	1857
	and	$W^{\pm}$	81 GeV	1968	1983
	quarks	Z^0	93 GeV	1968	1983
Quantum chromodynamics	Quarks	8 gluons	0	1974	—

Fig. 1.24.
Basic vertices of the colour force.

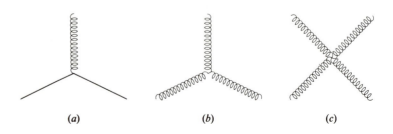

(*a*)   (*b*)   (*c*)

Further reading   Fuller accounts of the structure of matter at the elementary level of this chapter can be found in Einstein & Infeld 1938 (on the general structure of physical theory), Weinberg 1983 (on the first subatomic particles), Davies 1979, Polkinghorne 1979, Dodd 1984 (at a somewhat more advanced level), and Sutton 1985. See also O'Brien 1974. Accounts of quantum mechanics at this level can be found in Andrade e Silva & Lochak 1969, Zukav 1979 and Polkinghorne 1984. Heisenberg's own exposition (Heisenberg 1930) and Feynman's introduction (Feynman *et al.* 1965) are particularly recommended.

Problems on Chapter 1

1. In electrolysis metal ions in solution are attracted to a negatively charged plate (the **cathode**). Faraday's law states that the current is

$$I = \frac{FMv}{At}$$

where M is the mass of metal deposited on the cathode in time t, the integer v is the valency of the metal and A is its atomic weight, and F is a constant which has the value 9.65×10^7 coulomb kg^{-1} if I is measured in coulombs per second. Calculate Avogadro's number (the number of atoms of atomic weight A in a mass of A kg).

2. Calculate the mass of the electron in kg. [A charge of 1 coulomb falling through a potential difference of 1 volt acquires energy of 1 joule. The speed of light is 3×10^8 ms^{-1}. All other necessary information can be found in this chapter.]

3. An electron is suspended against gravity by an electric field between two plates 10 cm apart. Find the voltage between the plates.

4. Obtain the expression (1.8) for the energy of the electron in the hypothetical decay (1.4).

5. Show that an electron and a positron cannot annihilate to form a single real photon while conserving energy and momentum. Show that they can annihilate to form two real photons, and draw the Feynman diagram representing this process.

6. Show that the potential $V(\mathbf{r}) = qe^{-\mu r}/r$ satisfies $\nabla^2 V = \mu^2 V$ except at the origin, and that if R is any region containing the origin,

$$\int_{\partial R} \nabla V \cdot d\mathbf{S} = \mu^2 \int_R V \, d^3\mathbf{r} - 4\pi q$$

where ∂R is the boundary of R.

2

Quantum statics

In this chapter we will develop the mathematical apparatus which is needed to describe a physical system according to quantum mechanics. Our approach is basically deductive: we list the mathematical objects to be used and the properties assumed of them, and give precise statements of the fundamental postulates concerning physical systems and their relation to these mathematical objects. We then derive consequences of the postulates, in a form which can be compared with experimental results. The original postulates are justified to the extent that their consequences fit with the experimental facts – in other words, the only justification for the postulates is that they work. There can be no watertight argument leading from the experimental facts to the basic principles of the theory.

This order of presentation reflects the general logical structure of scientific theories, but it makes an uncomfortable situation for the student, who is asked to accept some pretty peculiar statements without being given any reason to do so at the time. I will try to soften the shock by leading up to the postulates with an indication of how they are suggested by experiment; but it must be remembered that these arguments are different in nature from those following the postulates. The arguments preceding the postulates are not proofs, and you have every right to find them unconvincing. On the other hand, the arguments following the postulates *are* intended to be proofs (and are labelled as such), and if you find them unconvincing then one or the other of us is at fault. The symbol ● announces a proposition which is to be proved, and ■ indicates that the proof is complete. Sometimes the proof of a proposition can be found in the discussion preceding the statement of it, and in that case the symbol ■ is placed immediately after the statement.

The arguments in the first section of this chapter are entirely of the first, heuristic, kind. The precise mathematical development starts in §2.2.

2.1. **Some examples** The description of particles and their behaviour given by quantum mechanics is fundamentally probabilistic. It does not offer definite predictions about what will happen in given physical circumstances, but can only state what events are possible and how probable each of them is. We have already seen a number of phenomena that seem to call for this statistical approach:

1. Radioactivity

As was described on p. 5, it is not possible to predict when a particular radioactive nucleus will decay. Two identical nuclei may survive for different lengths of time before decaying, and no features have been discovered which would distinguish the condition of the shorter-lived nucleus from that of the longer-lived. The only statement which can be made about a particular nucleus is to give the probability that it will decay in a given time interval.

2. Compton scattering

The response of an electron to an evenly spread plane wave of electromagnetic radiation is unpredictable: it may emerge from the encounter travelling in any direction. Note that this is what we would expect if the radiation is described as a collection of particle-like photons; not being able to predict the result of the collision is a result of not knowing the precise positions of the photons. But if the radiation is regarded as a field the probabilistic nature of the effect seems more fundamental.

3. Electron diffraction

Probabilistic ideas can resolve the clash between the concepts of a particle and a field, or wave. When a beam of electrons is passed through a crystal and observed on a photographic plate, the diffraction pattern that it creates is not truly continuous but is made up of a number of dots, just like a newspaper picture, each dot being caused by a single electron striking the plate. The pattern is a statistical effect of the large number of electrons; each individual electron moves like a particle, but its motion can only be predicted to the extent of a statement that it has a high probability of moving to the bright part of the diffraction pattern and a low probability of moving to the dark part. This means that the motion of the electron must be described by a probability distribution in space, i.e. there exists a probability density function $p(\mathbf{r})$ such that the probability that the electron will be found in a small volume dV at the point $\mathbf{r}$ is $p(\mathbf{r}) \, dV$.

Let us examine the phenomenon of interference to see what it suggests about the function $p(\mathbf{r})$. Consider any wave which has a definite frequency v. A **wave** does not necessarily involve the motion of a material medium (we speak of 'crime waves' and 'heat waves'); in general, any quantity that varies in space and time can constitute a wave. We will usually call it a 'disturbance'. To say that it has frequency v is to say that it oscillates in time with frequency v and with an amplitude A and phase ϕ which may vary from point to point. Thus at the point $\mathbf{r}$ the disturbance is given by

$$f(\mathbf{r}, t) = A(\mathbf{r}) \cos (\omega t + \phi(\mathbf{r})) \tag{2.1}$$

where $\omega = 2\pi v$. It is convenient to express this by means of complex numbers:

$$f(\mathbf{r}, t) = \text{Re}\,[\psi(\mathbf{r})e^{-i\omega t}] \tag{2.2}$$

where

$$\psi(\mathbf{r}) = A(\mathbf{r})e^{-i\phi(\mathbf{r})} \tag{2.3}$$

(the choice of $e^{-i\omega t}$ rather than $e^{i\omega t}$ in (2.2) is purely a matter of convention). In many types of wave phenomena the **intensity** of the wave at the point $\mathbf{r}$, i.e. the density of energy in the disturbance, is proportional to the square of the amplitude:

$$I(\mathbf{r}) = kA(\mathbf{r})^2 = k|\psi(\mathbf{r})|^2. \tag{2.4}$$

Now suppose two waves with the same frequency, but with different amplitudes $A_1(\mathbf{r})$ and $A_2(\mathbf{r})$ and phases $\phi_1(\mathbf{r})$ and $\phi_2(\mathbf{r})$, are superimposed in the same region of space. Then the total disturbance is

$$\begin{aligned} f(\mathbf{r}, t) &= A_1(\mathbf{r})\cos(\omega t + \phi_1(\mathbf{r})) + A_2(\mathbf{r})\cos(\omega t + \phi_2(\mathbf{r})) \\ &= \text{Re}\,[\{\psi_1(\mathbf{r}) + \psi_2(\mathbf{r})\}e^{-i\omega t}] \end{aligned} \tag{2.5}$$

where ψ_1 and ψ_2 are defined as in (2.3); and the total intensity is

$$I(\mathbf{r}) = k|\psi_1(\mathbf{r}) + \psi_2(\mathbf{r})|^2. \tag{2.6}$$

This is not equal to the sum of the intensities of the individual waves: it is larger than that when ψ_1 and ψ_2 have the same direction in the complex plane, so that the two oscillations are in phase, and smaller than either when they have opposite directions, so that the two oscillations are out of phase. Thus (2.6) is a mathematical description of interference.

The fact that electrons show interference patterns suggests that underlying the probability density $p(\mathbf{r})$ (which, like $I(\mathbf{r})$, is a positive function describing how much effect the wave has at $\mathbf{r}$) are amplitude and phase functions which can be put together into a complex function $\psi(\mathbf{r})$. Then the probability is given by

$$p(\mathbf{r}) = k|\psi(\mathbf{r})|^2. \tag{2.7}$$

ψ is called the **wave function** of the electron.

From (2.7) we see that the probability that an electron will be found somewhere in a region V is $k\int_V |\psi(\mathbf{r})|^2\,dV$. If we take the region V to be the whole of space then, since the electron must be somewhere, the probability must be 1: this determines the constant k. Thus (2.7) can be replaced by

$$p(\mathbf{r}) = \frac{|\psi(\mathbf{r})|^2}{\int |\psi|^2\,dV} \tag{2.8}$$

where the integral is taken over all space.

For (2.8) to make sense, the integral in the denominator must be finite, i.e. ψ must be **square-integrable**. In the mathematical development of quantum mechanics we will want to repeatedly differentiate ψ and multiply it by the coordinates of $\mathbf{r}$, and assume that the result has the same properties as ψ. We therefore impose the following conditions on ψ:

W1 ψ is square-integrable, i.e. $\int |\psi|^2 \, dV < \infty$.

W2 ψ has uniformly continuous partial derivatives of all orders, which are also square-integrable.

W3 If $f(\mathbf{r})$ is any polynomial in the coordinates of $\mathbf{r}$, then the product $f\psi$ is square-integrable.

By a **wave function** we will always mean a function ψ satisfying **W1–W3**. These conditions are not strictly necessary, but they make it possible to simplify the general discussion. In particular problems it is often convenient to relax these conditions; the justification for this will be discussed in §2.5.

Note that the probability density (2.8) is unchanged if the function $\psi(\mathbf{r})$ is multiplied by any complex constant c. Thus, as far as its position is concerned, the description of the particle given by the wave function $c\psi(\mathbf{r})$ is the same as that given by $\psi(\mathbf{r})$. This makes it permissible to assume that $\int |\psi|^2 \, dV = 1$, which is convenient because it simplifies (2.8). A wave function satisfying this condition is said to be **normalised**.

4. Polarisation of photons

As another example of probabilistic behaviour for which we can give a mathematical description, consider the photons in a beam of polarised light. The classical description of a light ray as an electromagnetic wave is that it consists of oscillating electric and magnetic fields at right angles to each other and to the direction of the ray. We can concentrate on the electric field vector; at every point on the ray, this lies in the plane perpendicular to the direction of the ray. If at each point the electric vector oscillates along a fixed line in this plane, the light is said to be **plane polarised** (because the lines along which the electric vector oscillates at different points on the ray are all parallel to each other and together make up a plane); the direction of the electric vector is called the direction of polarisation.

A **polaroid filter** is a sheet of material consisting of crystals aligned along a certain direction called the **axis** of the filter. The material allows polarised light to pass through if its direction of polarisation is parallel to the axis of the filter, but not if it is perpendicular to it. If the light is polarised along a direction which makes an angle θ with the axis of the polaroid, it destroys the component of the electric vector perpendicular to its axis. Thus if the electric vector of the light just in front of the polaroid oscillates as

$$\mathbf{E}(t) = \mathbf{E}_0 \cos \omega t, \tag{2.9}$$

and if $\mathbf{E}_0 = \mathbf{E}_1 + \mathbf{E}_2$ where $\mathbf{E}_1$ and $\mathbf{E}_2$ are parallel and perpendicular to the axis of the polaroid, then just behind the polaroid the light will have electric vector

$$\mathbf{E}(t) = \mathbf{E}_1 \cos \omega t. \tag{2.10}$$

Thus the light emerging from the polaroid is polarised parallel to its axis. Its intensity, which is proportional to the square of the amplitude of the oscillations of the electric vector, is determined by $|\mathbf{E}_1|^2$; hence (see Fig. 2.1)

$$\frac{\text{intensity of transmitted light}}{\text{intensity of incident light}} = \frac{|\mathbf{E}_1|^2}{|\mathbf{E}_0|^2} = \cos^2 \theta. \tag{2.11}$$

What of the photons in such a beam of light? Each individual photon is either absorbed by the polaroid or passes through it, and since all the photons are identical we cannot predict what will happen to a particular photon; we can only give the probability that it will pass through the polaroid. Since the intensity of the light is proportional to the number of photons in it, the proportion of the photons that pass through is $\cos^2 \theta$; this is therefore[†] the probability that an individual photon will pass through.

The photons which do pass through the polaroid make up a beam with a different direction of polarisation from the original beam. Thus passing through the polaroid changes the state of polarisation of the photons.

The most general state of polarisation of a coherent beam of monochromatic light is not plane polarisation but **elliptical polarisation**. If we choose x- and y-axes in the plane perpendicular to the direction of motion of the light, the x- and y-components of the electric vector at any point oscillate independently, and in general have different phases:

$$E_x(t) = E_{x0} \cos(\omega t + \phi_x), \quad E_y(t) = E_{y0} \cos(\omega t + \phi_y) \tag{2.12}$$

(the light is plane polarised when $\phi_x = \phi_y$). This means that the tip of the vector $\mathbf{E}(t)$ moves round an ellipse in the xy-plane. As in our previous discussion of waves, we can combine the amplitudes and phases to form complex numbers

$$c_1 = E_x e^{-i\phi_x}, \quad c_2 = E_y e^{-i\phi_y}. \tag{2.13}$$

† See the note on probability on p. 41.

Fig. 2.1   axis of polaroid

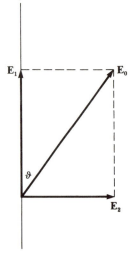

Then the electric vector is given by

$$\mathbf{E}(t) = \mathrm{Re}\,[\boldsymbol{\psi}e^{-i\omega t}] \tag{2.14}$$

where

$$\boldsymbol{\psi} = c_1\mathbf{i} + c_2\mathbf{j}, \tag{2.15}$$

and its intensity is proportional to

$$\boldsymbol{\psi}\cdot\bar{\boldsymbol{\psi}} = |c_1|^2 + |c_2|^2 \tag{2.16}$$

(the bar denotes complex conjugation). The condition of an individual photon in such a beam can be described only to the extent that the whole beam can be described; it is not possible to distinguish different photons. Thus the polarisation of a photon is described by the vector $\mathbf{E}(t)$, or equivalently (the frequency ω being given) by the constant vector $\boldsymbol{\psi}$. But if $\boldsymbol{\psi}$ is multiplied by a real number r the only effect on the beam is to change its intensity, i.e. the number of photons in the beam; thus a photon in the beam described by $r\boldsymbol{\psi}$ is in the same state of polarisation as one described by $\boldsymbol{\psi}$. Also, if $\boldsymbol{\psi}$ is multiplied by a complex number of the form $e^{i\theta}$ the only effect is to change the phase of the oscillation (i.e. the time at which $\mathbf{E}$ passes through a particular point of its ellipse) but not the shape of the ellipse. It is found that it is impossible to determine the phase of the oscillation for a single photon (there is an uncertainty relation between the phase and the number of photons in the beam): the polarisation of a photon is defined purely by the shape traced out by the electric vector. Thus $e^{i\theta}\boldsymbol{\psi}$ describes the same state of polarisation as $\boldsymbol{\psi}$.

We can summarise this mathematical description of the polarisation of a photon travelling in the z-direction as follows:

1. every state of polarisation corresponds to a vector $\boldsymbol{\psi}$ of the type (2.15) (we will call such a vector a **polarisation vector**);
2. if $\boldsymbol{\psi}$ is such a vector and c is any complex number, $c\boldsymbol{\psi}$ represents the same state of polarisation as $\boldsymbol{\psi}$.

A note on probability   Since probability is so fundamental in quantum mechanics, it may seem surprising that we have assumed that the reader already knows what it means, and have not explicitly defined it. It is not in fact possible to give a full definition of probability in elementary physical terms. As with other primitive terms in physics and mathematics, the most one can do is to give a partial, implicit definition by stating the properties that probability has, in axiomatic fashion.

We are concerned with the probability of a physical occurrence α, which we will describe as the result of an experiment E. The probability will depend on the conditions at the beginning of the experiment, i.e. on the state ψ of the system being investigated; it is denoted by $p_E(\alpha|\psi)$. Its properties are as follows:

> **P1** Suppose E is an experiment with possible results $\alpha_1, \ldots, \alpha_n$ which are exhaustive and exclusive, i.e. one and only one of them must

happen. The probability of the result α_i given the initial state ψ is a real number $p_E(\alpha_i | \psi)$ satisfying

(i) $0 \leqslant p_E(\alpha_i | \psi) \leqslant 1;$

(ii) $\sum_{i=1}^{n} p_E(\alpha_i | \psi) = 1.$

P2 If $\alpha_i, \alpha_j, \ldots, \alpha_r$ are different possible results of the experiment E, the probability that one of them will happen given the initial state ψ is

$$p_E(\alpha_i \text{ or } \alpha_j \text{ or } \cdots \text{ or } \alpha_r | \psi) = p_E(\alpha_i | \psi) + \cdots + p_E(\alpha_r | \psi). \tag{2.17}$$

P3 Suppose E and F are **independent** experiments, i.e. there is no causal influence of one of them on the other and no common causal influence on both of them. If $\alpha_1, \ldots, \alpha_m$ are the possible results of E and $\beta_1, \ldots, \beta_j$ are the possible results of F, the probability that the result of E (with initial state ψ) will be α_i and the result of F (with initial state ϕ) will be β_j is

$$p_{E \& F}(\alpha_i \text{ and } \beta_j | \psi \text{ and } \phi) = p_E(\alpha_i | \psi) p_F(\beta_j | \phi). \tag{2.18}$$

P4 Suppose the experiments E and F are linked in such a way that the initial state of F is determined by the result of E. Let ϕ_α be the initial state of F which follows from the result α of E. Then the initial states for E are initial states for the combined experiment $E \& F$, and the probability that E will have the result α and F will have the result β, given initial state ψ, is

$$p_{E \& F}(\alpha \text{ and } \beta | \psi) = p_E(\alpha | \psi) p_F(\beta | \phi_\alpha). \tag{2.19}$$

If the set of possible results is infinite, then the probability must be regarded as a measure on this set, i.e. a function of its subsets, in the way that is described in probability textbooks (e.g. Gnedenko 1968). Thus, for example, if E is the experiment of measuring the position of a particle and the initial state of the particle is specified by a normalised wave function $\psi(\mathbf{r})$, the basic probability statement is

$$p_E(\text{particle will be found in region } V | \psi) = \int_V |\psi(\mathbf{r})|^2 \, dV. \tag{2.20}$$

It follows from **P1–P3** that if the experiment E is repeated independently a large number of times with initial state ψ, the probability is close to 1 that the proportion of experiments which have the result α is close to $p_E(\alpha | \psi)$. In order to explain the quantitative notion of probability we need only add the qualitative statement that if the probability of an event is close to 1 then the event is very likely to happen. This also shows how the probability of a result is to be measured: the experiment is repeated a large number of times, and the probability of the result α is taken to be the proportion of times that this result occurred. The number so obtained is not certain to be exactly the probability being measured, but it is very likely to be close to it (which is as much as can be said of any experimental measurement).

It is often said that the notion of probability is only applicable to repeatable experiments (or that there are two kinds of probability, of which the kind that refers to single events is not relevant to physics). The motivation behind this distinction is the desire to accept only testable statements as meaningful, and as we have seen the only way to test a probability statement about the outcome of an experiment is to repeat the experiment a large number of times and observe the proportions of the various outcomes. But one can only repeat an experiment a finite number of times, and any finite collection of repetitions of the experiment can be regarded as a single experiment to which one must assign probabilities, in defiance of the above injunction. Thus the distinction between the two kinds of probability is logically shaky.

An alternative view of probability (not usually adopted in physics books) is that all probability is of the kind which is appropriate for single events: a statement about the probability of an event is a statement of the speaker's degree of belief that the event will occur, as shown by the odds one will accept in gambling on the event. This is logically unassailable, but is clearly not a suitable concept to use in framing objective laws of nature.

Our attitude in this book will be that a statement about the probability of an outcome of an experiment has consequences both for the degree of belief with which a rational being should believe in this outcome, and for the likely result of a series of repetitions of the experiment, but it cannot be reduced to either of these. It implies them but is not implied by them; they *explain* the probability statement but do not *define* it.

Attempts to define probability more explicitly than this are usually either circular (involving an appeal to 'likeliness') or mysterious (involving concepts like 'propensity' which are no more transparent than that of probability). This is not to say that the question of what probability means, or ought to mean, is not interesting and important; but the answer to that question, if there is one, will not affect the properties of probability that are set out here, and we can proceed without examining the concept any further.

2.2. **State space**

We now have two examples of properties of a quantum-mechanical system, for each of which we have given a mathematical description. In each case there is a mathematical object which describes the state of the properties we are interested in: the state of an electron as far as its position is concerned (ignoring any other properties it may have) is described by the wave function $\psi(\mathbf{r})$, while the state of polarisation of a photon (ignoring its position, and assuming it has a particular direction of motion) is described by the polarisation vector $\boldsymbol{\psi}$. (The word 'state', which has already appeared several times in this book, is to be understood for the time being in its ordinary-language sense, as when a mother complains to her teenage daughter 'Look at the *state* of your room.' It will eventually be given a precise technical meaning, but we are not yet in a position to define this.) The objects $\psi(\mathbf{r})$ and $\boldsymbol{\psi}$ are very different mathematically (as one might expect, position and polarisation being very different physical properties), but they share certain features which must be present in the description of any quantum system. We will now embark on a general description of these features, using the electron and the photon as examples. (Although we are ignoring some properties of each of these, we will regard them as complete physical systems. Thus for the purposes of this section an 'electron' is a particle with position but no internal properties – we will also call such a particle a **simple** particle – and a 'photon' is a particle with polarisation but no position and a fixed direction of motion. In §2.6 we will discuss how to put different properties together.)

Because we want to cover a number of different kinds of quantum system, the discussion will be mathematically abstract: we will not specify exactly what mathematical objects we are talking about, but will be content to say what properties they must have in order to be capable of describing a quantum system. These unspecified mathematical objects will be denoted by symbols like $|\psi\rangle$; the marks $|\rangle$ round the symbol indicate that it describes the state of a quantum system, rather as in mathematical handwriting a wavy line under a symbol indicates that it denotes a vector. Even when we are considering a definite system, so that we know exactly what the mathematical objects are, we will continue to use this general notation. This has the same sort of advantage as vector notation for three-dimensional vectors, in which one is not committed to using any particular set of coordinate axes.

The mathematical objects $|\psi\rangle$ are called **state vectors**. The mathematical properties that they are required to have can be summarised by saying that they form a **complex vector space with an inner product**. The statement that they form a complex vector space means that the following operations are possible:

S1   Scalar multiplication. If $|\psi\rangle$ is a state vector and c is a complex number, there is a state vector $c|\psi\rangle$.

S2   Addition. If $|\psi_1\rangle$ and $|\psi_2\rangle$ are any two state vectors, there is a state vector $|\psi_1\rangle + |\psi_2\rangle$.

These operations obey the rules

(i) $$|\psi_1\rangle + |\psi_2\rangle = |\psi_2\rangle + |\psi_1\rangle;$$ (2.21)

(ii) $$|\psi_1\rangle + (|\psi_2\rangle + |\psi_3\rangle) = (|\psi_1\rangle + |\psi_2\rangle) + |\psi_3\rangle;$$ (2.22)

(iii) $$c(|\psi_1\rangle + |\psi_2\rangle) = c|\psi_1\rangle + c|\psi_2\rangle;$$ (2.23)

(iv) $$(c_1 + c_2)|\psi\rangle = c_1|\psi\rangle + c_2|\psi\rangle;$$ (2.24)

(v) $$c_1(c_2|\psi\rangle) = (c_1 c_2)|\psi\rangle;$$ (2.25)

(vi) If $c = 0$, $c|\psi\rangle$ is always the same object, which is called the **zero vector** and denoted by 0.

The set $\mathscr{S}$ of all state vectors, together with the zero vector, is called **state space**. (Note that the zero vector is not written with the marks $|\rangle$ and is not regarded as a state vector; we will see the reason for this shortly.) The state space for the position states of an electron is the set of all wave functions $\psi(\mathbf{r})$ satisfying **W1**–**W3** (p. 39); the state space for the polarisation states of a photon is the set of all polarisation vectors of the form (2.15). It is left to the reader to verify that these satisfy **S1** and **S2** with the usual meanings for addition and scalar multiplication.

The property **S1** essentially says that state vectors involve complex numbers; we have seen in our two examples that the use of complex numbers reflects the fact that we are concerned with oscillatory phenomena. It is also true in both of these examples that if $|\psi\rangle$ is any state vector, then a scalar multiple $c|\psi\rangle$ describes the same physical state of the system as $|\psi\rangle$. We will assume that this is true for any system. (This may seem to make scalar multiplication unnecessary, but its significance becomes apparent when we consider addition: $|\psi_1\rangle + c|\psi_2\rangle$ does not describe the same state as $|\psi_1\rangle + |\psi_2\rangle$ if $c \neq 1$.) It now follows that the zero vector does not describe any physical state (for, if it did, it would describe *every* state, being a multiple of every state vector). Again, this is a feature of our examples: if the polarisation vector is zero there is no electric field and therefore no photon, and if the wave function is zero there is no probability of finding an electron anywhere, and therefore no electron.

Property **S2**, the possibility of addition, is at the heart of much specifically quantum-mechanical behaviour. It is the possibility of adding, or superimposing, two waves that gives rise to interference phenomena, for example. The statement that this addition is always possible is called the **principle of superposition**. It is closely connected with the probabilistic nature of quantum mechanics; for example, in the case of an electron, if $\psi_1(\mathbf{r})$ is a wave function which is localised in a region V_1 and $\psi_2(\mathbf{r})$ is localised in a separate region V_2, then $\psi_1 + \psi_2$ is a wave function describing a state of the electron in which it might be found in V_1 and might be found in V_2.

In general, $|\psi_1\rangle + |\psi_2\rangle$ describes a state of the system in which it might behave as if it was in the state described by the state vector $|\psi_1\rangle$, and it might behave as if it was in the state described by $|\psi_2\rangle$. By using complex coefficients we can form a state vector $c_1|\psi_1\rangle + c_2|\psi_2\rangle$ which describes a state in which the relative probability of these two forms of behaviour depends on the relative size of the coefficients c_1 and c_2. A precise statement of this will be given in the next section; it will require one further mathematical property of the state space, namely

> **S3 Inner product.** For any two state vectors $|\phi\rangle$ and $|\psi\rangle$ there is a complex number $\langle\phi|\psi\rangle$ called their **inner product** which has the properties

(i)
$$\text{if} \quad |\psi\rangle = c_1|\psi_1\rangle + c_2|\psi_2\rangle, \quad \text{then}$$
$$\langle\phi|\psi\rangle = c_1\langle\phi|\psi\rangle + c_2\langle\phi|\psi\rangle; \tag{2.26}$$

(ii)
$$\langle\psi|\phi\rangle = \overline{\langle\phi|\psi\rangle} \tag{2.27}$$

so that if $|\phi\rangle = c_1|\phi_1\rangle + c_2|\phi_2\rangle$, then

$$\langle\phi|\psi\rangle = \bar{c}_1\langle\phi_1|\psi\rangle + \bar{c}_2\langle\phi_2|\psi\rangle; \tag{2.28}$$

(iii)   for any $|\psi\rangle$, $\langle\psi|\psi\rangle \geqslant 0$; if $\langle\psi|\psi\rangle = 0$, then $|\psi\rangle = 0$. $\quad$ (2.29)

For the two systems we have been considering the inner product is defined as follows:

For the states of motion of an electron, for which the state vectors are wave functions, the inner product is

$$\langle\phi|\psi\rangle = \int \overline{\phi(\mathbf{r})}\psi(\mathbf{r})\, dV, \tag{2.30}$$

the integral being taken over all space. It can be shown (problem 2.1) that this integral is finite if ϕ and ψ are square-integrable.

For the polarisation states of a photon moving in the z-direction, for which the state vectors are polarisation vectors $\boldsymbol{\psi} = c_1\mathbf{i} + c_2\mathbf{j}$, the inner product is

$$\langle\phi|\psi\rangle = \bar{\boldsymbol{\phi}}\cdot\boldsymbol{\psi}. \tag{2.31}$$

From now on we will stick to the ket notation for polarisation states, and we will write the general polarisation state vector as

$$|\psi\rangle = c_1|\phi_x\rangle + c_2|\phi_y\rangle \tag{2.32}$$

where $|\phi_x\rangle$ and $|\phi_y\rangle$ describe the states with polarisation vectors $\mathbf{i}$ and $\mathbf{j}$. Eq. (2.32) has the direct physical meaning that the most general oscillation possible for the electric vector in light which is travelling in the z-direction is a superposition of oscillations along the x- and y-axes with different amplitudes and phases.

The statements in this section about state vectors constitute the first set of postulates of quantum mechanics. We collect them together as

> **Postulate I (the principle of superposition).** The state vectors of a quantum system belong to a complex vector space with an inner product, i.e. they satisfy **S1–S3**. Every non-zero state vector $|\psi\rangle$ describes a physical state of the system, and every non-zero scalar multiple of $|\psi\rangle$ describes the same state. Every state of the system is described by a non-zero state vector and its multiples, but by no other state vector.

It is often convenient to restrict the choice of a state vector to describe a particular state by requiring that it should satisfy $\langle\psi|\psi\rangle = 1$. Such a state vector is said to be **normalised**. This requirement does not determine the state vector uniquely, since it can still be multiplied by a complex number of the form $e^{i\phi}$ (which is known as a **phase factor**).

Two state vectors $|\phi\rangle$ and $|\psi\rangle$ are said to be **orthogonal** if $\langle\phi|\psi\rangle = 0$. A set of state vectors $|\psi_1\rangle$, $|\psi_2\rangle$, ... is **orthonormal** if

$$\langle\psi_i|\psi_j\rangle = \delta_{ij} = \begin{cases} 1 & \text{if } i=j, \\ 0 & \text{if } i\neq j. \end{cases} \tag{2.33}$$

A set of state vectors $|\psi_i\rangle$ is **complete** if any state vector $|\psi\rangle$ can be expressed in the form

$$|\psi\rangle = c_1|\psi_1\rangle + c_2|\psi_2\rangle + \cdots. \tag{2.34}$$

If the $|\psi_i\rangle$ are orthonormal, the coefficients c_i in (2.34) are given by

$$c_i = \langle \psi_i | \psi \rangle. \tag{2.35}$$

We will assume that an orthonormal complete set of state vectors exists for any state space. The number of elements in an orthonormal complete set of state vectors is called the **dimension** of the state space. If the dimension is infinite, so that the right-hand side of (2.34) is an infinite sum, questions of convergence will arise. We will not explore these questions in this book (see the note on Hilbert space at the end of §2.5). Our procedure in the rest of this chapter will be to give proofs which are valid if the state space is finite-dimensional, and then (in §2.5) state without proof what results are true for an infinite-dimensional state space.

2.3. The results of experiments

The second general principle of quantum mechanics gives a statement of the physical interpretation of the state vector by relating it to the properties of the system in the corresponding state. As we have seen, these properties cannot in general be known with certainty; we can only state the probability that an experiment to determine them will have a particular result.

Suppose a photon whose state of polarisation is described by the state vector $|\psi\rangle$ encounters a polaroid filter with its axis in the x-direction, and we ask whether it will pass through. It the photon is polarised parallel to the axis of the filter, so that $|\psi\rangle = |\phi_x\rangle$, then the answer will certainly be yes; if it is polarised perpendicular to the axis, so that $|\psi\rangle = |\phi_y\rangle$, the answer will certainly be no. In general, however, when $|\psi\rangle = c_1|\psi_x\rangle + c_2|\psi_y\rangle$, light passing through the polaroid will have its intensity reduced by a factor

$$p = \frac{|c_1|^2}{|c_1|^2 + |c_2|^2} \tag{2.36}$$

(cf. (2.11)), and so for an individual photon there is a probability p that it will pass through. If it does pass through, it will emerge polarised in the x-direction; this state is particularly associated with this experiment, and is called the 'eigenstate' corresponding to the result of passing through the polaroid. The eigenstate is related to the probability p of this result when the system is in its original state, described by the state vector $|\psi\rangle$; for the eigenstate is described by the state vector $|\phi_x\rangle$, and p is given by

$$p = \frac{|\langle \phi_x | \psi \rangle|^2}{\langle \psi | \psi \rangle}. \tag{2.37}$$

The eigenstate corresponding to the result of not passing through the polaroid is the state of being polarised parallel to the y-axis, described by state vector $|\phi_y\rangle$, and the probability of this result is determined by the inner product $\langle \phi_y | \psi \rangle$. This result, however, differs from the result in which the photon does pass through the polaroid, for if it happens the photon does not emerge in the associated eigenstate; instead it is destroyed by the experiment. This is a complication which we can do without at this stage; we wish to consider systems which can be observed in an experiment, and which, although they may be affected by the observation, remain intact after it. For this reason we will now switch our attention from a polaroid filter to a **doubly refracting** (or **birefringent**) **crystal**. This is a crystal like Iceland spar which has different indices of refraction for light polarised in the x- and y-directions (the x- and y-axes being inherent in the crystal as the axis of a polaroid is inherent in the polaroid). A narrow beam of light entering the crystal at an angle will therefore emerge split into two beams as in Fig. 2.2, one beam B_x polarised in the x-direction and the other B_y polarised in the y-direction. An individual photon travelling in the direction of the original beam will emerge along the path of B_x if its state vector is $|\phi_x\rangle$ and along the path of B_y if its state vector is $|\phi_y\rangle$; if its state vector is $|\psi\rangle$, there is a probability p (given by (2.37)) that it will emerge in the direction B_x, when its state vector will have become $|\phi_x\rangle$, and a probability $1 - p$ that it will emerge in the direction B_y, when its state vector will have become $|\phi_y\rangle$.

The polaroid filter and the doubly refracting crystal illustrate a distinction between two types of experiment which can be observed in classical physics as well as in quantum physics. In an **experiment of the first kind** the experiment determines some property of the system and leaves the system with the property which has been determined; thus if the experiment is repeated immediately afterwards it will give the same result. For example, a determination of the momentum of a particle by measuring its time of flight through a known distance is an experiment of the first kind; so is an experiment to determine whether a photon is polarised along the x-axis or the y-axis by passing it through a doubly refracting crystal. In an **experiment of the second kind** the property being determined is changed by the experiment; it may be possible to calculate the amount of this change, but the significant point is that if the experiment is immediately repeated on the same system, it will give a different result from when it was first performed. For example, a

Fig. 2.2.
Double refraction.

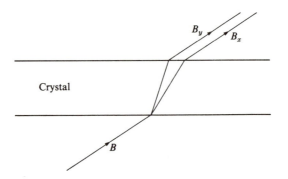

measurement of the momentum of a particle by observing its collision with a known mass is an experiment of the second kind; so is an experiment to determine the direction of polarisâtion of a photon by seeing if it passes through a polaroid filter. For simplicity, we will only consider experiments of the first kind.

In general, an **eigenstate** of an experiment on a quantum system is a state of the system in which the result of the experiment can be predicted with certainty. The eigenstate is **non-degenerate** if it is the only state in which this particular result will occur; in this case the result is also said to be non-degenerate.

We can now give a general statement of what predictions can be made about the result of an experiment.

> **Postulate II.** Let α be a non-degenerate result of an experiment E on a quantum system, and let $|\psi_\alpha\rangle$ be a normalised state vector describing the eigenstate associated with α. Then when the system is in a state $|\psi\rangle$, the probability that the experiment will have the result α is
>
> $$p_E(\alpha\,|\,\psi) = \frac{|\langle\psi_\alpha\,|\,\psi\rangle|^2}{\langle\psi\,|\,\psi\rangle}. \tag{2.38}$$

The following statement is a consequence of the definitions of 'eigenstate' and 'experiment of the first kind', but we state it as a postulate in anticipation of the general case:

> **Postulate III.** Suppose E is an experiment of the first kind. If the result of E is α, then immediately after the experiment the system will be in the eigenstate associated with α.

For a general statement, we must take account of degenerate results, i.e. results which are associated with more than one eigenstate. Let the state vectors $|\psi_1\rangle$ and $|\psi_2\rangle$ describe two eigenstates of an experiment E, in both of which the experiment would give the result α. According to the general considerations of the previous section, a superposition $c_1|\psi_1\rangle + c_2|\psi_2\rangle$ should describe a state which will have the characteristics of one of the two eigenstates, with probabilities determined by the coefficients c_1 and c_2. But as far as E is concerned, the characteristics of the eigenstates are the same, so the superposition state should also have them: it should also be an eigenstate of E with the result α. Thus the set $\mathscr{S}_\alpha$ of all state vectors which describe eigenstates of E with the result α has the property that if $|\psi_1\rangle$ and $|\psi_2\rangle$ are state vectors in $\mathscr{S}_\alpha$ so is $c_1|\psi_1\rangle + c_2|\psi_2\rangle$: $\mathscr{S}_\alpha$ is a **vector subspace** of the state space $\mathscr{S}$. It is called the **eigenspace** of E associated with the result α.

If $|\psi\rangle$ is any state vector, not contained in $\mathscr{S}_\alpha$, it can be written as a sum of a state vector in $\mathscr{S}_\alpha$ and one which is orthogonal to every vector in $\mathscr{S}_\alpha$. To prove this, let $|\psi_1\rangle, |\psi_2\rangle, \ldots$ be an orthonormal complete set of states for $\mathscr{S}_\alpha$, and write

$$|\psi\rangle = \sum_i c_i|\psi_i\rangle + |\psi'\rangle \tag{2.39}$$

where $c_i = \langle \psi_i | \psi \rangle$. Then $|\psi'\rangle$ is orthogonal to each $|\psi_i\rangle$ and therefore to every vector in $\mathscr{S}_\alpha$. The other term, which belongs to $\mathscr{S}_\alpha$, is called the **orthogonal projection** of $|\psi\rangle$ onto $\mathscr{S}_\alpha$ and denoted by $P_\alpha|\psi\rangle$:

$$P_\alpha|\psi\rangle = \sum |\psi_i\rangle\langle \psi_i|\psi\rangle. \tag{2.40}$$

The situation is pictured in Fig. 2.3 by taking $\mathscr{S}_\alpha$ to be a plane in the space of real three-dimensional vectors.

P_α, which is called the (orthogonal) **projection operator** onto $\mathscr{S}_\alpha$, is an example of a very important type of mathematical object. A **linear operator** A on state space is a rule which associates to each state vector $|\psi\rangle$ another state vector $A|\psi\rangle$ in such a way that

$$A(a|\psi\rangle + b|\psi\rangle) = aA|\psi\rangle + bA|\psi\rangle \tag{2.41}$$

for any two state vectors $|\psi\rangle, |\phi\rangle$ and any two complex numbers a, b. It is easy to see that P_α is a linear operator.

In this situation where there is not a unique eigenstate associated with the result α of the experiment E, we must expand on Postulate II by specifying just which element of the eigenspace is to be used for $|\psi_\alpha\rangle$ in (2.38). The state vector $P_\alpha|\psi\rangle$ is the nearest element of $\mathscr{S}_\alpha$ to $|\psi\rangle$ (in a certain precise sense: see problem 2.3), and it seems a reasonable choice. However, $P_\alpha|\psi\rangle$ is not normalised: its inner product with itself is $\sum |c_i|^2$, which is equal to $\langle \psi|P_\alpha|\psi\rangle$. Taking account of this, we arrive at the following general form of the postulate concerning the results of experiments:

> **Postulate II** (*continued*). In general, if the result α is degenerate, then the state vectors describing the associated eigenstates of E form a vector subspace $\mathscr{S}_\alpha$ of state space. The probability of the result α when the system is in the state described by the state vector $|\psi\rangle$ is
>
> $$p_E(\alpha|\psi) = \frac{\langle \psi|P_\alpha|\psi\rangle}{\langle \psi|\psi\rangle} \tag{2.42}$$
>
> where P_α is the orthogonal projection operator onto $\mathscr{S}_\alpha$.

Note that in terms of the orthonormal complete set of states $|\psi_1\rangle, |\psi_2\rangle, \ldots$ for $\mathscr{S}_\alpha$, (2.42) can be written

$$p_E(\alpha|\psi) = \sum_i \frac{|\langle \psi_i|\psi\rangle|^2}{\langle \psi|\psi\rangle}. \tag{2.43}$$

Fig. 2.3.
The projection operator.

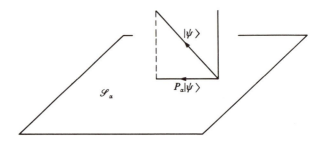

This makes it clear that (2.42) is a generalisation of (2.38) to the case where there are several eigenstates to be considered.

It is also necessary to postulate which of the possible eigenstates the system is in after the experiment, and again $P_\alpha|\psi\rangle$ is the obvious choice:

> **Postulate III** (*continued*). If the result of the experiment is α, then immediately after the experiment the system will be in the state described by the state vector $P_\alpha|\psi\rangle$.

This is called the **projection postulate**†.

From Postulates II and III we can deduce some properties of the state vectors which describe eigenstates of an experiment E (**eigenstate vectors** of E).

●**2.1** If the experiment E always gives an unambiguous result, then

(i) eigenstate vectors associated with different results are orthogonal;

(ii) the eigenstate vectors of E form a complete set.

Proof. (i) Let $|\psi_\alpha\rangle$ and $|\psi_\beta\rangle$ be eigenstate vectors associated with different results α and β. First suppose that α and β are non-degenerate; then when the system is in the eigenstate described by $|\psi_\beta\rangle$ the experiment will certainly give the result β and so the probability of the result α is zero; hence, by (2.38), $\langle\psi_\alpha|\psi_\beta\rangle = 0$.

If the result α is degenerate, then the conclusion of the first paragraph would, by (2.42), be $\langle\psi_\beta|P_\alpha|\psi_\beta\rangle = 0$. But, as we saw in the lines preceding (2.42),

$$\langle\psi_\beta|P_\alpha|\psi_\beta\rangle = \langle\phi|\phi\rangle \quad \text{where } |\phi\rangle = P_\alpha|\psi_\beta\rangle. \tag{2.44}$$

Hence, by the positivity of the inner product, (2.29), $\langle\psi_\beta|P_\alpha|\psi_\beta\rangle = 0$ implies that $P_\alpha|\psi_\beta\rangle = 0$; in other words, $|\psi_\beta\rangle$ is orthogonal to every state vector $|\psi_\alpha\rangle$ in $\mathscr{S}_\alpha$.

(ii) Now let $\alpha_1, \alpha_2, \ldots$ be all the possible results of E. If they are all non-degenerate, let $|\psi_1\rangle, |\psi_2\rangle, \ldots$ be normalised state vectors describing their eigenstates. Let $|\psi\rangle$ be any state vector, and write

$$|\psi'\rangle = |\psi\rangle - (c_1|\psi_1\rangle + c_2|\psi_2\rangle + \cdots) \tag{2.45}$$

where $c_i = \langle\psi_i|\psi\rangle$. Then if $|\psi'\rangle$ were non-zero it would describe a possible state of the system. But because the $|\psi_i\rangle$ are normalised and mutually orthogonal, $\langle\psi_i|\psi'\rangle = 0$ for each i, and so the probability of any result α_i would be zero when the system is in the state described by $|\psi'\rangle$. Since the experiment must give some result, it follows that $|\psi'\rangle$ does not describe a state of the system, and so $|\psi'\rangle = 0$ by Postulate I. Hence $|\psi\rangle$ can be expanded in terms of the eigenstate vectors $|\psi_i\rangle$.

In the general case we can use the projection operators $P_1, P_2, \ldots$ associated with the results $\alpha_1, \alpha_2, \ldots$ and write

$$|\psi'\rangle = |\psi\rangle - (P_1|\psi\rangle + P_2|\psi\rangle + \cdots). \tag{2.46}$$

By part (i), the $P_i|\psi\rangle$ are all orthogonal to each other; it follows that

$$P_i P_j|\psi\rangle = 0 \quad \text{if } i \neq j, \tag{2.47}$$

† Probably first formulated by Dirac, though it is often erroneously attributed to von Neumann.

while

$$P_i^2|\psi\rangle = P_i|\psi\rangle. \tag{2.48}$$

Using these in (2.46) gives $P_i|\psi\rangle = 0$ for each i. Hence, by (2.43), the probability of any result α_i is zero and so, as before, $|\psi'\rangle = 0$. Thus $|\psi\rangle$ can be written as a sum of eigenstate vectors, which therefore constitute a complete set. ∎

From now on we will follow common usage, which is careless of the distinction between states and state vectors. Thus we will talk of 'orthogonal states', 'a complete set of states', and say that a system is 'in the state $|\psi\rangle$'. In circumstances where this is likely to lead to confusion we will revert to the more pedantic style of the last two sections.

2.4. Observables

An **observable** is a physical quantity which can be measured by an experiment, of the type considered in §2.3, whose result is a real number (the measured value of the observable). The possible results are called **eigenvalues**; each is associated with an eigenstate (or possibly a set of eigenstates). From these eigenvalues and eigenstates we can construct for any observable A a corresponding linear operator $\hat{A}$; this will then describe the observable mathematically just as the state vectors describe the states of the system.

Let $|\psi_1\rangle, |\psi_2\rangle, \ldots$ be a complete set of eigenstates of the observable A, and let $\alpha_1, \alpha_2, \ldots$ be the corresponding eigenvalues. We define the operator $\hat{A}$ as follows:

$$\left. \begin{array}{ll} \text{If} & |\psi\rangle = c_1|\psi_1\rangle + c_2|\psi_2\rangle + \cdots, \\ \text{then} & \hat{A}|\psi\rangle = c_1\alpha_1|\psi_1\rangle + c_2\alpha_2|\psi_2\rangle + \cdots. \end{array} \right\} \tag{2.49}$$

Then $|\psi_i\rangle$ and α_i are eigenvectors and eigenvalues of $\hat{A}$ in the mathematical sense, i.e.

$$\hat{A}|\psi_i\rangle = \alpha_i|\psi_i\rangle. \tag{2.50}$$

A linear operator is called **hermitian** if

$$\langle\psi|\hat{A}|\phi\rangle = \overline{\langle\phi|\hat{A}|\psi\rangle} \tag{2.51}$$

for all states $|\phi\rangle$ and $|\psi\rangle$. We now prove that this is a property of the operator we have constructed.

●2.2   If A is any observable, the corresponding operator $\hat{A}$, defined by (2.49), is hermitian.

Proof. We can assume that the eigenstates $|\psi_i\rangle$ in (2.49) are all orthogonal to each other, by ●2.1. Take any two states $|\phi\rangle$ and $|\psi\rangle$, and write

$$|\phi\rangle = \sum b_i|\psi_i\rangle, \quad |\psi\rangle = \sum c_i|\psi_i\rangle. \tag{2.52}$$

Using the properties of the inner product, (2.26) and (2.28), we have

$$\langle\phi|\hat{A}|\psi\rangle = \sum \alpha_i \bar{b}_i c_i$$

and

$$\langle\psi|\hat{A}|\phi\rangle = \sum \alpha_i \bar{c}_i b_i. \tag{2.53}$$

Since the eigenvalues α_i are real, (2.51) follows. ∎

In a finite-dimensional vector space there is a converse to ●2.2: it is a theorem of linear algebra that in such a space every hermitian operator has real eigenvalues and a complete set of eigenstates, and therefore is qualified to describe an observable. (A similar theorem is valid if the vector space is infinite-dimensional, but it must be stated more carefully; see §2.5.)

As an example of an observable, suppose we define P_x for a beam of light to be the proportion of the light that will pass through a polaroid with its axis in the x-direction. Classically, P_x can take any value between 0 and 1; for plane polarised light whose direction of polarisation makes an angle θ with the x-axis, P_x has the value $\cos^2 \theta$. For a photon, however, which passes through the polaroid either entirely or not at all, the proportion can only be 1 or 0. Thus the eigenvalues of P_x are 1 and 0. The operator $\hat{P}$ is given, according to (2.49), by

$$P_x(c_1|\phi_x\rangle + c_2|\phi_y\rangle) = c_1|\phi_x\rangle. \tag{2.54}$$

Together with an observable A, we can also consider $f(A)$, where f is any function of a real variable; $f(A)$ is measured by measuring A and applying the function f to the result. Thus $f(A)$ has the same eigenstates as A, and eigenvalues $f(\alpha)$ where α is an eigenvalue of A. If $f(A) = A^n$, then the corresponding operator has the same effect on an eigenstate as applying $\hat{A}$ n times:

$$\widehat{A^n}|\psi_i\rangle = \alpha_i{}^n|\psi_i\rangle = \hat{A}^n|\psi_i\rangle. \tag{2.55}$$

Since the eigenstates are a complete set, it follows that $\widehat{A^n} = \hat{A}^n$. More generally, if f is a polynomial function we have

$$\widehat{f(A)}|\psi_i\rangle = f(\alpha_i)|\psi_i\rangle = f(\hat{A})|\psi_i\rangle \tag{2.56}$$

and therefore

$$\widehat{f(A)} = f(\hat{A}). \tag{2.57}$$

If f is not a polynomial, (2.57) serves to define $f(\hat{A})$ for any hermitian operator $\hat{A}$.

Observables and their corresponding operators are often called **q-numbers** to emphasise the fact that their algebra is different from that of ordinary complex numbers, which by contrast are called **c-numbers**. In particular, the product of two q-numbers may depend on the order in which they are multiplied. Two operators $\hat{A}$ and $\hat{B}$ are said to **commute** if this is not so, i.e. if

$$\hat{A}\hat{B} = \hat{B}\hat{A}. \tag{2.58}$$

A c-number c can be regarded as a particular kind of operator, whose effect is to multiply every state vector by c. This operator is a multiple of the identity operator; it commutes with all other operators.

Operators and matrices   It is often convenient to represent an operator on the state space of a photon by a 2×2 matrix. A state $|\psi\rangle = c_1|\phi_x\rangle + c_2|\phi_y\rangle$ corresponds to a two-component column vector

$$\mathbf{c} = \begin{pmatrix} c_1 \\ c_2 \end{pmatrix};$$

then if $\hat{A}|\psi\rangle = c_1'|\phi_x\rangle + c_2'|\phi_y\rangle$ corresponds to a column vector $\mathbf{c}'$, the operator $\hat{A}$ corresponds to a matrix $\mathbf{A}$ such that

$$\mathbf{c}' = \mathbf{Ac}. \tag{2.59}$$

For example, the observable P_x corresponds to the matrix

$$\mathbf{P_x} = \begin{pmatrix} 1 & 0 \\ 0 & 0 \end{pmatrix}. \tag{2.60}$$

In general, for any system with a finite-dimensional state space a choice of complete set $|\psi_1\rangle, \ldots, |\psi_n\rangle$ sets up a correspondence between states and column vectors: if $|\psi\rangle = c_1|\psi_1\rangle + \cdots + c_n|\psi_n\rangle$, the corresponding column vector has the coefficients c_i as its entries. Then any operator $\hat{A}$ corresponds to a matrix $\mathbf{A}$ which relates the column vectors corresponding to $|\psi\rangle$ and $\hat{A}|\psi\rangle$ as in (2.59). Alternatively, $\mathbf{A} = (a_{ij})$ can be defined by the equation

$$\hat{A}|\psi_j\rangle = \sum_{i=1}^{n} a_{ij}|\psi_i\rangle \quad (j = 1, \ldots, n). \tag{2.61}$$

If the complete set $|\psi_1\rangle, \ldots, |\psi_n\rangle$ is orthonormal, the matrix elements a_{ij} are given by

$$a_{ij} = \langle \psi_i|\hat{A}|\psi_j\rangle. \tag{2.62}$$

Eqs. (2.61) and (2.62) are valid even when the state space is not finite-dimensional. They show that the operator $\hat{A}$ is determined by the numbers $\langle \psi_i|\hat{A}|\psi_j\rangle$ where $|\psi_1\rangle, |\psi_2\rangle, \ldots$ is any orthonormal complete set. These numbers are called matrix elements of A. More generally, in quantum mechanics the phrase **matrix element of A** is used to denote any expression $\langle \phi|\hat{A}|\psi\rangle$ where $|\phi\rangle$ and $|\psi\rangle$ are any two states.

Hermitian conjugation   Given any linear operator X on state space, we define an operator $X^\dagger$ called the **hermitian conjugate** of X by

$$\langle \phi|X^\dagger|\psi\rangle = \overline{\langle \psi|X|\phi\rangle}. \tag{2.63}$$

An operator is determined by its matrix elements, so this is sufficient to define $X^\dagger$. Then

$$X \text{ is called } \textbf{hermitian} \quad \text{if } X = X^\dagger; \tag{2.64}$$

$$X \text{ is called } \textbf{unitary} \quad \text{if } X^\dagger X = 1. \tag{2.65}$$

Hermitian and unitary operators are analogous to real numbers and complex numbers of modulus 1, respectively; in particular their eigenvalues are such numbers. This is shown in an appendix to this chapter, where some other properties of these operators are collected.

We can also define a hermitian conjugate of a state vector $|\psi\rangle$; it is the mapping which takes any state vector $|\phi\rangle$ to the complex number $\langle \psi|\phi\rangle$. This mapping is denoted by $\langle \psi|$ and called a **bra** vector, because it is the left-hand part of a bracket; the ordinary state vector $|\psi\rangle$, on the other hand, is called a **ket** vector.

Any linear operator X on state space can be applied to a bra vector $\langle\psi|$ to produce another bra vector $\langle\psi|X$ which, like $\langle\psi|$, is defined by completing the bracket: $\langle\psi|X$ is the mapping which takes any state vector $|\phi\rangle$ to the c-number $\langle\psi|X|\phi\rangle$.

If we define the hermitian conjugate of a c-number to be its complex conjugate, we can give a general rule for the hermitian conjugate of any product of a number of objects, which may be c-numbers, ket vectors, bra vectors or operators. The rule is:

> The hermitian conjugate of the product of any number of objects is the product of their hermitian conjugates in the reverse order.

This rule incorporates (2.63) and all the following, in which c is any c-number, $|\psi\rangle$ is any state vector and X is any operator.

(a) $\quad (cX)^{\dagger}=\bar{c}X^{\dagger};$ (2.66a)

(b) $\quad (XY)^{\dagger}=Y^{\dagger}X^{\dagger};$ (2.66b)

(c) $\quad$ The hermitian conjugate of $c|\psi\rangle$ is $\bar{c}\langle\psi|;$ (2.66c)

(d) $\quad$ The hermitian conjugate of $X|\psi\rangle$ is $\langle\psi|X^{\dagger}.$ (2.66d)

Proof. (a) follows directly from (2.63) and (c) from (2.28). To prove (d), let $|\chi\rangle = X|\psi\rangle$; then

$$\langle\chi\,|\,\phi\rangle=\overline{\langle\phi\,|\,\chi\rangle}=\overline{\langle\phi|X|\psi\rangle}=\langle\psi|X^{\dagger}|\phi\rangle.$$

Since this is true for all $|\phi\rangle$,

$$\langle\chi|=\langle\psi|X^{\dagger};$$

and $\langle\chi|$ is the hermitian conjugate of $|\chi\rangle = X|\psi\rangle$. Finally, to prove (b), let $|\omega\rangle = Y|\phi\rangle$; then

$$\langle\phi|(XY)^{\dagger}|\psi\rangle=\overline{\langle\psi|XY|\phi\rangle}=\overline{\langle\psi|X|\omega\rangle}=\langle\omega|X^{\dagger}|\psi\rangle=\langle\phi|Y^{\dagger}X^{\dagger}|\psi\rangle,$$

(d) having been used in the last step. Since this is true for all $|\phi\rangle$ and $|\psi\rangle$, (b) follows. ∎

This can all be understood in terms of matrices. As we have seen, state vectors can be thought of as column vectors and operators as square matrices: then the product $X|\psi\rangle$ is given by matrix multiplication. Now think of a bra vector as a row vector, with the products $\langle\phi|X|\psi\rangle$ and $\langle\phi|\psi\rangle$ again given by matrix multiplication. Then hermitian conjugation is given by the combined operation of transposing the matrix and taking the complex conjugate of every entry. Thus we have the following list of correspondences between matrix notation and the notation of quantum mechanics:

Quantum mechanics	Matrices
State vector $\|\psi\rangle$	Column vector $\mathbf{a}$
Bra vector $\langle\psi\|$	Row vector $\bar{\mathbf{a}}^{\mathsf{T}}$
Inner product $\langle\phi\|\psi\rangle$	$\bar{\mathbf{b}}^{\mathsf{T}}\mathbf{a}$
Operator X	Square matrix $\mathbf{X}$
Hermitian conjugate $X^{\dagger}$	$\bar{\mathbf{X}}^{\mathsf{T}}$

The rule for the hermitian conjugate of a product in quantum mechanics now follows from the rule for the transpose of a product in matrix algebra.

Using bra vectors, we can write down a useful formula which describes the expansion of a state vector in terms of a complete set. Given a bra vector $\langle\phi|$ and a ket vector $|\psi\rangle$, we can form an operator $|\psi\rangle\langle\phi|$ which is defined by

$$(|\psi\rangle\langle\phi|)|\chi\rangle = \langle\phi\,|\,\chi\rangle|\psi\rangle. \tag{2.67}$$

Now let $|\psi_i\rangle$ be a complete orthonormal set of states. Then the fact that any state $|\psi\rangle$ can be expanded in terms of the $|\psi_i\rangle$, and the formula for the coefficients in this expansion (see (2.34) and (2.35)), can be expressed as

$$\sum_i |\psi_i\rangle\langle\psi_i| = 1 \tag{2.68}$$

where **1** denotes the identity operator.

Statistical properties of observables   When the system is in a state $|\psi\rangle$, the value obtained in a measurement of an observable A is a random variable with probability distribution given by (2.42), α now being understood as a value of A. The mean of this distribution, i.e. the average value obtained in a large number of measurements on identical systems in this state, is called the **expectation value** of A, and denoted by $\langle A\rangle$. The standard deviation, which is a measure of the spread of the results, is called the **uncertainty** in A and denoted by ΔA; it is defined to be the square root of the expectation value of $(A-\langle A\rangle)^2$.

The expectation value and the uncertainty can be expressed in terms of the operator $\hat{A}$ and the state vector $|\psi\rangle$ as follows:

●**2.3** If $|\psi\rangle$ is normalised, $\langle A\rangle$ and ΔA are given by

$$\langle A\rangle = \langle\psi|\hat{A}|\psi\rangle, \tag{2.69}$$
$$\Delta A^2 = \langle\psi|\hat{A}^2|\psi\rangle - \langle\psi|\hat{A}|\psi\rangle^2. \tag{2.70}$$

Proof. Let $|\psi_i\rangle$ be a complete set of eigenstates of A, with eigenvalues α_i, and expand $|\psi\rangle$ as a sum $\sum c_i|\psi_i\rangle$. Then $c_i = \langle\psi_i|\psi\rangle$, so the probability that a measurement of A will give the value α_i is $|c_i|^2$ (unless some of the α_i are equal, in which case the probability is the sum of the corresponding $|c_i|^2$). Thus the mean value of A is

$$\langle A\rangle = \sum_\alpha \alpha p_A(\alpha\,|\,\psi)$$

$$= \sum_i \alpha_i|c_i|^2.$$

Using (2.53) with $|\phi\rangle = |\psi\rangle$, we see that this is $\langle\psi|\hat{A}|\psi\rangle$.

To find ΔA, we must apply this result to find the expectation value of $(A-\langle A\rangle)^2$. This is a function of A, and the corresponding operator is

$(\hat{A} - \langle A \rangle)^2$; hence

$$\Delta A^2 = \langle \psi | (\hat{A} - \langle A \rangle)^2 | \psi \rangle$$
$$= \langle \psi | \hat{A}^2 | \psi \rangle - 2 \langle A \rangle \langle \psi | \hat{A} | \psi \rangle + \langle \hat{A} \rangle^2 \langle \psi | \psi \rangle$$
$$= \langle \psi | \hat{A}^2 | \psi \rangle - \langle A \rangle^2$$

since $| \psi \rangle$ is normalised. This proves (2.70). ∎

As an example of an expectation value, consider the observable P_x for a photon. When the photon is plane polarised at an angle θ to the x-axis, its state is $| \psi \rangle = \cos \theta | \phi_x \rangle + \sin \theta | \phi_y \rangle$, and then the expectation value of P_x is

$$\langle P_x \rangle = \langle \psi | P_x | \psi \rangle = \cos^2 \theta, \tag{2.71}$$

which is the classical value of P_x for light in this state.

Compatibility of observables

Because of the way that a measurement of an observable affects the system, it may not be possible to measure two different observables simultaneously: it may be necessary to use different experiments for the two observables, and measurement of one may change the value of the other. For example, consider the observable P_x of a photon, which has the value 1 if the photon is polarised parallel to the x-axis and 0 if it is polarised in the perpendicular direction; and let P_θ be the similar observable defined with the x-axis replaced by an axis at an angle θ to it. Then if the photon has the value 1 for P_x, a measurement of P_θ will result in the photon being polarised either parallel to or perpendicular to the inclined axis, and then a measurement of P_x will not necessarily give the value 1 again, for the photon will not necessarily pass through a polaroid with its axis parallel to the x-axis.

Two observables are said to be **compatible** if measurement of one does not affect the value of the other in this way. Thus the condition for A and B to be compatible is that if three measurements are performed, first of A, then of B, and then of A again, the second measurement of A will give the same value as the first; and vice versa.

The statement that two observables are compatible can be expressed very simply in terms of the corresponding operators:

⬤**2.4** A and B are compatible if and only if $\hat{A}$ and $\hat{B}$ commute.

Proof. We will show that both statements are equivalent to the statement that there is a complete set of states which are simultaneously eigenstates of A and eigenstates of B.

First, suppose that A and B are compatible and that three measurements are performed as above. After the first measurement of A, the system will be in an eigenstate $| \psi_\alpha \rangle$ of A with the measured eigenvalue α. After the measurement of B, the system will have jumped into an eigenstate of B; but the second measurement of A is certain to give the value α again if A and B are compatible, so the system is still in an eigenstate of A – not necessarily the same one as before, but one with the same eigenvalue α. Now let $| \phi_1 \rangle$, $| \phi_2 \rangle$, ... be a complete set of eigenstates of B; then we can expand $|\psi_\alpha\rangle$ as

$$|\psi_\alpha\rangle = c_1|\phi_1\rangle + c_2|\phi_2\rangle + \cdots, \tag{2.72}$$

and each c_i will be non-zero only if there is a chance that the measurement of B will leave the system in the corresponding state $|\phi_i\rangle$. In that case, as we have just seen, $|\phi_i\rangle$ must also be an eigenstate of A with eigenvalue α. Thus the subspace $\mathscr{S}_\alpha$ of eigenstates of A with eigenvalue α has a complete set of states consisting of states which are simultaneously eigenstates of B. Since any state can be written as a sum of states from the various subspaces $\mathscr{S}_\alpha$, it follows that the whole state space has a complete set of states consisting of simultaneous eigenstates of A and B.

Conversely, if there is a complete set of such states, then each subspace $\mathscr{S}_\alpha$ has a complete set of eigenstates of B. Hence any state in $\mathscr{S}_\alpha$ remains in $\mathscr{S}_\alpha$ after being projected onto a subspace of eigenstates of B. Thus if the first measurement of A gives the result α, putting the system in the subspace $\mathscr{S}_\alpha$, the measurement of B will leave it in the subspace $\mathscr{S}_\alpha$ and so the second measurement of A will also give the result α. So if such a complete set exists, A and B are compatible.

Now $\hat{A}\hat{B}$ and $\hat{B}\hat{A}$ have the same effect on any simultaneous eigenstate of A and B. Hence if there is a complete set of such states, $\hat{A}\hat{B} = \hat{B}\hat{A}$. Conversely, suppose $\hat{A}\hat{B} = \hat{B}\hat{A}$ and let $|\psi_\alpha\rangle$ be an eigenstate of A with eigenvalue α. Then

$$\hat{A}\hat{B}|\psi_\alpha\rangle = \hat{B}\hat{A}|\psi_\alpha\rangle = \alpha\hat{B}|\psi_\alpha\rangle;$$

so $\hat{B}|\psi_\alpha\rangle$ is also an eigenstate of A with eigenvalue α. Thus $\hat{B}$ acts as a hermitian operator inside the subspace $\mathscr{S}_\alpha$, and so $\mathscr{S}_\alpha$ has a complete set of eigenstates of B (if $\mathscr{S}_\alpha$ is finite-dimensional; the case of infinite-dimensional state spaces will be discussed in the next section). It follows, as before, that the whole state space has a complete set of simultaneous eigenstates of A and B. ■

Note the property of commuting operators which was used in this proof:

Suppose $\hat{A}$ and $\hat{B}$ commute. Then if $|\psi\rangle$ is an eigenstate of $\hat{A}$ with eigenvalue α, so is $\hat{B}|\psi\rangle$. Thus $\hat{B}$ acts as an operator on the eigenspace $\mathscr{S}_\alpha$. (2.73)

$\mathscr{S}_\alpha$ is said to be **invariant** under $\hat{B}$.

The proof of ●2.4 shows how degenerate eigenstates of an observable, i.e. eigenstates with the same eigenvalue, can be distinguished by means of the value of a second observable which is compatible with the first. If two simultaneous eigenstates of A and B have the same eigenvalues for both of them, then there must be a third observable C which is compatible with both of them, and which has different eigenvalues in these states or in some linear combinations of them (for if no experiment could distinguish between the states they would be the same physical state). If two states have the same eigenvalues for all three of A, B and C, there must be a fourth observable D which has different eigenvalues; and so on. A set of compatible observables is said to be **complete** if no two states have the same eigenvalues for all of them, so that knowledge of the eigenvalues of all the observables specifies a state uniquely. This can be used to give a precise definition: a **state** of a quantum system is a set of values for a complete set of compatible observables.

The **commutator** of two operators $\hat{A}$ and $\hat{B}$ is the operator

$$[\hat{A}, \hat{B}] = \hat{A}\hat{B} - \hat{B}\hat{A}. \tag{2.74}$$

It follows from ●2.3 that if $\hat{A}$ and $\hat{B}$ are hermitian, so is $i[\hat{A}, \hat{B}]$. We can therefore take this to describe an observable, which we denote by the same symbol without the hats, viz. $i[A, B]$. This observable is a measure of the extent to which A and B fail to be compatible, as is shown by the following relation.

●**2.5** **The generalised uncertainty relation.** In any state of the system,

$$\Delta A \cdot \Delta B \geqslant \tfrac{1}{2} |\langle i[A, B] \rangle|. \tag{2.75}$$

Proof. Let $|\psi\rangle$ be the state being considered, let $A_1 = A - \langle A \rangle$ and $B_1 = B - \langle B \rangle$, and let $|\phi\rangle$ be the state

$$|\phi\rangle = \hat{A}_1|\psi\rangle + ix\hat{B}_1|\psi\rangle \tag{2.76}$$

where x is an arbitrary real number. Then

$$\langle \phi| = \langle \psi|\hat{A}_1 - ix\langle \psi|\hat{B}_1 \tag{2.77}$$

since $\hat{A}_1$ and $\hat{B}_1$ are hermitian. Putting (2.76) and (2.77) together,

$$\langle \phi|\phi\rangle = \langle \psi|\hat{A}_1{}^2|\psi\rangle - x\langle \psi|i[\hat{A}_1, \hat{B}_1]|\psi\rangle + x^2\langle \psi|\hat{B}_1{}^2|\psi\rangle.$$

But $[\hat{A}_1, \hat{B}_1] = [\hat{A}, \hat{B}]$ since $\langle A \rangle$ and $\langle B \rangle$ are c-numbers and commute with all operators. Hence

$$\langle \phi|\phi\rangle = \langle A_1{}^2\rangle - x\langle i[A, B]\rangle + x^2\langle B_1{}^2\rangle$$
$$= \Delta A^2 - x\langle i[A, B]\rangle + x^2\Delta B^2, \tag{2.78}$$

using the definition of uncertainty. Now $\langle \phi|\phi\rangle \geqslant 0$ for all x, so the quadratic expression (2.76) has either no zeros or equal zeros; hence

$$\langle i[A, B]\rangle^2 \leqslant 4\Delta A^2 \Delta B^2$$

(i.e. '$b^2 \leqslant 4ac$'), which proves (2.75). ■

Knowledge of the commutators of observables is often sufficient to determine their properties. The commutator satisfies the following algebraic identities:

$$[B, A] = -[A, B]; \tag{2.79}$$

$$[A, BC] = [A, B]C + B[A, C]; \tag{2.80}$$

$$[A, [B, C]] + [B, [C, A]] + [C, [A, B]] = 0. \tag{2.81}$$

(2.81) is known as the **Jacobi identity**.

2.5. Observables of a particle moving in space

The previous section was expressed in terms appropriate to a finite-dimensional state space, and must be adapted if it is to apply to an infinite-dimensional space like the space of wave functions. To introduce the essential ideas, it is sufficient to consider functions of one variable, i.e. wave functions for a particle moving in one dimension. Thus we consider the space $\mathcal{W}$ of complex-valued functions ψ of one real variable x which satisfy one-dimensional versions of **W1**–**W3** (p. 39): ψ must be infinitely differentiable and square-integrable, i.e. $\int_{-\infty}^{\infty} |\psi(x)|^2 \, dx$ exists, and $x^n\psi(x)$ and $\psi^{(n)}(x)$ also have these properties for all n. The inner product in $\mathcal{W}$ is given by

$$\langle \phi \,|\, \psi \rangle = \int_{-\infty}^{\infty} \overline{\phi(x)}\,\psi(x) \, dx. \tag{2.82}$$

We define hermitian operators $\hat{X}$ and $\hat{K}$ by

$$(\hat{X}\psi)(x) = x\psi(x), \tag{2.83}$$

$$\hat{K}\psi = -i\frac{d\psi}{dx}. \tag{2.84}$$

It is clear that $\hat{X}$ is hermitian with respect to the inner product (2.82), and $\hat{K}$ can be seen to be hermitian by integrating by parts and using the fact that $\psi(x)$ must tend to 0 as $x \rightarrow \pm\infty$ if $\int |\psi(x)|^2 \, dx$ is to be finite. However, neither of these operators has a complete set of eigenvectors; in fact, they do not have so much as a single eigenvector. It is, of course, possible to find a function $\varepsilon_k(x)$ such that $\hat{K}\varepsilon_k = k\varepsilon_k$, namely $\varepsilon_k(x) = e^{ikx}$, but this is not relevant in this context because it does not belong to the space $\mathcal{W}$; and there are no continuous functions at all which satisfy $\hat{X}\psi = a\psi$ for any number a.

The way round this difficulty is to regard the function $\varepsilon_k(x)$ not as a state vector but as a bra vector, namely as the map $\langle \varepsilon_k |$ which takes any wave function ψ to the complex number

$$\langle \varepsilon_k | \psi \rangle = \int_{-\infty}^{\infty} \overline{\varepsilon_k(x)}\,\psi(x) \, dx. \tag{2.85}$$

This map exists, i.e. the integral converges, for all real k. Now we have

$$\langle \varepsilon_k | \hat{K} = k\langle \varepsilon_k | \tag{2.86}$$

in the sense that

$$\langle \varepsilon_k | \hat{K} | \psi \rangle = k\langle \varepsilon_k \,|\, \psi \rangle \quad \text{for all } |\psi\rangle \in \mathcal{W}.$$

Because of (2.85), $\langle \varepsilon_k |$ can be called an **eigenbra** of $\hat{K}$. We can also find eigenbras of $\hat{X}$, namely $\langle \delta_a |$ defined by

$$\langle \delta_a \,|\, \psi \rangle = \psi(a). \tag{2.87}$$

Then $\langle \delta_a |$ is defined for all real a, and

$$\langle \delta_a | \hat{X} = a\langle \delta_a |. \tag{2.88}$$

These eigenbras of $\hat{X}$ form a complete set, in the following sense. Because they are labelled by a continuous eigenvalue, the sum over eigenvalues which was appropriate in the finite-dimensional case must be replaced by an integral. Thus the appropriate expansion is

$$\langle \psi | = \int_{-\infty}^{\infty} \bar{c}_a \langle \delta_a |\, da. \tag{2.89}$$

This means

$$\langle\psi|\phi\rangle = \int_{-\infty}^{\infty} \bar{c}_a\langle\delta_a|\phi\rangle \, da \quad \text{for all } |\phi\rangle \in \mathscr{W}. \tag{2.90}$$

Now if $\langle\psi|$ is the conjugate bra vector of a ket vector $|\psi\rangle \in \mathscr{W}$, and if the coefficients c_a are given by

$$c_a = \psi(a), \tag{2.91}$$

then (2.90) follows from the definition (2.82) of the inner product. Thus the expansion is possible for all such $\langle\psi|$, and so the eigenbras $\langle\delta_a|$ form a complete set for them.

The eigenbras $\langle\varepsilon_k|$ form a complete set in a similar way. To show this we will need the following facts from Fourier analysis:

Fourier inversion theorem. Let $\tilde{\psi}(k)$ be the Fourier transform of ψ, defined by

$$\tilde{\psi}(k) = \frac{1}{\sqrt{(2\pi)}} \int_{-\infty}^{\infty} \psi(x) e^{-ikx} \, dx. \tag{2.92}$$

Then $\tilde{\psi}(k)$ exists if $\psi(x)$ belongs to $\mathscr{W}$, and

$$\psi(x) = \frac{1}{\sqrt{(2\pi)}} \int_{-\infty}^{\infty} \tilde{\psi}(k) e^{ikx} \, dx. \tag{2.93}$$

For comparison with (2.89), we write

$$\psi(x) = \int_{-\infty}^{\infty} c_k e^{ikx} \, dk \quad \text{where } c_k = \frac{1}{\sqrt{(2\pi)}} \tilde{\psi}(k). \tag{2.94}$$

Then for all $|\phi\rangle \in \mathscr{W}$ we have

$$\langle\psi|\phi\rangle = \int_{-\infty}^{\infty} \bar{c}_k \int_{-\infty}^{\infty} e^{-ikx} \phi(x) \, dx \, dk$$

$$= \int_{-\infty}^{\infty} \bar{c}_k \langle\varepsilon_k|\phi\rangle \, dk. \tag{2.95}$$

Hence

$$\langle\psi| = \int_{-\infty}^{\infty} \bar{c}_k \langle\varepsilon_k| \, dk. \tag{2.96}$$

Note that not every bra vector (i.e. linear map from $\mathscr{W}$ to $\mathbb{C}$) is of the form $\langle\psi|$ where $|\psi\rangle$ is an element of $\mathscr{W}$, and not every bra vector can be expanded in terms of the sets $\{\langle\delta_a|\}$ or $\{\langle\varepsilon_k|\}$. These are complete sets only for the expansion of bra vectors $\langle\psi|$ which are the conjugates of ket vectors.

In general, a hermitian operator may have both ordinary eigenvalues associated with eigenvalues as in the previous section, and eigenvalues of the generalised sort, associated with eigenbras, that we have just been considering. These are called **discrete** and **continuous** eigenvalues respectively; for provided the state space satisfies certain technical conditions† it can be shown that there

---

† It must be a dense subset of a separable Hilbert space.

is a countable number of the first sort of eigenvalue, and the second sort make up a continuous set of real numbers, i.e. a collection of (possibly infinite) intervals. The set of all eigenvalues, of both sorts, is called the **spectrum** of the operator. Suppose for the moment that all the eigenvalues are non-degenerate. Let $|\psi_i\rangle$ $(i = 1, 2, \ldots)$ be eigenvectors associated with the discrete eigenvalues α_i, and let $\langle \psi_\alpha |$ be an eigenbra associated with the continuous eigenvalue α; then, suitably normalised, these form a complete set in the sense that if $|\psi\rangle$ is any vector in $\mathscr{S}$, the corresponding bra can be written as

$$\langle \psi | = \sum_i \bar{c}_i \langle \psi_i | + \int \bar{c}_\alpha \langle \psi_\alpha | \, d\alpha \tag{2.97}$$

with the coefficients c_i and c_α given by

$$c_i = \langle \psi_i | \psi \rangle, \quad c_\alpha = \langle \psi_\alpha | \psi \rangle. \tag{2.98}$$

If the eigenvalues are not all non-degenerate, we must consider a complete set of commuting operators, all of which may have both discrete and continuous eigenvalues.

Eqs. (2.97) and (2.98) define the normalisation of the eigenbras $\langle \psi_\alpha |$. This normalisation is not an intrinsic property of the $\langle \psi_\alpha |$ as bra vectors, but relates to the operator $\hat{A}$ of which they are eigenbras: for if $\hat{B} = f(\hat{A})$ is any function of $\hat{A}$, $\langle \psi_\alpha |$ is also an eigenbra of $\hat{B}$ with eigenvalue $\beta = f(\alpha)$. If this relation can be inverted to write α as a function of β, (2.97) can be written

$$\langle \psi | = \sum_i \bar{c}_i \langle \psi_i | + \int \overline{c_{\alpha(\beta)}} \langle \psi_{\alpha(\beta)} | \rho(\beta) \, d\beta \tag{2.99}$$

where $\rho(\beta) = d\alpha/d\beta$. This can be restored to the form of (2.97) with respect to β, i.e.

$$\langle \psi | = \sum_i \bar{c}_i \langle \psi_i | + \int \overline{c'_\beta} \langle \psi'_\beta | \, d\beta, \tag{2.100}$$

with $\overline{c'_\beta} = \langle \psi'_\beta | \psi \rangle$, if we define

$$\langle \psi'_\beta | = g(\beta) \langle \psi_{\alpha(\beta)} |$$

where $\hspace{8cm}$ (2.101)

$$|g(\beta)|^2 = \rho(\beta).$$

We will say that the eigenbras $\langle \psi_\alpha |$ are **normalised relative to** $\hat{A}$ (the $\langle \psi'_\beta |$, correspondingly, being normalised relative to $\hat{B}$). The function $\rho(\beta)$ is called the **density of states** (in full, the density of eigenstates of $\hat{A}$ relative to eigenstates of $\hat{B}$).

Now suppose the operator we are considering represents an observable A. Then the continuous eigenvalues, like the discrete ones, have the physical meaning that they are possible values of A; but the probability statement (2.42) must be adapted to cater for the continuous variable. Thus Postulate II must be supplemented by

> **Postulate II** (*continued*). Let α be a non-degenerate continuous eigenvalue of an observable A, and let $\langle \psi_\alpha |$ be the corresponding

eigenbra, normalised relative to $\hat{A}$. Then when the system is in the state $|\psi\rangle$, the probability that a measurement of A will give a result between α and $\alpha + d\alpha$ is $p_A(\alpha \mid \psi)\, d\alpha$, where

$$p_A(\alpha \mid \psi) = \frac{|\langle \psi_\alpha | \psi \rangle|^2}{\langle \psi | \psi \rangle}. \tag{2.102}$$

Note that if we apply Postulate II to an operator $\hat{B} = f(\hat{A})$ and express the probability in terms of eigenbras $\langle \psi_\alpha|$ normalised relative to $\hat{A}$, the result is

$$p_B(\beta \mid \psi) = \frac{|\langle \psi_{\alpha(\beta)} | \psi \rangle|^2}{\langle \psi | \psi \rangle}\, \rho(\beta) \tag{2.103}$$

where ρ is the density of states.

Because of (2.97) and (2.98), it is still true that the expectation value of A in the state $|\psi\rangle$ is $\langle \psi | \hat{A} | \psi \rangle$, and therefore that the uncertainty is given by (2.70).

The postulate about the state of the system immediately after the measurement becomes a little more complicated if precisely stated:

> **Postulate III** (*continued*). If the measurement gives a result lying between α_1 and α_2, then immediately after the measurement the system is in the state which is obtained from $|\psi\rangle$ by orthogonal projection onto the subspace of states which are orthogonal to all states $|\psi'\rangle$ which satisfy
>
> $$\int_{\alpha_1}^{\alpha_2} \langle \psi_\alpha | \psi' \rangle\, d\alpha = 0,$$

but we will see shortly that there is a non-rigorous version which is essentially the same as our original statement.

Now let us return to the operators $\hat{X}$ and $\hat{K}$ defined in (2.83) and (2.84), which act on the space of functions of one real variable. Suppose these correspond to observables X and K relating to a particle moving along a line. As we have seen, each of them has a continuous spectrum consisting of all real numbers, with eigenbras $\langle \delta_a|$ and $\langle \varepsilon_k|$ respectively. According to (2.102), when the system is in the normalised state $|\psi\rangle$ the probability density of the value of X is

$$p(a) = |\langle \delta_a | \psi \rangle|^2 = |\psi(a)|^2. \tag{2.104}$$

Thus by comparison with (2.7), X can be identified as the position of the particle.

To identify K, note that its eigenbras $\langle \varepsilon_k|$ do correspond to functions, namely $\varepsilon_k(x) = e^{ikx}$, even though these functions do not belong to our state space. The reason for excluding them is that it is impossible to multiply $\varepsilon_k(x)$ by a constant to obtain a function ψ for which $\int |\psi|^2\, dx = 1$, and so for any such function $|\psi(x)|^2$ cannot be a probability density. However, as long as it is possible to integrate $|\psi|^2$ over any finite interval, it can be interpreted as a *relative* probability density in the sense that

$$\frac{\text{Probability that } a_1 < X < a_2}{\text{Probability that } a_3 < X < a_4} = \frac{\int_{a_1}^{a_2} |\psi|^2 \, dx}{\int_{a_3}^{a_4} |\psi|^2 \, dx}. \tag{2.105}$$

Thus $\psi(x) = e^{ikx}$ can be taken as the wave function of a particle for which all positions are equally likely. This function actually describes a periodic wave with the wavelength $\lambda = 2\pi/k$; hence according to de Broglie's relation (1.24) the particle associated with this wave has momentum

$$p = \frac{h}{\lambda} = \frac{h}{2\pi} k = \hbar k. \tag{2.106}$$

But k is the eigenvalue of $\hat{K}$; so the observable K is proportional to the *momentum* of the particle.

Thus by extending our notion of state, we have found an eigenstate of K. It is not described by a state vector according to our original conception, but by a bra vector; nevertheless, this bra vector is associated with a function which can be interpreted in a similar way to the wave functions which constitute true state vectors.

By making a further extension of the notion of a state, we can do a similar thing for X. The 'wave function' $\varepsilon_k(x)$ for the eigenstate of K was suggested by the representation of $\langle \varepsilon_k | \psi \rangle$ as an integral in (2.85); to find a similar function for X we would need to represent $\langle \delta_a | \psi \rangle$ also in an integral form, with a function $\delta_a(x)$ such that

$$\psi(a) = \langle \delta_a | \psi \rangle = \int_{-\infty}^{\infty} \overline{\delta_a(x)} \psi(x) \, dx. \tag{2.107}$$

This is not possible; there is no such function. But it is convenient to pretend that it is possible by taking

$$\delta_a(x) = \delta(x - a)$$

where δ is the **Dirac δ-function**, which is supposed to have the properties

$$\delta(x) = 0 \qquad \text{if } x \neq 0; \tag{2.108}$$

$$\int_{-a}^{b} \delta(x) \, dx = 1 \qquad \text{if } -a < 0 < b; \tag{2.109}$$

$$\int_{-\infty}^{\infty} f(x) \, \delta(x) \, dx = f(0) \quad \text{for any function } f. \tag{2.110}$$

Clearly these are impossible properties for a true function; nevertheless, the δ-function can be useful as a shorthand device for writing equations involving integrals. Any equation in which it occurs must be integrated in order to make sense. As an example of the use of the δ-function, the Fourier inversion theorem (2.93) can be written as

$$\int_{-\infty}^{\infty} e^{ik(x-y)} \, dk = 2\pi \, \delta(x - y). \tag{2.111}$$

In physical terms, $\delta(x)$ can be thought of as the line density of a distribution of matter consisting of a point particle with unit mass, situated at the origin. In quantum mechanics, $\delta(x-a)$ can be regarded as a wave function describing a particle which will certainly be found at $x=a$.

On physical grounds as well as mathematical ones, neither $\varepsilon_k(x)=e^{ikx}$ nor $\delta_a(x)=\delta(x-a)$ should be taken too literally as wave functions. The first would describe a particle which was equally likely to be found anywhere in the universe; the second, one whose position is known with infinite accuracy. But they are convenient idealisations, corresponding to the fiction of infinitely precise measurement. Such an ideal measurement can be regarded as the limit of a sequence of increasingly accurate actual measurements; in the same way, the states described by the 'wave functions' ε_k or δ_a can be regarded as limits of sequences of genuine states. The δ-function can also be understood as a limit of a sequence of genuine functions; for an explanation of it in these terms, see Lighthill 1958.

In terms of the ideal states described by such generalised wave functions, the postulate concerning the results of measurements of a continuous quantity can be phrased in the same way as for a discrete quantity: for momentum, for example, we postulate that if the result of the measurement is p then after the measurement the particle is in the state with wave function e^{ikx} where $p=\hbar k$.

For a particle moving in three dimensions, the basic observables are its Cartesian coordinates (x, y, z), which we will also denote by x_i $(i=1, 2, 3)$, and the components of momentum (p_x, p_y, p_z) or p_i $(i=1, 2, 3)$. By extension from the one-dimensional case, the corresponding operators, which act on wave functions $\psi(\mathbf{r})$, are given by

$$(\hat{x}_i\psi)(x_1, x_2, x_3)=x_i\psi(x_1, x_2, x_3), \tag{2.112}$$

$$\hat{p}_i\psi=-i\hbar\,\frac{\partial\psi}{\partial x_i}. \tag{2.113}$$

These are components of the *vector operators* $\hat{\mathbf{r}}$ and $\hat{\mathbf{p}}=-i\hbar\nabla$.

As a complete set of compatible observables we can take (x_1, x_2, x_3). The corresponding operators $\hat{x}_i$ clearly commute with each other; their possible simultaneous eigenvalues are any three real numbers (a_1, a_2, a_3), i.e. the components of any vector $\mathbf{a}$. The simultaneous eigenbra with these eigenvalues is $\langle\delta_\mathbf{a}|$, defined by

$$\langle\delta_\mathbf{a}|\psi\rangle=\psi(\mathbf{a})=\int\psi(\mathbf{r})\,\delta(\mathbf{r}-\mathbf{a})\,dV \tag{2.114}$$

where the second equality defines the three-dimensional δ-function.

An alternative complete set of compatible observables is provided by (p_1, p_2, p_3). The possible simultaneous eigenvalues are the components of any vector $\hbar\mathbf{k}$; the corresponding simultaneous eigenbra is $\langle\varepsilon_\mathbf{k}|$ where

$$\langle\varepsilon_\mathbf{k}|\psi\rangle=\int\psi(\mathbf{r})e^{-i\mathbf{k}\cdot\mathbf{r}}\,dV. \tag{2.115}$$

The probability distribution of **p** is then given by the density function $f(\mathbf{p}) = |\langle \varepsilon_\mathbf{k} | \psi \rangle|^2$ where $\mathbf{p} = \hbar \mathbf{k}$. Thus the state can be described by the function of momentum $\phi(\mathbf{p}) = \langle \varepsilon_\mathbf{k} | \psi \rangle$ just as well as by the wave function $\psi(\mathbf{r})$; these are said to be alternative **representations** of the same state vector $|\psi\rangle$. Their relation to each other is similar to the relation of different sets of coordinates (with respect to different bases) of a vector in a finite-dimensional space. The function $\phi(\mathbf{p})$ (essentially the Fourier transform of $\psi(\mathbf{r})$) is often called the **wave function in momentum space**. We see here an indication of a fundamental equality of status between position and momentum in quantum mechanics.

In order to recognise this equal status, we take as the basis of the theory of a single particle not the specification of the operators $\hat{x}_i$ and $\hat{p}_i$ in (2.112) and (2.113), but the commutation relations which follow from them:

> **Postulate IV.** The components of the position and momentum vectors of a particle in space are described by operators $\hat{x}_i$ and $\hat{p}_i$ which satisfy
>
> $$[\hat{x}_i, \hat{x}_j] = 0 = [\hat{p}_i, \hat{p}_j], \tag{2.116}$$
>
> $$[\hat{x}_i, \hat{p}_j] = i\hbar \delta_{ij}. \tag{2.117}$$
>
> A particle with no internal properties has no observables which are compatible with all the x_i and p_j.

Eqs. (2.116)–(2.117) are called the **canonical commutation relations**. We define a **simple** particle as one which has observables x_i and p_j satisfying these commutation relations, and no observables which are compatible with all the x_i and p_j.

Postulate IV is actually equivalent to (2.112)–(2.113), for it has been shown that Postulate III implies that the state space is (essentially) isomorphic to the space of wave functions and that the operators $\hat{x}_i$ and $\hat{p}_j$ are given by (2.112)–(2.113). This is the **Stone/von Neumann theorem** (see Jauch 1968, p. 201). Most of the arguments in this book will be based directly on the commutation relations (2.116)–(2.117) and do not use the specific forms (2.112)–(2.113), but of course the latter are indispensable in most of the applications of quantum mechanics.

From the commutator (2.117) we have

$$\langle i[x, p_x] \rangle = \langle -\hbar \rangle = -\hbar. \tag{2.118}$$

Thus the uncertainty relation (2.75) becomes

$$\Delta x \cdot \Delta p_x \geqslant \tfrac{1}{2}\hbar, \tag{2.119}$$

which is the original Heisenberg uncertainty relation (1.25).

Eq. (2.119) can be understood as describing the relationship between the wave function $\psi(\mathbf{r})$ and the wave function in momentum space, $\phi(\mathbf{p})$. An example is shown in Fig. 2.4. This shows the x-dependence of ψ, so it is concerned with a one-dimensional function which can be written as in (2.94):

$$\psi(x) = \int_{-\infty}^{\infty} c_k e^{ikx} \, dk \quad \text{where} \quad k = \frac{p_x}{\hbar} \quad \text{and} \quad c_k = \frac{\phi(p_x)}{2\pi}. \tag{2.120}$$

Thus $\psi(x)$ is a superposition of the oscillatory functions e^{ikx}. In the example the coefficients are taken to be real and to be concentrated in a range of k of length Δk. Then the different oscillations e^{ikx} in $\psi(x)$ cancel each other outside a range of x of length Δx, where $\Delta x \cdot \Delta k \simeq \frac{1}{2}$, but reinforce each other inside this region to produce a localised wave function. A wave function like this is called a **wave packet**.

From now on we will not need to be pedantic about the distinction between observables and operators. We will apply the same terms to both (referring to observables as 'commuting', for example) and we will use the same symbol for an observable and its operator (i.e. we will drop the circumflex on operators).

A note on Hilbert space   As we noted on p. 39, the demand that a wave function should satisfy **W1**–**W3** is unnecessarily restrictive. The reason for imposing these conditions is that they provide us with a state space on which $\hat{x}_i$ and $\hat{p}_i$, and all products of them, are well-defined operators. However, if we continued to work with this space we would soon find that it had disadvantages: in particular, there are Cauchy sequences of functions in the space that do not converge to a limit in the space. It is usual to work with a space which is **complete**, i.e. which does include all limits of Cauchy sequences. Such a space is called a **Hilbert space**. The smallest Hilbert space containing our space of wave functions is $L^2(\mathbb{R}^3)$, the space of all square-integrable functions: this is often taken to be the true state space of a

Fig. 2.4.
A wave packet.

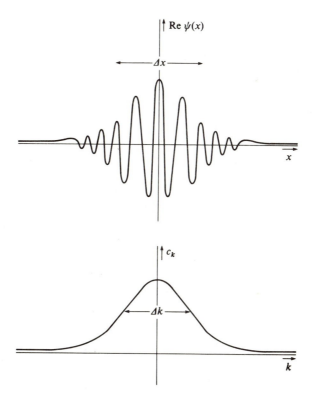

single particle. It has the disadvantage that many interesting operators are not now defined on the whole space; it is necessary to take into account the domain of definition of every operator considered. This puts quite a lot of grit in the way of the smooth running of the machinery. An account of the relevant parts of Hilbert space theory can be found in Jauch 1968.

Our approach can be formalised as the theory of a **rigged Hilbert space**, which consists of a chain of spaces $\mathscr{S} \subset \mathscr{H} \subset \mathscr{S}^*$, where $\mathscr{H}$ is a Hilbert space, $\mathscr{S}$ is a dense subspace of $\mathscr{H}$, and $\mathscr{S}^*$ is the space of continuous linear functionals on $\mathscr{S}$ ($\mathscr{S}$ is our space of wave functions and $\mathscr{S}^*$ is our space of bras). This theory then accounts for generalised states like eigenstates of position and momentum in the way we have described. An account of rigged Hilbert spaces and their use in quantum mechanics can be found in Bogolubov, Logunov and Todorov 1975 or Böhm 1978.

2.6. Combined systems

Suppose a system is made up of two or more parts, each of which can be considered as a system on its own; for example, the system of several particles moving in space, or a beam of light consisting of many photons moving in the z-direction, or both of these put together. We want to describe the states of the whole system in terms of the states of its component parts.

We will start by considering a system with two parts S and T; we will call the combined system ST. The two parts can be put together with S in any state $|\phi\rangle$ and T in any state $|\psi\rangle$; we will denote this state of the whole system by $|\phi\rangle|\psi\rangle$. Then if we consider a fixed state $|\phi\rangle$, the states $|\phi\rangle|\psi\rangle$ will correspond to all the possible states of subsystem T. We will assume that the relations between these states of ST are the same as those between the corresponding states of T; if $|\psi\rangle$ is a combination of $|\psi_1\rangle$ and $|\psi_2\rangle$, then $|\phi\rangle|\psi\rangle$ is the same combination of $|\phi\rangle|\psi_1\rangle$ and $|\phi\rangle|\psi_2\rangle$, and (if $|\phi\rangle$ is normalised) the inner product between $|\phi\rangle|\psi_1\rangle$ and $|\phi\rangle|\psi_2\rangle$ is the same as that between $|\psi_1\rangle$ and $|\psi_2\rangle$. Similar considerations apply to the states of S combined with a fixed state of T. Thus

$$|\phi\rangle(c_1|\psi_1\rangle + c_2|\psi_2\rangle) = c_1|\phi\rangle|\psi_1\rangle + c_2|\phi\rangle|\psi_2\rangle, \qquad (2.121)$$

$$(c_1|\phi_1\rangle + c_2|\phi_2\rangle)|\psi\rangle = c_1|\phi_1\rangle|\psi\rangle + c_2|\phi_2\rangle|\psi\rangle, \qquad (2.122)$$

and

$$(\langle\phi|\langle\psi_1|)(|\phi\rangle|\psi_2\rangle) = \langle\phi|\phi\rangle\langle\psi_1|\psi_2\rangle \qquad (2.123)$$

where the left-hand side of (2.123) denotes the inner product between $|\phi\rangle|\psi_1\rangle$ and $|\phi\rangle|\psi_2\rangle$. From this it can be deduced (problem 2.17) that

$$(\langle\phi_1|\langle\psi_1|)(|\phi_2\rangle|\psi_2\rangle) = \langle\phi_1|\phi_2\rangle\langle\psi_1|\psi_2\rangle. \qquad (2.124)$$

Eqs. (2.121)–(2.122) show that $|\phi\rangle|\psi\rangle$ can be regarded as a kind of product of $|\phi\rangle$ and $|\psi\rangle$. In physical terms, they mean that any experiment which can be performed on one subsystem can be performed in the presence of the other without affecting or being affected by the state of the second subsystem.

Not every state of the combined system can be expressed as a product $|\phi\rangle|\psi\rangle$, for by the principle of superposition there must be states of the form $|\phi\rangle|\psi\rangle + |\phi'\rangle|\psi'\rangle$, and these cannot be written as a single product. We will assume that there are no states of ST other than those demanded by the principle of superposition; then every state of ST is of the form

$$|\Psi\rangle = |\phi\rangle|\psi\rangle + |\phi'\rangle|\psi'\rangle + \cdots. \tag{2.125}$$

Let $|\phi_1\rangle, |\phi_2\rangle, \ldots$ be a complete set of states for S and $|\psi_1\rangle, |\psi_2\rangle, \ldots$ a complete set of states for T, and write

$$\begin{aligned} |\phi\rangle = \sum a_i |\phi_i\rangle, \quad |\phi'\rangle = \sum a_i' |\phi_i\rangle \\ |\psi\rangle = \sum b_i |\psi_i\rangle, \quad |\psi'\rangle = \sum b_i' |\psi_i\rangle \end{aligned} \tag{2.126}$$

Then

$$|\Psi\rangle = \sum_{ij} (a_i b_j + a_i' b_j' + \cdots)|\phi_i\rangle|\psi_j\rangle. \tag{2.127}$$

Thus the states $|\phi_i\rangle|\psi_j\rangle$ constitute a complete set of states for ST.

Let $\mathscr{S}$ denote the state space of S and $\mathscr{T}$ that of T. The state space of ST is denoted by $\mathscr{S} \otimes \mathscr{T}$ and called the **tensor product** of $\mathscr{S}$ and $\mathscr{T}$. If $\mathscr{S}$ and $\mathscr{T}$ are finite-dimensional, the dimension of $\mathscr{S} \otimes \mathscr{T}$ is the product of the dimensions of $\mathscr{S}$ and $\mathscr{T}$.

Operators on the individual state spaces $\mathscr{S}$ and $\mathscr{T}$ can act on $\mathscr{S} \otimes \mathscr{T}$ in the obvious way: if A is an operator on $\mathscr{S}$ and B is an operator on $\mathscr{T}$, we define†

$$\begin{aligned} A(|\phi\rangle|\psi\rangle) = (A|\phi\rangle)|\psi\rangle \\ B(|\phi\rangle|\psi\rangle) = |\phi\rangle(B|\psi\rangle) \end{aligned} \tag{2.128}$$

Then all the operators on $\mathscr{S}$ commute with all operators on $\mathscr{T}$. In the case of operators representing observables, this reflects the fact that experiments on S are not affected by the state of T. A complete set of compatible observables for the combined system ST consists of the union of a complete set for S and a complete set for T.

Let us see how this applies to a system of two particles moving in space. Since a complete set of compatible observables for one particle consists of the components of its position vector $\mathbf{r}$, a complete set for two particles will be given by their two position vectors $\mathbf{r}_1, \mathbf{r}_2$. Hence a state of two particles is completely specified by its values for all eigenbras $\langle \delta_{\mathbf{a}1}|, \langle \delta_{\mathbf{a}2}|$; in other (plainer) words, it is given by a function of two position variables, $\psi(\mathbf{r}_1, \mathbf{r}_2)$. This has the interpretation that one would expect by analogy with the case of one particle: $|\psi(\mathbf{r}_1, \mathbf{r}_2)|^2 \, dV_1 \, dV_2$ is the probability that particle 1 will be found in the volume dV_1 at $\mathbf{r}_1$, and at the same time particle 2 will be found in the volume dV_2 at $\mathbf{r}_2$.

Thus in this case $\mathscr{S} \otimes \mathscr{T}$ consists of functions $\psi(\mathbf{r}_1, \mathbf{r}_2)$ satisfying similar conditions to those specified in §2 for the wave functions of one particle. Let $\psi_1, \psi_2, \ldots$ be a complete set of one-particle wave functions. By expanding $\psi(\mathbf{r}_1, \mathbf{r}_2)$ in a series of $\psi_i(\mathbf{r}_1)$ with coefficients which depend on $\mathbf{r}_2$, and then expanding these coefficients in a series of $\psi_j(\mathbf{r}_2)$, we find that $\psi(\mathbf{r}_1, \mathbf{r}_2)$ can be expanded as

---

† Strictly speaking, we should use different symbols for the operators on the two sides of (2.128), since they act on different spaces; the operators on the left-hand sides should be called $A \otimes 1$ and $1 \otimes B$. However, the notation used here is convenient and will not cause confusion.

$$\psi(\mathbf{r}_1, \mathbf{r}_2) = \sum_{ij} c_{ij} \psi_i(\mathbf{r}_1) \psi_j(\mathbf{r}_2). \tag{2.129}$$

Thus the functions $\psi_i(\mathbf{r}_1)\psi_j(\mathbf{r}_2)$ constitute a complete set for $\mathscr{S} \otimes \mathscr{T}$, corresponding to the complete set $|\phi_i\rangle |\psi_j\rangle$ in the general case: for wave functions the tensor product is formed by ordinary multiplication.

The extension to a system consisting of several subsystems is straightforward: one can define a tensor product of n state spaces, $\mathscr{S}_1 \otimes \cdots \otimes \mathscr{S}_n$, whose elements are linear combinations of products $|\psi_1\rangle |\psi_2\rangle \cdots |\psi_n\rangle$ with $|\psi_i\rangle$ taken from $\mathscr{S}_i$. For a system of n particles moving in space, the state space consists of wave functions $\psi(\mathbf{r}_1, \ldots, \mathbf{r}_n)$.

This formalism applies not only to a system consisting of separate objects, but also to a system consisting of one object which has a number of separate aspects. For example, to give a full description of a photon we should consider not only its state of polarisation but also its state of motion in space; thus its full state space is $\mathscr{S} \otimes \mathscr{T}$ where $\mathscr{S}$ is the two-dimensional space of polarisation states and $\mathscr{T}$ is the space of wave functions for a particle moving in space. Since $\mathscr{S}$ has a complete set consisting of two states $|\phi_x\rangle$ and $|\phi_y\rangle$, the general state in $\mathscr{S} \otimes \mathscr{T}$ is

$$|\Psi\rangle = \sum_i (a_i |\phi_x\rangle |\psi_i\rangle + b_i |\phi_y\rangle |\psi_i\rangle) \tag{2.130}$$

where $|\psi_i\rangle$ represents a complete set of wave functions. This can be written as a two-component wave function

$$\Psi(\mathbf{r}) = \begin{bmatrix} \psi_1(\mathbf{r}) \\ \psi_2(\mathbf{r}) \end{bmatrix} \tag{2.131}$$

where $\psi_1(\mathbf{r}) = \sum a_i \psi_i(\mathbf{r})$ and $\psi_2(\mathbf{r}) = \sum b_i \psi_i(\mathbf{r})$.

Identical particles   If the two subsystems are the same, as in the case of a system of two particles of the same type, then the state space of the combined system, viz. $\mathscr{S} \otimes \mathscr{S}$ where $\mathscr{S}$ is the state space of one particle, has an **exchange operator** X which exchanges the states of the two particles:

$$X(|\phi\rangle |\psi\rangle) = |\psi\rangle |\phi\rangle. \tag{2.132}$$

Not every state of the two-particle system is of the form $|\phi\rangle |\psi\rangle$, but (2.132) is sufficient to define X since every state is a sum of states of this form. Now if the two systems are really of the same type, then it will be impossible to distinguish the state $|\phi\rangle |\psi\rangle$ from $|\psi\rangle |\phi\rangle$; for in order to tell whether it is the first particle or the second that is in the state $|\phi\rangle$ one would need some way of distinguishing the two particles, and such a distinguishing mark would mean that they were not after all of exactly the same type. Hence if $|\Psi\rangle$ is any state vector in $\mathscr{S} \otimes \mathscr{S}$, $X|\Psi\rangle$ describes an indistinguishable physical state, and therefore the same one. It follows, by Postulate I, that $X|\Psi\rangle$ is a multiple of $|\Psi\rangle$:

$$X|\Psi\rangle = \varepsilon |\Psi\rangle. \tag{2.133}$$

But $X^2 = 1$, so $\varepsilon^2 = 1$; hence the only possibilities for ε are ± 1.

A state for which $\varepsilon = 1$ is called **symmetric**; it must be a sum of states of the form

$$|\phi\rangle|\psi\rangle + |\psi\rangle|\phi\rangle. \qquad (2.134)$$

A state for which $\varepsilon = -1$ is called **antisymmetric**; it must be a sum of states of the form

$$|\phi\rangle|\psi\rangle - |\psi\rangle|\phi\rangle. \qquad (2.135)$$

The state vectors (2.134)–(2.135) describe states in which the two particles are in the states $|\phi\rangle$ and $|\psi\rangle$, but it is not possible to ask which particle is in which state. This is the right sort of description for two truly identical particles.

We have argued that any physical state must be either symmetric or antisymmetric; it cannot be a superposition of a symmetric state and an antisymmetric state. But if both symmetric and antisymmetric states were possible this would contradict the principle of superposition. It follows that, for a particular type of particle, either all states are symmetric or all states are antisymmetric; in other words, the state space for the two-particle system is not $\mathscr{S} \otimes \mathscr{S}$ but either the subspace of symmetric states or the subspace of antisymmetric states. Which of these two possibilities occurs depends on the type of particle: the 'particles of matter' described in Chapter 1(A) (leptons, baryons and quarks), which are called **fermions**, always have antisymmetric two-particle states; the 'particles of force' described in Chapter 1(B) (photons, W and Z particles, and gluons), which are called **bosons**, always have symmetric two-particle states.

This extends naturally to a system of n particles. The n-fold tensor product $\otimes^n \mathscr{S} = \mathscr{S} \otimes \cdots \otimes \mathscr{S}$ has, for each pair (i, j), an exchange operator X_{ij} which exchanges the states of the particles numbered i and j:

$$X_{ij}(|\psi_1\rangle \cdots |\psi_i\rangle \cdots |\psi_j\rangle \cdots |\psi_n\rangle) = |\psi_1\rangle \cdots |\psi_j\rangle \cdots |\psi_i\rangle \cdots |\psi_n\rangle. \qquad (2.136)$$

Then if $|\Psi\rangle \in \otimes^n \mathscr{S}$ is a state of n bosons,

$$X_{ij}|\Psi\rangle = |\Psi\rangle \quad \text{for all pairs } (i, j). \qquad (2.137)$$

Such a state vector is called **totally symmetric**. If $|\Psi\rangle$ is a state of n fermions,

$$X_{ij}|\Psi\rangle = -|\Psi\rangle \quad \text{for all pairs } (i, j). \qquad (2.138)$$

Such a state vector is called **totally antisymmetric**.

By performing a number of exchanges in succession, we can put the states $|\psi_1\rangle, \ldots, |\psi_2\rangle$ into any order $|\psi_{\rho(1)}\rangle, \ldots, |\psi_{\rho(n)}\rangle$ where ρ is a permutation of $(1, \ldots, n)$. Thus we have an operator X_ρ for the permutation ρ:

$$X_\rho(|\psi_1\rangle \cdots |\psi_n\rangle) = |\psi_{\rho(1)}\rangle \cdots |\psi_{\rho(n)}\rangle. \qquad (2.139)$$

The permutation ρ is called **even** or **odd** according as the number of exchanges it requires is even or odd. The **signature** of the permutation, denoted by $\varepsilon(\rho)$, is $+1$ if ρ is even and -1 if it is odd. Now (2.137)–(2.138) give:

if $|\Psi\rangle$ is totally symmetric, $\qquad X_\rho|\Psi\rangle = |\Psi\rangle, \qquad (2.140)$

if $|\Psi\rangle$ is totally antisymmetric, $\quad X_\rho|\Psi\rangle = \varepsilon(\rho)|\Psi\rangle, \qquad (2.141)$

for all permutations ρ. Any totally symmetric state must be a sum of states of the form

$$S(|\psi_1\rangle \cdots |\psi_n\rangle) \quad \text{where } S = \sum_\rho X_\rho; \tag{2.142}$$

any totally antisymmetric state must be a sum of states of the form

$$A(|\psi_1\rangle \cdots |\psi_n\rangle) \quad \text{where } A = \sum_\rho \varepsilon(\rho) X_\rho. \tag{2.143}$$

If the state vectors $|\psi_i\rangle$ are wave functions, then (2.143) gives an n-particle wave function

$$\Psi(\mathbf{r}_1, \ldots, \mathbf{r}_n) = \sum_\rho \varepsilon(\rho) \psi_{\rho(1)}(\mathbf{r}_1) \cdots \psi_{\rho(n)}(\mathbf{r}_n)$$

$$= \begin{vmatrix} \psi_1(\mathbf{r}_1) & \cdots & \psi_1(\mathbf{r}_n) \\ \vdots & & \vdots \\ \psi_n(\mathbf{r}_1) & \cdots & \psi_n(\mathbf{r}_n) \end{vmatrix}. \tag{2.144}$$

This is known as a **Slater determinant**.

The principles introduced in this section can be summarised as follows:

> **Postulate V.** If two systems S and T are combined to form a system ST, the state space of ST is the tensor product of the state spaces of S and T.
>
> Every elementary particle is either a fermion or a boson. A state of many identical particles is totally antisymmetric if they are fermions, totally symmetric if they are bosons.

Fermions and bosons obey different probability laws from each other and from classical particles. Consider the elementary probability problem of two particles which are placed in two differently coloured boxes at random. What are the probabilities that (a) both particles are in the yellow box, (b) both are in the blue box, (c) there is one particle in each box? In the quantum version of this problem there are two particles which can occupy two orthogonal states $|\phi\rangle$ or $|\psi\rangle$ (i.e. each particle has a two-dimensional state space $\mathscr{S}$ in which $|\phi\rangle$ and $|\psi\rangle$ form a complete set of states), and it is assumed that all two-particle states are equally likely. With $|\phi\rangle$ as the yellow box and $|\psi\rangle$ as the blue box, we ask the same question. It can be shown (see problem 5.1) that the probabilities are proportional to the dimensions of the corresponding subspaces – in other words, despite the possibility of superposition one simply counts the numbers of independent states in the same way as one counts the number of ways of putting two beads into two boxes.

The state space we first thought of, namely $\mathscr{S} \otimes \mathscr{S}$, which would be appropriate if the particles were distinguishable, has four mutually orthogonal two-particle states:

$$|\phi\rangle|\phi\rangle, \quad |\psi\rangle|\psi\rangle, \quad |\phi\rangle|\psi\rangle, \quad |\psi\rangle|\phi\rangle. \tag{2.145}$$

If the particles are bosons, there are three orthogonal states:

$$|\phi\rangle|\phi\rangle, \quad |\psi\rangle|\psi\rangle, \quad |\phi\rangle|\psi\rangle + |\psi\rangle|\phi\rangle. \tag{2.146}$$

And if the particles are fermions, there is only state:

$$|\phi\rangle|\psi\rangle - |\psi\rangle|\phi\rangle. \tag{2.147}$$

Hence the probabilities we are looking for, with (2.145) corresponding to the classical answer, are:

	$\lvert\cdot\cdot\rvert\ \rvert$	$\rvert\ \lvert\cdot\cdot\rvert$	$\rvert\cdot\lvert\cdot\rvert$
Classical	$\frac{1}{4}$	$\frac{1}{4}$	$\frac{1}{2}$
Bosons	$\frac{1}{3}$	$\frac{1}{3}$	$\frac{1}{3}$
Fermions	0	0	1

Thus bosons are more likely than classical particles to occupy the same state; they appear to attract each other. Fermions, on the other hand, force each other into different states. This property of fermions is

> ●**2.6   Pauli's exclusion principle.** Two identical fermions cannot occupy the same state. ■

This plays a crucial role in the explanation of the structure of atoms which gives rise to their different chemical properties.

Fermions are said to obey **Fermi–Dirac statistics**; bosons obey **Bose–Einstein statistics**.

If a particle is composite, like an atom or a nucleus, then its statistics (i.e. whether it is a fermion or a boson) are determined by those of its constituents. Consider first a particle P with two constituents A and B. A state of two Ps is a state of two As combined with two Bs, and the operation X_P of interchanging the Ps is the product of the (commuting) operations X_A and X_B of interchanging the As and interchanging the Bs. Now X_A multiplies the state by ε_A, where

$$\varepsilon_A = \begin{cases} +1 & \text{if A is a boson} \\ -1 & \text{if A is a fermion,} \end{cases} \tag{2.148}$$

and similarly X_B multiplies the state by ε_B. Hence

$$\varepsilon_P = \varepsilon_A \varepsilon_B, \tag{2.149}$$

i.e. P is a boson if A and B are both bosons or both fermions, a fermion if one is a boson and the other a fermion. The general statement is

> ●**2.7   A composite system of m bosons and n fermions is a boson if n is even, a fermion if n is odd. ■

Thus, since quarks and antiquarks are fermions, mesons (made of a quark and an antiquark, i.e. two fermions) are bosons: baryons (made of three quarks) are fermions. The α-particle (made of two protons and two neutrons, i.e. four fermions) is a boson.

**Appendix: properties of
hermitian and unitary
operators**

1. The eigenvalues of a hermitian operator are real.
2. The eigenvalues of a unitary operator have modulus 1.
3. Let $|\phi\rangle$ and $|\psi\rangle$ be eigenvectors of an operator A corresponding to different eigenvalues. If A is either hermitian or unitary, $|\phi\rangle$ and $|\psi\rangle$ are orthogonal.
4. If H is hermitian, e^{iH} is unitary.

Proofs: 1. Let λ be an eigenvalue of the hermitian operator H, and let $|\psi\rangle$ be a (non-zero) eigenvector. Then

$$H|\psi\rangle = \lambda|\psi\rangle. \tag{2.A1}$$

Taking the hermitian conjugate of this equation,

$$\langle\psi|H = \bar{\lambda}\langle\psi|. \tag{2.A2}$$

Multiplying (2.A1) on the left by $\langle\psi|$ and (2.A2) on the right by $|\psi\rangle$,

$$\langle\psi|H|\psi\rangle = \lambda\langle\psi|\psi\rangle = \bar{\lambda}\langle\psi|\psi\rangle. \tag{2.A3}$$

Since $|\psi\rangle$ is non-zero, it follows that $\lambda = \bar{\lambda}$, i.e. λ is real. ∎

2. Let λ be an eigenvalue and $|\psi\rangle$ an eigenvector of the unitary operator U. Then

$$\left.\begin{array}{l} U|\psi\rangle = \lambda|\psi\rangle \\ \langle\psi|U = \bar{\lambda}\langle\psi| \end{array}\right\}. \tag{2.A4}$$

Multiplying these equations together,

$$\langle\psi|U^\dagger U|\psi\rangle = \lambda\bar{\lambda}\langle\psi|\psi\rangle. \tag{2.A5}$$

Since $U^\dagger U = 1$ and $|\psi\rangle$ is non-zero, it follows that $|\lambda| = 1$. ∎

3. First suppose $A = H$ where H is hermitian. Then

$$H|\phi\rangle = \lambda|\phi\rangle, \quad H|\psi\rangle = \mu|\psi\rangle \tag{2.A6}$$

where λ and μ are real. Multiplying by $\langle\phi|$ and $\langle\psi|$ respectively,

$$\langle\psi|H|\phi\rangle = \lambda\langle\psi|\phi\rangle; \tag{2.A7}$$

$$\langle\phi|H|\psi\rangle = \mu\langle\phi|\psi\rangle; \tag{2.A8}$$

$$\therefore \ \langle\psi|H|\phi\rangle = \mu\langle\psi|\phi\rangle. \tag{2.A9}$$

Since $\lambda \neq \mu$, it follows from (2.A7) and (2.A 9) that $\langle\psi|\phi\rangle = 0$.

Now suppose $A = U$ where U is unitary. Then

$$U|\phi\rangle = e^{i\alpha}|\phi\rangle, \tag{2.A10}$$

$$U|\psi\rangle = e^{i\beta}|\psi\rangle \tag{2.A11}$$

where α and β are real. Taking the hermitian conjugate of (2A.11),

$$\langle\psi|U^\dagger = e^{-i\beta}\langle\psi|. \tag{2.A12}$$

Multiplying (2.A11) and (2.A13) together and using $U^\dagger U = 1$,

$$\langle\psi|\phi\rangle = e^{i(\alpha-\beta)}\langle\psi|\phi\rangle.$$

Since $e^{i\alpha} \neq e^{i\beta}$, it follows that $\langle\psi|\phi\rangle = 0$. ∎

4. Let $|\psi_1\rangle, |\psi_2\rangle, \ldots$ be a complete set of eigenstates of H with eigenvalues $\lambda_1, \lambda_2, \ldots$, and let $U = e^{iH}$. Then

$$\langle \psi_m | U | \psi_n \rangle = e^{i\lambda_n} \delta_{mn},$$
$$\therefore \ \langle \psi_n | U^\dagger | \psi_m \rangle = e^{-i\lambda_m} \delta_{mn} = \langle \psi_n | e^{-i\lambda_m} | \psi_m \rangle.$$

Hence

$$U^\dagger | \psi_m \rangle = e^{-i\lambda_m} | \psi_m \rangle$$

and so

$$U^\dagger U | \psi_m \rangle = | \psi_m \rangle.$$

Since the $|\psi_m\rangle$ form a complete set, it follows that $U^\dagger U = 1$. ∎

Bones of Chapter 2

Postulate I. Principle of superposition 46

Postulate II. Results of experiments 49, 50, 62

Postulate III. Projection postulate 49, 51, 63

Postulate IV. Position and momentum of a particle 66

Postulate V. Combined systems 72

● 2.1   Eigenstates are orthogonal and complete 51

● 2.2   The operator of an observable is hermitian 52

● 2.3   Expectation value and uncertainty 56

● 2.4   Compatible observables ⇔ commuting operators 57

● 2.5   Generalised uncertainty relation 59

● 2.6   Exclusion principle 73

● 2.7   Statistics of a composite particle 73

Further reading

The classic exposition of quantum mechanics, which remains unrivalled for the elegance of its presentation of the general structure of the theory (despite – or perhaps because of – its cavalier attitude to mathematical niceties) is Dirac 1930. Other excellent textbooks which are recommended for their careful treatment of conceptual matters are Pauli 1933, Bohm 1951 and Gottfried 1966. Feynman *et al.* 1965 (Vol. III) is also highly recommended.

Mathematically rigorous accounts of quantum mechanics can be found in von Neumann 1932 and Jauch 1968. These follow a different line from the approach via rigged Hilbert spaces which has been sketched in this chapter; for this see Böhm 1978 or Bogolubov *et al.* 1975.

Problems on Chapter 2

1. Prove that the set of all wave functions satisfying W1–W3 (p. 39) forms a complex vector space and that (2.30) defines an inner product.

2. Let P_α be the projection operator associated with the result α of an experiment E. Show that

(i) $P_\alpha^2 = P_\alpha$;

(ii)   if α and β are different results of E, $P_\alpha P_\beta = 0$;

(iii)   if α is non-degenerate, with eigenstate $|\psi_\alpha\rangle$,

$$P_\alpha | \psi \rangle = \langle \psi_\alpha | \psi \rangle | \psi_\alpha \rangle;$$

(iv) $\sum_\alpha P_\alpha = 1$ where the sum is over all results of E.

3. Let $|\psi\rangle$ be any state vector, let $|\phi\rangle$ be any element of the eigenspace $\mathscr{S}_\alpha$, and let

$$|\theta_0\rangle = |\psi\rangle - P_\alpha|\psi\rangle, \quad |\theta\rangle = |\psi\rangle - |\phi\rangle$$

where P_α is the projection operator onto $\mathscr{S}_\alpha$. Show that

$$\langle\theta|\theta\rangle \geqslant \langle\theta_0|\theta_0\rangle.$$

[This shows that $P_\alpha|\psi\rangle$ is the nearest element of $\mathscr{S}_\alpha$ to $|\psi\rangle$.]

4. A quantum system can exist in two states $|a_0\rangle$ and $|a_1\rangle$, which are normalised eigenstates of the observable A with eigenvalues 0 and 1. A second observable B corresponds to the operator $\hat{B}$ defined by

$$\hat{B}|a_0\rangle = 7|a_0\rangle - 24i|a_1\rangle, \quad \hat{B}|a_1\rangle = 24i|a_0\rangle - 7|a_1\rangle.$$

Find the eigenvalues and eigenstates of B.

The system is in the state $|a_0\rangle$ when B is measured, and immediately afterwards A is measured. Find the probability that the measurement of A gives the result 0.

5. Let A and B be two observables on a system with a two-dimensional state space, and suppose measurements are made of A, B and A again in quick succession. Show that the probability that the second measurement of A gives the same result as the first is independent of the initial state of the system.

6. If A is any operator, show that $A^\dagger A$ has non-negative eigenvalues.

7. If H is a hermitian operator and $|\psi\rangle$ is an eigenvector of H with eigenvalue E, show that $\langle\psi|e^{iH} = e^{iE}\langle\psi|$.

8. Show that the normalised state $|\psi\rangle$ is an eigenstate of an observable A if and only if $\langle\psi|\hat{A}^2|\psi\rangle = \langle\psi|\hat{A}|\psi\rangle^2$.

9. Find the uncertainty in the polarisation observable P_x when the photon is polarised at an angle θ to the x-axis.

10. Let P_θ be the photon polarisation observable for an axis at an angle θ to the x-axis (so that $P_\theta = P_x$ if $\theta = 0$). Show that P_θ and P_ϕ are compatible if and only if $\theta - \phi = \frac{1}{2}n\pi$ for some integer n.

11. Prove that

$$\Delta A^2 \Delta B^2 \geqslant |\langle\tfrac{1}{2}i[A, B]\rangle|^2 + (\langle\tfrac{1}{2}\{A, B\}\rangle^2 - \langle A\rangle\langle B\rangle)^2$$

where A and B are any two observables and $\{A, B\} = AB + BA$.

12. Check that the formulae (2.69)–(2.70) for expectation value and uncertainty remain valid if A has continuous eigenvalues.

13. Use formal manipulations with the δ-function to obtain Plancherel's formula

$$\int |\tilde{\psi}(k)|^2 \, dk = \int |\psi(x)|^2 \, dx$$

where $\tilde{\psi}(k)$ is the Fourier transform of $\psi(x)$.

14. Show that

$$\delta(f(x)) = \sum_i \frac{\delta(x - x_i)}{|f'(x_i)|}$$

where the sum is over all zeros x_i of $f(x)$.

15. For a particle moving in one dimension, find the uncertainty Δx when the wave function is the Gaussian $\psi(x) = (2/\pi a^2)^{\frac{1}{4}} \exp(-x^2/a^2)$. Show that the Fourier transform of this function has the same form, and deduce that $\Delta x \cdot \Delta p$ has its minimum value in this state.

16. From the commutation relations between $\hat{x}$ and $\hat{p}$, show that

    $$(i\hbar)^{-1}[\hat{p}, \hat{x}^n] = n\hat{x}^{n-1}.$$

    Hence show that if f is a polynomial in x_1, x_2, x_3, the commutation relations (2.116)–(2.117) imply $[\hat{p}_i, f] = -i\hbar \, \partial f/\partial x_i$.

17. Let $D = \mathbf{r} \cdot \mathbf{p}$. If f is a product of m position operators and n momentum operators, show that

    $$[D, f] = -i\hbar(m-n)f.$$

18. Deduce (2.124) from (2.121)–(2.123).

19. Let X be a fermion with an n-dimensional state space. What is the dimension of the state space of a system of r X particles? What if X is a boson?

3

Quantum dynamics

The previous chapter was concerned with describing the state of a system at one instant of time, and with the results and effects of an instantaneous experiment. In this chapter we discuss how the system changes between experiments, in response to forces which operate over an extended period. The discussion is quite general, and applies to systems described by any kind of state vector (e.g. polarisation vectors, wave functions, etc.).

3.1. The equations of motion

The time development of a quantum system is specified by saying how the state vector changes from one time to another; this is the quantum counterpart of the classical equations of motion. (We will use 'equation of motion' to denote any equation describing a change of state; the term does not imply the existence of motion in the classical sense of 'change of position'.) An indication of what the quantum equation of motion should be is provided by Planck's fundamental equation $E = h\nu$, which can be applied to any system as a relation between its energy E and the frequency ν of an oscillation associated with the system. We saw in Chapter 2, when we were describing how the amplitude and phase of an oscillation could be represented by a complex number, that the oscillatory time dependence is given by a factor $e^{-2\pi i\nu t}$. Thus Planck's equation implies that if a system has energy E, its state vector $|\psi(t)\rangle$ at time t should contain a factor $e^{-2\pi i\nu t} = e^{-iEt/\hbar}$, i.e.

$$|\psi(t)\rangle = e^{-iEt/\hbar}|\psi(0)\rangle. \tag{3.1}$$

Now energy is an observable, so for a system to have a definite energy E it must be in an eigenstate of this observable. Eq. (3.1) says that if this is so, the state vector at time t differs from that at $t=0$ only by a c-number factor, and therefore describes the same physical state. For this reason an eigenstate of energy is called a **stationary state**.

In order to say how a general state evolves in time we make the further assumption that as long as the system is undisturbed by an experiment the evolution of states is linear, i.e. if $|\psi(0)\rangle = c_1|\psi_1\rangle + c_2|\psi_2\rangle$ and if in time t the states $|\psi_1\rangle$ and $|\psi_2\rangle$ would evolve to $|\psi_1(t)\rangle$ and $|\psi_2(t)\rangle$, then $|\psi(0)\rangle$ evolves to

$|\psi(t)\rangle = c_1|\psi_1(t)\rangle + c_2|\psi_2(t)\rangle$. In other words,

$$|\psi(t)\rangle = U(t)|\psi(0)\rangle \tag{3.2}$$

where $U(t)$ is a linear operator. Since (3.1) holds whenever $|\psi(0)\rangle$ is an eigenstate of energy, we can identify $U(t)$ as a function of the operator H representing energy:

$$U(t) = e^{-iHt/\hbar}. \tag{3.3}$$

It is usual to state this law of evolution in the form of a differential equation:

> **Postulate VI.** Let $|\psi(t)\rangle$ be the state of the system at time t. Then as long as the system is not disturbed by any experiments, $|\psi(t)\rangle$ satisfies
>
> $$i\hbar \frac{d}{dt}|\psi(t)\rangle = H|\psi(t)\rangle \tag{3.4}$$
>
> where H is the operator describing the total energy of the system.

This is a general statement prescribing the form that the equation of motion must have in quantum mechanics, as Newton's second law of motion does in classical mechanics. The equation of motion (3.4) is called the (time-dependent) **Schrödinger equation**. The operator H is called the **Hamiltonian** of the system; it corresponds to the force in Newtonian mechanics – in fact, since the total energy includes the potential energy, knowledge of the total energy is equivalent to knowledge of the force.

The Hamiltonian takes its name from Hamilton's formulation of Newtonian mechanics, which is suitable for a direct comparison with quantum mechanics. In Hamiltonian mechanics the configuration and velocity of a mechanical system are described by a number of coordinates $(q_1, \ldots, q_n)$ and corresponding momenta $(p_1, \ldots, p_n)$; the motion of the system is determined by a function called the Hamiltonian function $H(q_1, \ldots, q_n, p_1, \ldots, p_n)$, whose value is the total energy of the system, by means of **Hamilton's equations**

$$\begin{aligned} \frac{dq_i}{dt} &= \frac{\partial H}{\partial p_i}, \\ \frac{dp_i}{dt} &= -\frac{\partial H}{\partial q_i}. \end{aligned} \tag{3.5}$$

Thus the state of the classical mechanical system is specified by the values of $(q_1, \ldots, q_n, p_1, \ldots, p_n)$, just as the state of a quantum system is specified by the state vector $|\psi\rangle$; an observable property of the classical system is a function $f(q_1, \ldots, q_n, p_1, \ldots, p_n)$, while an observable of a quantum system is an operator on state space; and in both cases the behaviour of the system is governed by a first-order differential equation giving the rate of change of the state in terms of the particular observable H.

For a single particle moving in space, the classical state is given by the position and momentum vectors $(\mathbf{r}, \mathbf{p})$; if the particle is acted on by a force $\mathbf{F}(\mathbf{r})$ which is derived from a potential $V(\mathbf{r})$, so that $\mathbf{F} = -\nabla V$, the Hamiltonian function is

$$H(\mathbf{r}, \mathbf{p}) = \frac{\mathbf{p}^2}{2m} + V(\mathbf{r}).$$
(3.6)

The operator corresponding to this observable in quantum mechanics is obtained by substituting the operators $\hat{\mathbf{r}}$ and $\hat{\mathbf{p}}$ (see after (2.112)–(2.113)) for $\mathbf{r}$ and $\mathbf{p}$. The state vector $|\psi(t)\rangle$, being a wave function which varies with time, is a function $\psi(\mathbf{r}, t)$; thus in this context eq. (3.6) becomes the partial differential equation

$$i\hbar \frac{\partial \psi}{\partial t} = -\frac{\hbar^2}{2m} \nabla^2 \psi + V(\mathbf{r})\psi.$$
(3.7)

If n particles are moving under the action of forces which are derived from a potential $V(\mathbf{r}_1, \ldots, \mathbf{r}_n)$, so that the force on the kth particle is $\mathbf{F}_k = -\nabla_k V$ (where $\nabla_k = \partial/\partial \mathbf{r}_k$), then the equation satisfied by the wave function $\psi(\mathbf{r}_1, \ldots, \mathbf{r}_k)$ is

$$i\hbar \frac{\partial \psi}{\partial t} = -\frac{\hbar^2}{2m_1} \nabla_1{}^2 \psi - \cdots - \frac{\hbar^2}{2m_n} \nabla_n{}^2 \psi + V\psi.$$
(3.8)

Now suppose the Hamiltonian H has a purely discrete spectrum, so that it has a complete set of eigenstates $|\psi_1\rangle, |\psi_2\rangle, \ldots$ with corresponding eigenvalues $E_1, E_2, \ldots$. Then $|\psi(t)\rangle$ can be expanded in terms of this complete set with coefficients which will depend on t:

$$|\psi(t)\rangle = \sum_m c_m(t)|\psi_m\rangle$$
(3.9)

and eq. (3.4) gives

$$i\hbar \sum_m \frac{dc_m}{dt} |\psi_m\rangle = \sum_m E_m c_m(t)|\psi_m\rangle.$$

Hence, equating coefficients of $|\psi_m\rangle$,

$$i\hbar \frac{dc_m}{dt} = E_m c_m(t);$$

$$\therefore \quad c_m(t) = e^{-iE_m t/\hbar} c_m(0).$$
(3.10)

Thus

$$|\psi(t)\rangle = \sum e^{-iE_m t/\hbar} c_m(0)|\psi_m\rangle.$$
(3.11)

This is the same as (3.2) and (3.3), and shows that they follow from Postulate VI.

As an example of the effect of time development on observables, suppose a system is initially in an eigenstate $|\psi_0\rangle$ of an observable A, where $|\psi_0\rangle$ is not a stationary state but a superposition of two different stationary states:

$$|\psi_0\rangle = c_1|\psi_1\rangle + c_2|\psi_2\rangle.$$
(3.12)

Assume that $|\psi_0\rangle$, $|\psi_1\rangle$ and $|\psi_2\rangle$ are all normalised, so that

$$|c_1|^2 + |c_2|^2 = 1.$$
(3.13)

Then the probability that a measurement of A at time t will show A to have the same value as it had initially is

$$\left|\langle\psi_0|\psi(t)\rangle\right|^2 = \left|\langle\psi_0|(c_1 e^{-iE_1 t/\hbar}|\psi_1\rangle + c_2 e^{-iE_2 t/\hbar}|\psi_2\rangle)\right|^2$$
$$= \left||c_1|^2 e^{-iE_1 t/\hbar} + |c_2|^2 e^{-iE_2 t/\hbar}\right|^2 \tag{3.14}$$

since $\langle\psi_0|\psi_1\rangle = \bar{c}_1$ and $\langle\psi_0|\psi_2\rangle = \bar{c}_2$. Using (3.13), this can be written as

$$\left|\langle\psi_0|\psi(t)\rangle\right|^2 = 1 - 4|c_1|^2|c_2|^2 \sin^2\left[\tfrac{1}{2}(E_1 - E_2)t/\hbar\right]. \tag{3.15}$$

Thus the value of A oscillates with frequency

$$v = \frac{E_1 - E_2}{h}. \tag{3.16}$$

Note that if the energy of the system had been measured at $t=0$, when the state was given by (3.12), then the measurement would have changed the state of the system in one of two different ways, and there would have been two alternative courses of development. With probability $|c_1|^2$, the measurement would have given the value E_1, the state of the system would have become $|\psi_1\rangle$, and at time t this would have evolved to $e^{-iE_1 t/\hbar}|\psi_1\rangle$. With probability $|c_2|^2$, the measurement would have given the value E_2, the state would have become $|\psi_2\rangle$, and at time t it would be $e^{-iE_2 t/\hbar}|\psi_2\rangle$. Thus the state at time t would be

$$\begin{aligned} e^{-iE_1 t/\hbar}|\psi_1\rangle \quad &\text{with probability } |c_1|^2, \\ e^{-iE_2 t/\hbar}|\psi_2\rangle \quad &\text{with probability } |c_2|^2 \end{aligned} \tag{3.17}$$

instead of

$$c_1 e^{-iE_1 t/\hbar}|\psi_1\rangle + c_2 e^{-iE_2 t/\hbar}|\psi_2\rangle. \tag{3.18}$$

If A is measured at time t, the probability that it has the same value as it had initially is $|\langle\psi_0|\psi_1\rangle|^2 = |c_1|^2$ in the first case of (3.17), $|\langle\psi_0|\psi_2\rangle|^2$ in the second case. Hence the total probability is

$$|c_1|^2 \cdot |c_1|^2 + |c_2|^2 \cdot |c_2|^2 = |c_1|^4 + |c_2|^4. \tag{3.19}$$

This should be compared with (3.14) and (3.15). We see that the oscillation in (3.15) has entirely disappeared. This oscillation is the result of interference between the two states $e^{-iE_1 t/\hbar}|\psi_1\rangle$ and $e^{-iE_2 t/\hbar}|\psi_2\rangle$; it occurs when the states are added **coherently**, as in (3.18). In (3.17), on the other hand, the states are added **incoherently**. The mathematical difference between coherent and incoherent addition is shown in the contrast between (3.14) and (3.19).

In the case of a system of particles governed by the time-dependent Schrödinger equation (3.8), the solution (3.11) can be written

$$\psi(\mathbf{r}_1, \ldots, \mathbf{r}_n, t) = \sum_m e^{-iE_m t/\hbar} c_m(0)\psi_m(\mathbf{r}_1, \ldots, \mathbf{r}_n) \tag{3.20}$$

where ψ_m satisfies the equation

$$-\frac{\hbar^2}{2m_1}\nabla_1^2\psi_m - \cdots - \frac{\hbar^2}{2m_n}\nabla_n^2\psi_m + V(\mathbf{r}_1, \ldots, \mathbf{r}_n)\psi_m = E_m\psi_m, \tag{3.21}$$

which is known as the **time-independent Schrödinger equation**. This is the solution that would be obtained by the method of separation of variables (see problem 3.4). If the Hamiltonian has a continuous spectrum, the sum in (3.20) must be replaced by an integral to give

$$\psi(\mathbf{r}_1, \ldots, \mathbf{r}_n, t) = \int e^{-iEt/\hbar} \psi_E(\mathbf{r}_1, \ldots, \mathbf{r}_n) \, dE. \tag{3.22}$$

Write $E = \omega\hbar$, $\phi(\mathbf{r}_1, \ldots, \mathbf{r}_n, \omega) = \hbar\psi_E(\mathbf{r}_1, \ldots, \mathbf{r}_n)$; then

$$\psi(\mathbf{r}_1, \ldots, \mathbf{r}_n, t) = \int e^{-i\omega t} \phi(\mathbf{r}_1, \ldots, \mathbf{r}_n, \omega) \, d\omega \tag{3.23}$$

so in this case (3.11) corresponds to solving the differential equation by taking the Fourier transform with respect to t.

The simplest system of this type is that of a single particle moving in space under no forces, so that the Hamiltonian is $H = \mathbf{p}^2/2m$. Since this is a function of momentum, any eigenstate of momentum will also be an eigenstate of H. As we saw in the previous chapter, there are no true eigenstates of momentum, but there is an expression for the wave function corresponding to an expansion in momentum eigenstates, namely the Fourier transform with respect to $\mathbf{r}$:

$$\psi(\mathbf{r}, t) = \int \tilde{\psi}(\mathbf{p}', t) e^{i\mathbf{p}' \cdot \mathbf{r}/\hbar} \, d^3\mathbf{p}' \tag{3.24}$$

(we use $\mathbf{p}'$ for the variable of integration to avoid confusion with the momentum operator). The solution of the Schrödinger equation that corresponds to this as (3.11) corresponds to (3.9) is

$$\psi(\mathbf{r}, t) = \int \tilde{\psi}(\mathbf{p}') \exp\left\{ \frac{i}{\hbar}\left(\mathbf{p}' \cdot \mathbf{r} - \frac{\mathbf{p}'^2}{2m} t \right) \right\} d^3\mathbf{p}'. \tag{3.25}$$

Rate of change of expectation value

● **3.1**   Let A be any observable. As the state $|\psi(t)\rangle$ changes according to Postulate VI, the expectation value of A in this state changes according to

$$\frac{d}{dt}\langle A \rangle = \left\langle \frac{i}{\hbar}[H, A] \right\rangle. \tag{3.26}$$

Proof. The expectation value at time t is

$$\langle A \rangle = \langle \psi(t)|A|\psi(t) \rangle. \tag{3.27}$$

We have

$$i\hbar \frac{d}{dt}|\psi(t)\rangle = H|\psi(t)\rangle.$$

Taking the hermitian conjugate of this equation,

$$-i\hbar \frac{d}{dt}\langle \psi(t)| = \langle \psi(t)|H$$

since H is hermitian. Now (3.27) gives

$$\frac{d}{dt}\langle A\rangle = \left[\frac{d}{dt}\langle\psi(t)|\right]A|\psi(t)\rangle + \langle\psi(t)|A\left[\frac{d}{dt}|\psi(t)\rangle\right]$$

$$= \left[\frac{i}{\hbar}\langle\psi(t)|H\right]A|\psi(t)\rangle + \langle\psi(t)|A\left[-\frac{i}{\hbar}H|\psi(t)\rangle\right]$$

$$= \langle\psi(t)|\frac{i}{\hbar}[H, A]|\psi(t)\rangle$$

$$= \left\langle\frac{i}{\hbar}[H, A]\right\rangle. \quad\blacksquare$$

Let us apply this to the case of a simple particle moving in space in a potential $V(\mathbf{r})$. For this system

$$H = \frac{p_j p_j}{2m} + V(\mathbf{r}) \tag{3.28}$$

(here, and everywhere else, summation over repeated indices is understood; see Appendix A). Using the commutator rule (2.80) and the form of the momentum operators, (2.113), we have

$$[H, x_i] = \frac{1}{2m}(p_j[p_j, x_i] + [p_j, x_i]p_j) + [V, x_i]$$

$$= -\frac{i\hbar}{m}p_i. \tag{3.29}$$

Also,

$$[V, p_i]\psi = V\left[-i\hbar\frac{\partial\psi}{\partial x_i}\right] + i\hbar\frac{\partial}{\partial x_i}(V\psi)$$

$$= i\hbar\frac{\partial V}{\partial x_i}\psi,$$

so

$$[H, p_i] = [V, p_i] = i\hbar\frac{\partial V}{\partial x_i}. \tag{3.30}$$

Putting (3.29) and (3.30) in turn into (3.26), we find

$$\frac{d}{dt}\langle\mathbf{r}\rangle = \frac{\langle\mathbf{p}\rangle}{m}, \tag{3.31}$$

$$\frac{d}{dt}\langle\mathbf{p}\rangle = -\langle\nabla V\rangle = \langle\mathbf{F}\rangle \tag{3.32}$$

where $\mathbf{F}$ is the classical force on the particle. Thus the equations for the motion of the expectation values are obtained by taking expectation values of all terms in the classical equations of motion.

From eqs. (3.31) and (3.32) the classical equations of motion can be deduced as approximations which hold if the particle can be described approximately as a point particle, i.e. when its wave function is a localised wave packet. The precise statement is

● **3.2   Ehrenfest's theorem.** If the wave function of a particle vanishes outside a convex region V in which the force $\mathbf{F}$ is approximately constant, then the particle behaves like a classical particle with position $\langle \mathbf{r} \rangle$ obeying the classical equation of motion

$$m \frac{d^2}{dt^2} \langle \mathbf{r} \rangle = \mathbf{F}(\langle \mathbf{r} \rangle). \tag{3.33}$$

Proof. From (3.31) and (3.32) we have

$$m \frac{d^2}{dt^2} \langle \mathbf{r} \rangle = \langle \mathbf{F}(\mathbf{r}) \rangle. \tag{3.34}$$

If the normalised wave function of the particle is ψ, the formula (2.69) for the expectation value gives

$$\langle \mathbf{F}(\mathbf{r}) \rangle = \int_V |\psi(\mathbf{r})|^2 \mathbf{F}(\mathbf{r}) \, d^3\mathbf{r}. \tag{3.35}$$

Now $\mathbf{F}$ can be regarded as constant in the region V, which contains $\langle \mathbf{r} \rangle$ since V is convex, so we can write

$$\langle \mathbf{F}(\mathbf{r}) \rangle \simeq \mathbf{F}(\langle \mathbf{r} \rangle) \int_V |\psi(\mathbf{r})|^2 \, d^3\mathbf{r} = \mathbf{F}(\langle \mathbf{r} \rangle) \tag{3.36}$$

since ψ is normalised. ∎

The conditions of Ehrenfest's theorem will not in general remain satisfied at all times, as we can see by studying the change of the uncertainty in position. As an example let us take the case of a free particle ($V = 0$), so that by (3.32) $\langle \mathbf{p} \rangle$ is constant. We will look at the 1-components x_1, p_1 of $\mathbf{r}$ and $\mathbf{p}$. From the formula (2.70) for the uncertainty, together with (3.26), we obtain

$$\frac{d}{dt} (\Delta x_1{}^2) = \left\langle \frac{i}{\hbar} [H, x_1{}^2] \right\rangle - 2 \langle x_1 \rangle \frac{d}{dt} \langle x_1 \rangle, \tag{3.37}$$

$$\frac{d}{dt} (\Delta p_1{}^2) = \left\langle \frac{i}{\hbar} [H, p_1{}^2] \right\rangle - 2 \langle p_1 \rangle \frac{d}{dt} \langle p_1 \rangle. \tag{3.38}$$

Since H commutes with p_1, (3.38) shows that $\Delta p_1{}^2$ is constant. Now a calculation similar to that leading to (3.29) gives

$$[H, x_1{}^2] = -\frac{i\hbar}{m} (x_1 p_1 + p_1 x_1),$$

$$\therefore \quad \frac{d}{dt} (\Delta x_1{}^2) = \frac{1}{m} \langle x_1 p_1 + p_1 x_1 \rangle - \frac{2}{m} \langle x_1 \rangle \langle p_1 \rangle \tag{3.39}$$

$$\therefore \quad \frac{d^2}{dt^2} (\Delta x_1{}^2) = \frac{1}{m} \left\langle \frac{i}{\hbar} [H, x_1 p_1 + p_1 x_1] \right\rangle - \frac{2}{m} \langle p_1 \rangle^2$$

$$= \frac{2}{m} \langle p_1{}^2 \rangle - \frac{2}{m} \langle p_1 \rangle^2$$

$$= \frac{2}{m} (\Delta p_1)^2. \tag{3.40}$$

Thus the expectation values and uncertainties in the components of momentum are constant, but the uncertainties in the coordinates will in general increase with time and the wave packet will spread out.

The probability current The time-dependent Schrödinger equation for a single particle moving in a potential $V(\mathbf{r})$ is

$$ih\frac{\partial\psi}{\partial t}=-\frac{h^2}{2m}\nabla^2\psi+V(\mathbf{r})\psi. \tag{3.7}$$

Multiplying this by $\bar{\psi}$ and taking the imaginary part gives

$$\frac{1}{2}h\left[\bar{\psi}\frac{\partial\psi}{\partial t}+\psi\frac{\partial\bar{\psi}}{\partial t}\right]=-\frac{h^2}{4mi}\,[\bar{\psi}\,\nabla^2\psi-\psi\,\nabla^2\bar{\psi}], \tag{3.41}$$

which can be written as

$$\frac{\partial\rho}{\partial t}+\boldsymbol{\nabla}\cdot\mathbf{j}=0 \tag{3.42}$$

where

$$\rho=|\psi|^2$$

and

$$\mathbf{j}=\frac{\hbar}{2mi}[\bar{\psi}\nabla\psi-\psi\nabla\bar{\psi}]=\frac{\hbar}{m}\,\mathrm{Im}\,[\bar{\psi}\,\boldsymbol{\nabla}\psi]. \tag{3.43}$$

Now ρ is the probability density: if a large number of particles were moving in the potential V simultaneously but independently, forming a cloud of dust, ρ would (almost certainly) be the density of the dust (taking the total mass of the cloud to be 1). Eq. (3.42), which is known in fluid mechanics as the **equation of continuity**, suggests that the vector $\mathbf{j}$ should be interpreted as the **current density** describing the flow of the quantity whose density is ρ. The reason for this can be seen by integrating (3.41) over a region V and using the divergence theorem; this gives

$$\frac{d}{dt}\int_V\rho\,dV=-\int_{\partial V}\mathbf{j}\cdot d\mathbf{S} \tag{3.44}$$

where ∂V is the boundary of the region V. Thus the total probability that the particle is in V decreases at a rate given by the flux of the vector $\mathbf{j}$ across the boundary of V; we can imagine the probability being carried by the vector $\mathbf{j}$. (For a classical cloud of dust we would have $\mathbf{j}=\rho\mathbf{v}$ where $\mathbf{v}$ is the velocity of the dust at the point in question.) For this reason $\mathbf{j}$ is known as the **probability current**.

Boundary conditions for the Schrödinger equation We have seen (in eqs. (3.20)–(3.21) that the time development of the wave function of a set of particles is obtained by first solving the time-independent Schrödinger equation (3.21), i.e. by finding the eigenfunctions and eigenvalues of the Hamiltonian. In the next chapter we will tackle this eigenvalue problem for the most interesting physical systems by algebraic methods; in general, however, the problem must be treated as a differential equation. Since this book is concerned with principles rather than techniques, we refer to other books (e.g. Schiff 1968, Bohm 1951) for details of methods of solving the Schrödinger equation. Here we will simply complete the specification of the problem by finding the boundary conditions that the wave function must satisfy.

The simplest case to consider is that of a single particle moving in one dimension, for which the time-independent Schrödinger equation becomes

$$H\psi \equiv -\frac{\hbar^2}{2m}\frac{d^2\psi}{dx^2} + V(x)\psi = E\psi. \tag{3.45}$$

As an element of the state space $\mathcal{W}$ defined by **W1**–**W3** (p. 39), ψ is square-integrable and uniformly continuous, and therefore satisfies

$$\psi(x) \to 0 \quad \text{as } x \to \pm\infty. \tag{3.46}$$

This is sufficient to act as a boundary condition for the differential equation (3.45). It is sometimes too restrictive: as we saw in §2.5, a hermitian operator may have no eigenfunctions in the space defined by **W1**–**W3**, so that it becomes necessary to look for eigenbras. These will also be given by functions ψ satisfying (3.45), but with the conditions

$$\int_{-\infty}^{\infty} \overline{\psi(x)}\,\phi(x)\,dx < \infty \quad \text{for all } \phi \in \mathcal{W}. \tag{3.47}$$

In particular,

$$\frac{\psi(x)}{x} \to 0 \quad \text{as } x \to \pm\infty. \tag{3.48}$$

If the Hamiltonian H is to be an operator on the state space $\mathcal{W}$, taking smooth functions to smooth functions, then the potential $V(x)$ must also be a smooth function of x. Physically, this is a reasonable condition. However, it is often convenient to consider a discontinuous potential function; such a potential can be a good idealisation of a physical situation, and it can give a Schrödinger equation which is easy to solve. In order to admit discontinuous functions as operators we must use a different space of wave functions; the conditions to be imposed on the wave functions will depend on the nature of the discontinuities in the potential. (This does not involve any departures from the basic physical principles: remember that we are dealing with a simplification of an actual physical situation in which the potential function is smooth and the wave function belongs to our standard state space $\mathcal{W}$.)

We consider only potentials which have simple jump discontinuities at a finite number of points $x_1, \ldots, x_n$ and are smooth everywhere else. We want to extend the space $\mathcal{W}$ to a space $\mathcal{S}$ on which the operator

$$H = -\frac{\hbar^2}{2m}\frac{d^2}{dx^2} + V(x) \tag{3.49}$$

acts and is hermitian. Clearly the only modifications necessary refer to the points of discontinuity of $V(x)$; thus we assume that the functions in $\mathscr{S}$ are smooth everywhere except at $x_1, \ldots, x_n$. Then if $\phi(x)$ and $\psi(x)$ are any two functions in $\mathscr{S}$, regarded as state vectors $|\phi\rangle$, $|\psi\rangle$, we have

$$\langle\phi|H|\psi\rangle = \sum_{i=0}^{n} \int_{x_i}^{x_{i+1}} \phi(x)\left\{-\frac{\hbar^2}{2m}\frac{d^2\psi}{dx^2} + V(x)\psi(x)\right\} dx \tag{3.50}$$

(with $x_0 = -\infty$, $x_{n+1} = \infty$). Integrating the first term by parts and comparing with $\overline{\langle\psi|H|\phi\rangle}$, we find

$$\langle\phi|H|\psi\rangle - \overline{\langle\psi|H|\phi\rangle} = \frac{\hbar^2}{2m}\sum_{i=0}^{n}\Delta_i\left[\bar{\phi}\frac{d\psi}{dx} - \psi\frac{d\bar{\phi}}{dx}\right] \tag{3.51}$$

where $\Delta_i f$ denotes the discontinuity in the function f at x_i:

$$\Delta_i f = f(x_i+) - f(x_i-). \tag{3.52}$$

If H is to be hermitian, the right-hand side of (3.51) must vanish for all ϕ, ψ in $\mathscr{S}$. The most natural way to ensure this is to demand that all functions in $\mathscr{S}$ are continuous and have continuous derivatives. Then to ensure that functions in $\mathscr{S}$ stay in $\mathscr{S}$ after H has operated, we need to impose a condition on the second derivative: it must have a discontinuity to balance the discontinuity in V. Thus the space $\mathscr{S}$ consists of all functions ψ satisfying

    1. ψ and ψ' are continuous at x_i, $\qquad\qquad$ (3.53)

    2. $\dfrac{\hbar^2}{2m}\Delta_i\psi'' = (\Delta_i V)\psi,$ $\qquad\qquad\qquad\qquad$ (3.54)

where the dashes denote differentiation with respect to x. (3.54) is only the first in a chain of conditions giving the discontinuities in the higher derivatives of ψ; these conditions are all automatically satisfied if ψ is a superposition of solutions of the Schrödinger equation (3.45) satisfying the basic continuity conditions (3.53).

If it is necessary to look for eigenbras the problem becomes that of finding a function ψ which satisfies

$$\int_{-\infty}^{\infty} \overline{\psi(x)}\left[-\frac{\hbar^2}{2m}\frac{d^2\phi}{dx^2} + V\phi\right] dx = E\int_{-\infty}^{\infty} \bar{\psi}\phi\, dx \quad \text{for all} \quad \phi \in \mathscr{S}.$$

$$\tag{3.55}$$

In this case the same boundary conditions (3.53) arise as the condition for (3.55) to be equivalent to the Schrödinger equation (3.45) (see problem 3.12).

For example, let us take $V(x)$ to be a simple step function with a discontinuity at $x = 0$:

$$V(x) = \begin{cases} 0 & \text{if } x < 0, \\ V_0 & \text{if } x \geqslant 0. \end{cases}$$

Then if $E > V_0$ the general solution of the Schrödinger equation (3.45) is

    For $x < 0$, $\quad \psi(x) = Ae^{ikx} + Be^{-ikx} \quad$ where $k^2 = 2mE/\hbar^2$;

    for $x > 0$, $\quad \psi(x) = Ce^{iKx} + De^{-iKx} \quad$ where $K^2 = 2m(E - V_0)/\hbar^2$

and the continuity conditions impose the following relations on the constants

A, B, C, D:

$$\psi \text{ continuous at } x=0 \quad \Rightarrow \quad A+B=C+D;$$

$$\psi' \text{ continuous at } x=0 \quad \Rightarrow \quad ikA-ikB=iKC-iKD.$$

A further idealisation which it is sometimes convenient to make is to suppose that the particle can only be found in a certain region R, so that $\psi=0$ outside R (this is often expressed by saying that '$V=\infty$ outside R'). Let us take R to be an interval $[a, b]$. Then the state space is a space of smooth functions on $[a, b]$, with inner product

$$\langle \phi | \psi \rangle = \int_a^b \overline{\phi(x)}\,\psi(x)\,dx. \tag{3.56}$$

Boundary conditions at a and b are needed to make the Hamiltonian (3.49) hermitian, even when $V(x)$ is a smooth function on $[a, b]$. The condition for H to be hermitian is

$$\left[\overline{\phi}\frac{d\psi}{dx} - \frac{d\overline{\phi}}{dx}\,\psi \right]_a^b = 0. \tag{3.57}$$

This can be satisfied by imposing on all functions ψ in $\mathscr{S}$ the condition

$$\psi(a)=\psi(b)=0; \tag{3.58}$$

there is no need to impose a condition on the derivative of ψ. ((3.56) could be satisfied by imposing the condition $\psi'(a)=\psi'(b)=0$ instead of (3.58); the latter is chosen because it makes the momentum operator $p=-i\hbar\,d/dx$ hermitian.)

For a particle which is confined to the interval $[a, b]$ but is otherwise free, the Hamiltonian is $H=p^2/2m$ and the Schrödinger equation (3.45) becomes

$$-\frac{\hbar^2}{2m}\frac{d^2\psi}{dx^2} = E\psi \tag{3.59}$$

with boundary conditions (3.58). The solutions of this differential equation are

$$\psi = A \sin k(x-a) + B \cos k(x-a) \tag{3.60}$$

where

$$k^2 = 2mE/\hbar^2;$$

the boundary conditions then require

$$B=0, \quad k(b-a)=n\pi \tag{3.61}$$

where n is an integer. Thus E can only take the values

$$E=\frac{n^2\pi^2\hbar^2}{2m(b-a)^2}, \quad n\in\mathbb{Z}, \tag{3.62}$$

which form a discrete set. This discreteness is the basic quantum effect; it is characteristic of a particle which is confined to a bounded region.

3.2. Invariances and constants of the motion

Suppose the physical system we are considering can exist anywhere in space. Then we can consider the effect of moving every part of the system through the same displacement $\mathbf{a}$; such an operation is called a **translation**. To every state

$|\psi\rangle$ of the system there will correspond another state $|\psi'\rangle$ in the new position; $|\psi'\rangle$ is obtained by doing whatever was done to obtain $|\psi\rangle$ but with the apparatus all displaced through the vector **a**. Now if the basic laws of physics are the same in all places, it seems reasonable to suppose that relations of superposition between states will not be affected by the translation – that is,

$$|\psi\rangle=c_1|\psi_1\rangle+c_2|\psi_2\rangle \quad \Rightarrow \quad |\psi'\rangle=c_1|\psi_1'\rangle+c_2|\psi_2'\rangle \tag{3.63}$$

– because whatever modification to the apparatus in one place is needed to produce $c_1|\psi_1\rangle+c_2|\psi_2\rangle$ rather than $|\psi_1\rangle$ or $|\psi_2\rangle$, the same modification in the new place will produce $c_1|\psi_1'\rangle+c_2|\psi_2'\rangle$. (A more general possibility will be considered later; see p. 96.) Thus the correspondence between $|\psi\rangle$ and $|\psi'\rangle$ is linear:

$$|\psi'\rangle=U(T_a)|\psi\rangle \tag{3.64}$$

where $U(T_a)$ is a linear operator on state space. Similarly, the relations between states expressed by their inner products will be the same after the translation, so

$$\langle\phi'|\psi'\rangle=\langle\phi|\psi\rangle. \tag{3.65}$$

Taking the hermitian conjugate of (3.64),

$$\langle\psi'|=\langle\psi|U(T_a)^\dagger. \tag{3.66}$$

Hence (3.65) gives

$$\langle\phi|U(T_a)^\dagger U(T_a)|\psi\rangle=\langle\phi|\psi\rangle. \tag{3.67}$$

Since this holds for any states $|\phi\rangle$ and $|\psi\rangle$, we must have

$$U(T_a)^\dagger U(T_a)=1. \tag{3.68}$$

Thus $U(T_a)$ is a unitary operator.

The linearity and unitarity of $U(T_a)$ are consequences of the assumption that the relations between states at one time are not affected by the translation, or in other words that the general formalism for describing the system is invariant under translations. We will say that translations are **unitaristic** operations.

In general, we will distinguish between an **operation**, which is performed on the physical system itself, and an **operator**, which acts mathematically on state vectors. (The distinction runs parallel to that between states and state vectors.) An operation is unitaristic if it does not affect the static laws of quantum mechanics for the system, and is represented by a unitary operator on the state space of the system. Our notation incorporates the distinction between operations and operators: T_a denotes the operation of translating the system through the vector **a**, while $U(T_a)$ denotes the corresponding operator.

Now suppose that the evolution in time of the system is also invariant under translations, so that the relations between states at different times remain the same after translation. Then if $|\psi(t)\rangle$ is the sequence of states to which $|\psi(0)\rangle$ evolves, the translated state $U(T_a)|\psi(0)\rangle$ will evolve to the sequence $U(T_a)|\psi(t)\rangle$. Thus if $|\psi(t)\rangle$ satisfies the Schrödinger equation, so does $U(T_a)|\psi(t)\rangle$:

$$i\hbar\frac{d}{dt}[U(T_a)|\psi(t)\rangle]=HU(T_a)|\psi(t)\rangle. \tag{3.69}$$

But

$$ih\frac{d}{dt}\left[U(T_a)|\psi(t)\rangle\right] = U(T_a)\cdot ih\frac{d}{dt}|\psi(t)\rangle = U(T_a)H|\psi(t)\rangle \qquad (3.70)$$

if $|\psi(t)\rangle$ satisfies the Schrödinger equation. In particular this applies when $t=0$; and since $|\psi(0)\rangle$ could be any state, we can conclude that

$$HU(T_a) = U(T_a)H. \qquad (3.71)$$

Thus the condition for the behaviour of a system to be invariant under translations is that the translation operators should commute with the Hamiltonian.

Whether this invariance is true for a particular system depends on how widely the system is defined. In classical mechanics, for example, the system of a particle moving in the earth's gravitational field is not invariant under translations; the particle will behave differently if it is moved away from the earth, because the field there is weaker. However, if the system is widened to include the earth, then this wider system is invariant under translations: if the particle and the earth are translated together to any position in the universe, the forces between them remain the same because the distance between them does. It is generally, and strongly, believed that if one includes all relevant matter in the system, this invariance will always apply: the laws of physics do not vary from place to place. This is the lasting legacy of the Copernican revolution: the firm belief that there is no centre of the universe, or any other special point in space.

When $\mathbf{a}=\mathbf{0}$ the translation has no effect on the system, so $T(\mathbf{0})$ is the identity operator. Now assume that $U(T_a)$ is a differentiable function of $\mathbf{a}$, and put

$$D_i = \frac{\partial}{\partial a_i}U(T_a)\Big|_{\mathbf{a}=\mathbf{0}}. \qquad (3.72)$$

D_i is called an **infinitesimal translation operator**. Differentiating (3.68) and putting $\mathbf{a}=\mathbf{0}$, we find

$$D_i + D_i^\dagger = 0. \qquad (3.73)$$

Thus D_i is antihermitian. Let $P_i = ihD_i$; then P_i is hermitian, and therefore is qualified to describe an observable.

We will identify the observable P_i for the case of a single particle in space by determining the operators $U(T_a)$ and D_i. The effect of $U(T_a)$ on a wave function ψ is to produce a wave function ψ' whose value at $\mathbf{r}+\mathbf{a}$ is the same as the value of ψ at $\mathbf{r}$ (this is illustrated in Fig. 3.1, in which the value of ψ at a point is given by the density of ink at that point). Thus we have

$$\psi'(\mathbf{r}+\mathbf{a}) = \psi(\mathbf{r}) \qquad (3.74)$$

or, writing $\psi' = U(T_a)\psi$,

$$[U(T_a)\psi](\mathbf{r}) = \psi(\mathbf{r}-\mathbf{a}). \qquad (3.75)$$

Hence, from (3.72),

$$D_i\psi(\mathbf{r}) = \frac{\partial}{\partial a_i}\left[U(T_\mathbf{a})\psi\right](\mathbf{r})\big|_{\mathbf{a}=0} = \frac{\partial}{\partial a_i}\left[\psi(\mathbf{r}-\mathbf{a})\right]_{\mathbf{a}=0}$$

$$= -\frac{\partial\psi}{\partial x_i}(\mathbf{r}); \tag{3.76}$$

$$\therefore\ P_i = ihD_i = -ih\frac{\partial}{\partial x_i}. \tag{3.77}$$

Thus the observables P_i are the components of momentum.

Now suppose the behaviour of the system is invariant under translations, so that (3.71) holds. Differentiating, we find

$$HP_i = P_iH, \tag{3.78}$$

i.e. P_i commutes with the Hamiltonian.

An observable which commutes with the Hamiltonian is called a **conserved quantity** or a **constant of the motion**. It can be seen from ●3.1 that such an observable has a constant expectation value; in fact it has the following stronger property:

●**3.3** If the observable A commutes with the Hamiltonian, the probability that A takes a particular value α is constant.

Proof. Let $\mathscr{S}_\alpha$ be the subspace of eigenstates of A with the given eigenvalue α, and let P_α be the orthogonal projection onto $\mathscr{S}_\alpha$. We will show that H commutes with P_α. First note that if $|\psi_\alpha\rangle$ belongs to $\mathscr{S}_\alpha$, so that $A|\psi_\alpha\rangle = \alpha|\psi_\alpha\rangle$, then

$$AH|\psi_\alpha\rangle = HA|\psi_\alpha\rangle = \alpha H|\psi_\alpha\rangle, \tag{3.79}$$

i.e. $H|\psi_\alpha\rangle$ also belongs to $\mathscr{S}_\alpha$. Now any state $|\psi\rangle$ can be written as

$$|\psi\rangle = |\psi_\alpha\rangle + |\psi_\perp\rangle \tag{3.80}$$

where $|\psi_\alpha\rangle = P_\alpha|\psi\rangle$ belongs to $\mathscr{S}_\alpha$ and $|\psi_\perp\rangle$ is orthogonal to $\mathscr{S}_\alpha$. Then $H|\psi_\alpha\rangle$ belongs to $\mathscr{S}_\alpha$; also $H|\psi_\perp\rangle$ is orthogonal to $\mathscr{S}_\alpha$, for if $|\phi\rangle$ is any vector in $\mathscr{S}_\alpha$ then $H|\phi\rangle$ is also in $\mathscr{S}_\alpha$ and so

$$\langle\phi|H|\psi_\perp\rangle = \overline{\langle\psi_\perp|H|\phi\rangle} = 0. \tag{3.81}$$

Fig. 3.1.
The effect of translation on a wave packet.

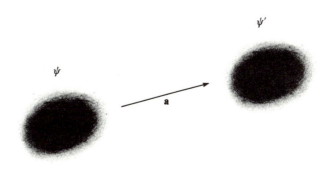

Thus $H|\psi_\alpha\rangle$ is the orthogonal projection of $H|\psi\rangle$ onto $\mathcal{S}_\alpha$, i.e.

$$HP_\alpha|\psi\rangle = P_\alpha H|\psi\rangle. \tag{3.82}$$

Since $|\psi\rangle$ is any state vector, it follows that $HP_\alpha = P_\alpha H$.

Now using the form (2.42) for the probability that A has the value α, and ●3.1 for the rate of change of an expectation value, we find

$$\frac{d}{dt}p_A(\alpha\,|\,\psi) = \frac{d}{dt}\langle\psi|P_\alpha|\psi\rangle = \frac{i}{\hbar}\langle\psi|[H, P_\alpha]|\psi\rangle = 0. \quad\blacksquare$$

We have now seen the property of commuting with the Hamiltonian being significant for two different types of operator: unitary operators, like those which describe the effect of translations; and hermitian operators, which represent observables. The significance is different in the two cases; the general situation can be summarised as

> ●**3.4** If a unitary operator commutes with the Hamiltonian, it describes a physical operation on the system which leaves its behaviour invariant.
>
> If a hermitian operator commutes with the Hamiltonian, it describes an observable which is a conserved quantity. ■

And we can easily generalise the reasoning which, in the case of translations and momentum, yielded a connection between unitary operators and hermitian ones, to prove

> ●**3.5** If $U(\lambda)$ is a family of unitary operators depending differentiably on a real parameter λ, with $U(0) = 1$, then
>
> $$X = i\hbar\frac{dU}{d\lambda}\bigg|_{\lambda=0} \tag{3.83}$$
>
> is a hermitian operator. If the $U(\lambda)$ describe invariances Q_λ of the system, then X describes a conserved observable. ■

In many examples of physical interest, the operations Q_λ form a *group* in which the composition of operations is given by adding the parameters, i.e.

$$Q_\lambda \circ Q_\mu = Q_{\lambda+\mu} \tag{3.84}$$

where $Q_\lambda \circ Q_\mu$ is the combined operation of Q_μ followed by Q_λ. In this case the unitary operators $U(Q_\lambda)$ can be expressed in terms of the hermitian operator X, as follows:

> ●**3.6** Let Q_λ be a family of operations on a physical system which are labelled by a real parameter and satisfy (3.84), and whose effect on the states of the system is described by unitary operators. Then these operators can be chosen to be of the form
>
> $$U(Q_\lambda) = e^{-i\lambda X/\hbar} \tag{3.85}$$
>
> where X is a hermitian operator.

Proof. To simplify the notation we will write $U(Q_\lambda) = U(\lambda)$. Suppose that we start off by associating the operation Q_λ with the unitary operator $U_1(\lambda)$. Then the state vector $U_1(\lambda)U_1(\mu)|\psi\rangle$ describes the state of the system obtained by applying first Q_μ and then Q_λ to the system when it is in the state $|\psi\rangle$. This state results from applying $Q_{\lambda+\mu}$, so it is also described by $U_1(\lambda+\mu)|\psi\rangle$. Hence

$$U_1(\lambda+\mu)|\psi\rangle = \omega(\lambda, \mu)U_1(\lambda)U_1(\mu)|\psi\rangle \tag{3.86}$$

where $\omega(\lambda, \mu)$ is a complex number with unit modulus; it is independent of $|\psi\rangle$ since $U_1(\lambda+\mu)$ has to be linear. This holds for all $|\psi\rangle$, so

$$U_1(\lambda+\mu) = \omega(\lambda, \mu)U_1(\lambda)U_1(\mu). \tag{3.87}$$

We can assume that $U_1(0) = 1$; this gives

$$\omega(\lambda, 0) = \omega(0, \mu) = 1. \tag{3.88}$$

Now

$$\frac{dU_1}{d\lambda} = \lim_{\delta\lambda \to 0}\left[\frac{U_1(\lambda+\delta\lambda) - U_1(\lambda)}{\delta\lambda}\right] = \lim_{\delta\lambda \to 0}\left[\frac{\omega(\lambda, \delta\lambda)U_1(\delta\lambda)U_1(\lambda) - U_1(\lambda)}{\delta\lambda}\right]$$

$$= \lim_{\delta\lambda \to 0}\left[\frac{\omega(\lambda, \delta\lambda)U_1(\delta\lambda) - U_1(0)}{\delta\lambda}\right]U_1(\lambda) = \frac{\partial}{\partial\mu}\left[\omega(\lambda, \mu)U_1(\mu)\right]_{\mu=0}U_1(\lambda).$$

Let $(\partial\omega/\partial\mu)(\lambda, 0) = \alpha(\lambda)$ and $i\hbar(dU_1/d\mu) = X$; then

$$\frac{dU_1}{d\lambda} = \left[\alpha(\lambda) + \frac{X}{i\hbar}\right]U_1(\lambda) \tag{3.89}$$

since $U_1(0) = 1$ and $\omega(\lambda, 0) = 1$. Let

$$U(\lambda) = \exp\left[-\int_0^\lambda \alpha(\lambda')\, d\lambda'\right]U_1(\lambda); \tag{3.90}$$

then

$$\frac{dU}{d\lambda} = \frac{X}{i\hbar}U(\lambda). \tag{3.91}$$

The solution of this differential equation with $U(0) = 1$ is

$$U(\lambda) = e^{-i\lambda X/\hbar} \tag{3.92}$$

as can be seen by writing $|\psi(\lambda)\rangle = U(\lambda)|\psi_0\rangle$, so that

$$i\hbar\frac{d}{d\lambda}|\psi(\lambda)\rangle = X|\psi(\lambda)\rangle; \tag{3.93}$$

this is the same as the Schrödinger equation (3.4) with X instead of H and λ instead of t, so its solution is the same as (3.3) with these changes.

Since $\omega(\lambda, \mu)$ has modulus 1 for all λ and μ, $\alpha(\lambda)$ is purely imaginary and so the exponential in (3.90) has modulus 1. Thus $U(\lambda)$ differs from $U_1(\lambda)$ only by a phase factor and so it describes the same physical operation Q_λ. ∎

As an example of (3.85), take $Q_\lambda = T_{\lambda i}$ to be a translation in the x-direction, so that $U(Q_\lambda)$ acts on wave functions by (3.74). Take $X = -i\hbar\,\partial/\partial x$; then by expanding the exponential as a power series, (3.85) can be recognised as Taylor's theorem.

The hermitian operator X is called the **hermitian generator** of the unitary family $U(Q_\lambda)$. Note that in general X is not uniquely determined by the unitaristic operations Q_λ, because of the freedom to multiply the unitary operators $U(Q_\lambda)$ by a phase factor $\omega(\lambda)$. This could have the effect of adding a multiple of the identity to X. However, this possibility is often eliminated if there are several families of Q_λ which combine to form a larger group (see §3.3).

As we noted in the course of proving ●3.6, the hermitian generator X is related to the unitary operators $U(Q_\lambda)$ in the same way as the Hamiltonian H is related to the time development operators $U(t)$ of §3.1. Now $U(t)$ represents the operation of taking the state at time t_0 to the state at time $t_0 + t$, which operation is to time as translations are to space. Thus the Hamiltonian can be seen as the hermitian generator of **time translations.**

In classical mechanics as well as in quantum mechanics a continuous family of invariances is associated with a conserved quantity (see Goldstein 1980, p. 411). There, too, invariance under translations is associated with conservation of momentum. Another important example is invariance under rotations (about a given point, say the origin); the associated conserved quantity is angular momentum (about that point). These two examples are of fundamental importance, and need to be formulated as one of the basic postulates:

> **Postulate VII.** Translations and rotations are unitaristic operations: i.e. on the state space of any physical system in space there are unitary operators $U(T_\mathbf{a})$ and $U(R(\mathbf{n}, \theta))$ representing the effects of a translation through the vector $\mathbf{a}$ and a rotation about the axis $\mathbf{n}$ through the angle θ. For fixed $\mathbf{n}$, the observable associated with the family $U(\lambda) = U(T_{\lambda\mathbf{a}})$ is $\mathbf{P} \cdot \mathbf{n}$, where $\mathbf{P}$ is the total momentum of the system, and the observable associated with the family $U(\lambda) = U(R(\mathbf{n}, \lambda))$ is $\mathbf{J} \cdot \mathbf{n}$, where $\mathbf{J}$ is the total angular momentum of the system.

From this postulate we can deduce the form of the operators representing the components of angular momentum of a single particle described by a wave function, using an argument similar to that which gave the components of linear momentum in (3.77). A rotation R is an operation upon three-dimensional vectors; if we regard these as three-component column vectors, R can be identified with a 3×3 matrix. If R is a rotation about the z-axis through an angle θ, for example,

$$R = R(\mathbf{k}, \theta) = \begin{bmatrix} \cos\theta & -\sin\theta & 0 \\ \sin\theta & \cos\theta & 0 \\ 0 & 0 & 1 \end{bmatrix}. \tag{3.94}$$

The unitary operator $U(R)$ acts on wave functions in a similar way to the translation operator $U(T_\mathbf{a})$, as shown in Fig. 3.1; $U(R)$ acts on a wave function ψ to produce a new function ψ' whose value at $R\mathbf{r}$ is the same as the value of ψ at $\mathbf{r}$. Hence

$$[U(R)\psi](\mathbf{r}) = \psi(R^{-1}\mathbf{r}) \tag{3.95}$$

(cf. (3.75)). In particular,

$$[U(R(\mathbf{k}, \theta))\psi](x, y, z) = \psi(x\cos\theta + y\sin\theta, -x\sin\theta + y\cos\theta, z). \tag{3.96}$$

By ●3.5 and Postulate VII, the z-component of angular momentum is represented by the operator

$$J_z = i\hbar \frac{d}{d\theta} U(R(\mathbf{k}, \theta))|_{\theta=0}. \tag{3.97}$$

Hence, by differentiating (3.96) and putting $\theta = 0$, we find

$$\frac{1}{i\hbar} [J_z \psi](x, y, z) = y \frac{\partial \psi}{\partial x}(x, y, z) - x \frac{\partial \psi}{\partial y}(x, y, z),$$

i.e.

$$J_z = i\hbar \left(y \frac{\partial}{\partial x} - x \frac{\partial}{\partial y} \right)$$

$$= xp_y - yp_x \tag{3.98}$$

using the usual operators (2.112)–(2.113) for the components of momentum. (3.98) is the z-component of the vector equation

$$\mathbf{J} = \mathbf{r} \times \mathbf{p}, \tag{3.99}$$

which is the classical definition of the angular momentum of the particle. Similar arguments show that the x- and y-components of (3.99) are also true in the quantum case.

Postulate VII also enables us to determine the angular momentum operator for a photon. Consider a photon travelling in the z-direction, with its polarisation states $|\phi_x\rangle$ and $|\phi_y\rangle$. The effect of a rotation about the z-axis through angle θ is to take the state $|\phi_x\rangle$ to a state with polarisation vector at an angle θ to the x-axis:

$$U(R(\mathbf{k}, \theta))|\phi_x\rangle = \cos\theta|\phi_x\rangle + \sin\theta|\phi_y\rangle$$

Similarly

$$U(R(\mathbf{k}, \theta))|\phi_y\rangle = -\sin\theta|\phi_x\rangle + \cos\theta|\phi_y\rangle$$
$$\left. \right\} . \tag{3.100}$$

Thus the z-component of angular momentum is given (as an operator) by

$$J_z|\phi_x\rangle = i\hbar \frac{d}{d\theta}[\cos\theta|\phi_x\rangle + \sin\theta|\phi_y\rangle]_{\theta=0} = i\hbar|\phi_y\rangle$$
$$J_z|\phi_y\rangle = -i\hbar|\phi_x\rangle$$
$$\left. \right\} . \tag{3.101}$$

The other components of angular momentum will involve states with different directions of motion for the photon, since other rotations will take a photon moving in the z-direction to one moving in some other direction.

Parity A third kind of geometrical operation which assumes dynamical significance in quantum mechanics is **space inversion**. Like the operations of rotation, it is defined with respect to a particular point, which we take to be the origin; it consists of taking the point $\mathbf{r}$ to the point $-\mathbf{r}$. An operation which is somewhat easier to visualise can be obtained by combining this with a rotation through π about an axis $\mathbf{n}$; the result is mirror reflection in the plane perpendicular to $\mathbf{n}$. Neither of these operations can be performed on a physical object;

nevertheless they constitute physical operations in the sense that for any physical system one can construct another system which is related to the first by space inversion (if the first system had a particle at the point **r**, the second has one at $-\mathbf{r}$).

This operation changes a left hand into a right hand; thus to say that it is unitaristic is to say that there is no particular handedness built into the laws of physics. This seems a reasonable supposition. (The laws of economics may discriminate against left-handed people who want to buy an appropriate pair of scissors, but the laws of physics do not make the use of their scissors, once they have found some, any more difficult than for right-handed people.) If it is true, there is a unitary operator P which represents the effect of space inversion on state vectors: on wave functions, for example, it acts by

$$[P\psi](\mathbf{r}) = \psi(-\mathbf{r}). \tag{3.102}$$

If space inversion is applied to a system twice, it brings it back to its original state; thus $P^2|\psi\rangle$ must be a multiple of $|\psi\rangle$. By redefining P with a phase factor if necessary, we can arrange that

$$P^2 = 1. \tag{3.103}$$

Since there is no continuous family of operators in this case, there is no hermitian operator associated with inversion as in ●3.5, and so we would not expect there to be a conserved quantity associated with invariance under inversion. In classical mechanics indeed there is no such conserved quantity. In quantum mechanics, however, (3.103) and the fact that P is unitary give

$$P = P^{-1} = P^\dagger, \tag{3.104}$$

i.e. P is hermitian as well. Thus P itself represents an observable. This observable is called **parity**; because of (3.103) its eigenvalues can only be ± 1 (these eigenvalues are also called **even** and **odd** respectively). If the operation of space inversion is an invariance of the system, then P (as a unitary operator) commutes with the Hamiltonian and therefore (as an observable) is a conserved quantity.

Time reversal   So far we have assumed that a physical operation must be represented on state space by a *linear* operator. The grounds for this assumption were that an equation of the form

$$|\psi\rangle = c_1|\psi_1\rangle + c_2|\psi_2\rangle \tag{3.105}$$

expresses a relation between the states $|\psi\rangle$, $|\psi_1\rangle$ and $|\psi_2\rangle$ which must be preserved by the operation. However, this argument is careless of the distinction between states and state vectors. The coefficients c_1, c_2 describe a relation between state vectors, but it is only their squared moduli $|c_1|^2, |c_2|^2$ that are significant in the relations between physical states. Their phases are significant, but only to the extent that they enter into probabilities, which are given by expressions like $|\langle\phi|\psi\rangle|^2$. Thus the only property which is needed to describe the effect of an operation which does not affect the static laws of quantum mechanics is that the state vectors $|T\phi\rangle, |T\psi\rangle$ which describe the states $|\phi\rangle, |\psi\rangle$ after the operation should satisfy

$$|\langle T\phi | T\psi\rangle| = |\langle \phi | \psi\rangle| \quad \text{for all } |\phi\rangle, |\psi\rangle. \tag{3.106}$$

Wigner's theorem (whose proof is sketched in problem 3.23) states that if T is a mapping of state vectors which satisfies (3.106), then for each $|\psi\rangle$ we can find a complex number ω, with $|\omega| = 1$, so that the map U defined by

$$U|\psi\rangle = \omega |T\psi\rangle \tag{3.107}$$

either is linear and unitary or satisfies

$$U(c_1|\psi_1\rangle + c_2|\psi_2\rangle) = \bar{c}_1 U|\psi_1\rangle + \bar{c}_2 U|\psi_2\rangle. \tag{3.108}$$

In the latter case U is said to be **antilinear**. As a consequence of (3.106), if U is antilinear it must also satisfy

$$\langle \phi' | \psi'\rangle = \overline{\langle \phi | \psi\rangle} \quad \text{where } |\phi'\rangle = U|\phi\rangle, |\psi'\rangle = U|\psi\rangle : \tag{3.109}$$

U is said to be **antiunitary**.

An example of an antilinear operator is the operator K of complex conjugation with respect to a complete set $|\psi_1\rangle, |\psi_2\rangle, \ldots$, which is defined as follows:

$$\text{If} \quad |\psi\rangle = \sum c_i |\psi_i\rangle, \quad K|\psi\rangle = \bar{c}_i |\psi_i\rangle. \tag{3.110}$$

(Note that the notion of complex conjugation has no absolute meaning in a vector space, but depends on a choice of complete set.) More generally, the operator of complex conjugation with respect to a complete set of bras is defined by

$$\langle \psi_i | K | \psi\rangle = \overline{\langle \psi_i | \psi\rangle}. \tag{3.111}$$

Given such a complex conjugation operator K, any antiunitary operator U can be written as $U = KV$ where V is linear and unitary.

The only physically significant operation which is represented by an antilinear operator is that of **time reversal**. This is the operation of leaving all the parts of a system in the same positions but reversing all their momenta and angular momenta. For a system of a single simple particle time reversal is represented by the operator T of complex conjugation with respect to the eigenbras of position, which simply has the effect of complex-conjugating the wave function:

$$T\psi(\mathbf{r}) = \overline{\psi(\mathbf{r})}. \tag{3.112}$$

This leaves the particle in the same position, in the sense that it does not change the probability that the particle will be found in a given position; it reverses the momentum of the particle, for if $\psi = e^{i\mathbf{k}\cdot\mathbf{r}}$ is an eigenfunction of momentum with eigenvalue $\hbar\mathbf{k}$ then $T\psi = e^{-i\mathbf{k}\cdot\mathbf{r}}$ is an eigenfunction of momentum with eigenvalue $-\hbar\mathbf{k}$ (more correctly: the probability that the particle will be found to have momentum $\mathbf{p}$ when it is in state $|\psi\rangle$ equals the probability that it will be found to have momentum $-\mathbf{p}$ when it is in the state $T|\psi\rangle$).

A system is **invariant under time reversal** if $T|\psi(-t)\rangle$ is a possible sequence of states (i.e. a solution of the Schrödinger equation) whenever $|\psi(t)\rangle$ is a possible sequence of states. The condition for this invariance is the same as the condition for invariance under other operations:

●**3.7**   A system is invariant under time reversal if the Hamiltonian commutes with the antilinear operator T.

In particular, a system of several simple particles moving in a real potential $V(\mathbf{r}_1, \ldots, \mathbf{r}_n)$ is invariant under time reversal.

Proof. If $|\psi(t)\rangle$ is a possible sequence of states,

$$i\hbar \frac{d}{dt} |\psi(t)\rangle = H|\psi(t)\rangle. \tag{3.113}$$

Let $|\phi(t)\rangle = T|\psi(-t)\rangle$; then

$$
\begin{aligned}
i\hbar \frac{d}{dt} |\phi(t)\rangle &= i\hbar T \frac{d}{dt} |\psi(-t)\rangle \\
&= T\left[-i\hbar \frac{d}{dt} |\psi(-t)\rangle \right] \quad \text{since } T \text{ is antilinear} \\
&= TH|\psi(-t)\rangle \\
&= HT|\psi(-t)\rangle \quad \text{if } T \text{ commutes with } H \\
&= H|\phi(t)\rangle.
\end{aligned}
$$

Thus $|\phi(t)\rangle$ is also a possible sequence of states and so the system is invariant under time reversal.

For a system of several simple particles moving in a real potential the Hamiltonian is

$$H = -\frac{\hbar^2}{2m_1} \nabla_1{}^2 - \cdots - \frac{\hbar^2}{2m_n} \nabla_n{}^2 + V(\mathbf{r}_1, \ldots, \mathbf{r}_n).$$

This is a real operator and clearly commutes with complex conjugation of the wave function:

$$H\bar{\psi} = \overline{H\psi}.$$

i.e. H commutes with T. Hence the system is invariant under time reversal. ∎

Combined systems   If two systems S_1 and S_2 are put together to form a combined system $S_1 S_2$, as in §2.6, then the effect of a physical operation on the combined system can be obtained by performing the operation on both S_1 and S_2. Let $\mathscr{S}_1$ and $\mathscr{S}_2$ be the state spaces of S_1 and S_2, and suppose the operation Q is represented by a unitary operator $U_1(Q)$ on $\mathscr{S}_1$ and by a unitary operator $U_2(Q)$ on $\mathscr{S}_2$. Then if the combined system $S_1 S_2$ is in the state $|\phi\rangle|\psi\rangle$, i.e. S_1 is in state $|\phi\rangle$ and S_2 is in state $|\psi\rangle$, the result of the operation will be to put $S_1 S_2$ in the state in which S_1 is in state $U_1(Q)|\phi\rangle$ and S_2 is in state $U_2(Q)|\psi\rangle$. Denoting the corresponding unitary operator on $\mathscr{S}_1 \otimes \mathscr{S}_2$ by $U(Q)$, we have

$$U(Q)[|\phi\rangle|\psi\rangle] = [U_1(Q)|\phi\rangle][U_2(Q)|\psi\rangle]. \tag{3.114}$$

$U(Q)$ is called the tensor product of the operators $U_1(Q)$ and $U_2(Q)$; we write

$$U(Q) = U_1(Q) \otimes U_2(Q). \tag{3.115}$$

If we identify $U_1(Q)$ with $U_1(Q) \otimes 1$, as was discussed in §2.6 (in the footnote to (2.128)), we can write

$$U(Q) = U_1(Q)U_2(Q). \tag{3.116}$$

Now suppose we have a sequence of operations Q_λ, with Q_0 being the identity as usual. Then according to ●3.5 there are observables K_1 and K_2 of the systems S_1 and S_2 whose operators are the hermitian generators of the families $U_1(Q_\lambda)$ and $U_2(Q_\lambda)$; these operators act on the combined state space $\mathscr{S}_1 \otimes \mathscr{S}_2$ as in (2.128). The hermitian generator K of the family $U(Q)$ is given by

$$\begin{aligned} K|\phi\rangle|\psi\rangle &= \frac{d}{d\lambda} [U(Q_\lambda)|\phi\rangle|\psi\rangle]_{\lambda=0} \\ &= \frac{d}{d\lambda} [U_1(Q_\lambda)|\phi\rangle U_2(Q_\lambda)|\psi\rangle]_{\lambda=0} \\ &= [K_1|\phi\rangle]|\psi\rangle + |\phi\rangle[K_2|\psi\rangle], \end{aligned} \tag{3.117}$$

using the product rule for differentiation and the fact that $U_1(Q_0) = U_2(Q_0) = 1$. Hence

$$K = K_1 + K_2: \tag{3.118}$$

the observable K for the combined system is the sum of the corresponding observables for the constituent systems. An observable with this property is called an **additive quantum number**. We have just proved

●**3.8** If an observable is the hermitian generator of a family of operations, it is an additive quantum number. ■

Thus momentum and angular momentum, as we would expect from classical mechanics, are additive (vector) quantities. Parity, on the other hand, is not. Since the parity operator is itself the unitary operator representing an operation, (3.116) gives for a combined system

$$P = P_1 P_2. \tag{3.119}$$

This means that the eigenvalues of parity in the combined system are the products of its eigenvalues in the constituent systems. Such an observable is called a **multiplicative quantum number**.

Transformation of observables A physical operation like a translation or a rotation has an effect on the observables of a system as well as its states. If A is any observable, we can define an observable A' which is measured by the same experiment as A but with the apparatus all translated through a vector **a**. The operator representing A can then be expressed in terms of A and the unitary operator

$U(T_a)$. The general statement is

●**3.9** Let Q be a unitaristic operation on a quantum system, represented by the unitary operator $U(Q)$. Let A be an observable of the system, measured by an experiment E, and let A' be the observable which is measured by the same experiment after the operation Q has been applied to the apparatus. Then A' is described by the operator

$$A' = U(Q) \cdot A \cdot U(Q)^{-1}. \tag{3.120}$$

If Q_λ is a continuous family of operations associated with the hermitian operator X by (3.83), then

$$i\hbar \left. \frac{dA'}{d\lambda} \right|_{\lambda=0} = [X, A]. \tag{3.121}$$

Proof. Let $|\psi_\alpha\rangle$ be an eigenstate of A with eigenvalue α, so that the experiment E certainly gives the result α. If the experiment is performed after applying the operation Q then, since Q does not affect the static laws of physics, the result will again certainly be α. This second experiment measures A'; the state of the system is $U(Q)|\psi_\alpha\rangle$; so $U(Q)|\psi_\alpha\rangle$ is an eigenstate of A' with eigenvalue α. Hence

$$A'U(Q)|\psi_\alpha\rangle = \alpha U(Q)|\psi_\alpha\rangle = U(Q)A|\psi_\alpha\rangle. \tag{3.122}$$

Since the eigenstates $|\psi_\alpha\rangle$ form a complete set of states, we can deduce that

$$A'U(Q) = U(Q)A, \tag{3.123}$$

which is the same as (3.120).

In the case of a continuous family of operations we must replace Q in (3.123) by Q_λ. Differentiating with respect to λ gives

$$\frac{dA'}{d\lambda} U(Q_\lambda) + A' \frac{d}{d\lambda} U(Q_\lambda) = \frac{d}{d\lambda} U(Q_\lambda)A.$$

When $\lambda = 0$, $U(Q_\lambda) = 1$ and $A' = A$; hence, using (3.83),

$$i\hbar \left. \frac{dA'}{d\lambda} \right|_{\lambda=0} = XA - AX. \quad ■$$

This has important applications when the operations are translations and rotations. First let us take Q to be a translation through $\mathbf{a}$, and let $A = f(\mathbf{r})$ be any function of the position vector $\mathbf{r}$. If the position-measuring apparatus is translated through a vector $\mathbf{a}$, it will give the result $\mathbf{r}'$ when the particle is at $\mathbf{r} = \mathbf{r}' + \mathbf{a}$. Hence

$$A' = f(\mathbf{r} - \mathbf{a}). \tag{3.124}$$

Now put $\mathbf{a} = \lambda\mathbf{n}$; then we have the family of unitary operators $U(T_{\lambda\mathbf{n}})$. According to Postulate VII, the associated hermitian operator is $X = \mathbf{P} \cdot \mathbf{n}$. Taking $\mathbf{n}$ to be a unit vector along the ith coordinate axis gives

$$i\hbar \frac{dA'}{d\lambda} = -i\hbar \frac{\partial f}{\partial x_i} = [P_i, f(\mathbf{r})] \tag{3.125}$$

by (3.121). Taking $f(\mathbf{r}) = x_j$ shows that the canonical commutation relation (2.117) is a special case of (3.125).

Secondly, take Q to be a rotation R, and let $A = \mathbf{m} \cdot \mathbf{V}$ be the component in the direction $\mathbf{m}$ of the vector observable $\mathbf{V}$. Then when the apparatus is tilted by the rotation R, it will measure the component of $\mathbf{V}$ in the rotated direction $R\mathbf{m}$; thus

$$A' = (R\mathbf{m}) \cdot \mathbf{V} = \mathbf{m} \cdot (R^{-1}\mathbf{V}). \tag{3.126}$$

If $U(R)$ is the unitary operator representing R, (3.120) gives

$$U(R)\mathbf{V}U(R)^{-1} = R^{-1}\mathbf{V}, \tag{3.127}$$

i.e.

$$U(R)V_i U(R)^{-1} = (R^{-1})_{ij}V_j = R_{ji}V_j \tag{3.128}$$

where R_{ij} are the matrix elements of the orthogonal 3×3 matrix R. Now consider the family of rotations $R(\mathbf{n}, \lambda)$. According to Postulate VII, the associated hermitian operator is $X = \mathbf{J} \cdot \mathbf{n}$. In (3.121) $dA'/d\lambda$ is a component of the vector

$$\frac{d}{d\lambda}(R^{-1}\mathbf{V}) = \frac{d}{d\lambda}[R(\mathbf{n}, -\lambda)\mathbf{V}]_{\lambda=0} = -\mathbf{n} \times \mathbf{V}, \tag{3.129}$$

which is the velocity of the tip of the vector $\mathbf{V}$ when it rotates about the vector $\mathbf{n}$ with unit angular velocity (see Fig. 3.2). Thus (3.121) gives the vector equation

$$-i\hbar\mathbf{n} \times \mathbf{V} = [\mathbf{J} \cdot \mathbf{n}, \mathbf{V}]. \tag{3.130}$$

Since $\mathbf{n}$ is any vector, this can also be written

$$[J_i, V_j] = i\hbar\varepsilon_{ijk}V_k \tag{3.131}$$

(see Appendix A for the ε_{ijk} notation).

Finally, let S be a scalar observable. Then S is not related to any particular direction in space, and so it is unaffected by a rotation of the apparatus; thus $S' = S$. Hence

$$U(R)SU(R)^{-1} = S \tag{3.132}$$

and

$$[J_i, S] = 0.$$

In particular, this applies whenever S is a scalar product of two vector observables or the magnitude of a vector observable (e.g. the distance $r = |\mathbf{r}|$).

Fig. 3.2.
Explanation of (3.129).

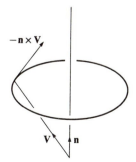

To summarise, we have shown that Postulate VII implies the following:

●3.10 The basic commutation relations

$$[P_i, f] = -i\hbar \frac{\partial f}{\partial x_i},$$ (3.133)

$$[J_i, V_j] = i\hbar \varepsilon_{ijk} V_k,$$ (3.134)

$$[J_i, S] = 0,$$ (3.135)

where $\mathbf{P}$ is the total momentum of the system, $\mathbf{J}$ is its total angular momentum, $f(\mathbf{r})$ is any observable function of the position of a point of the system, $\mathbf{V}$ is any vector observable, and S is any scalar observable.

Active and passive transformations

In this section we have considered operations which are actually performed on a physical system, so that we have two states of the system to consider; the state before and the state after the operation. Given a specific way of associating states with definite mathematical objects (the state vectors), we then have two state vectors, and these are related by the unitary operator $U(Q)$. This is called the **active** interpretation of $U(Q)$. A unitary operator can also be given a **passive** interpretation when there is *one* state but two ways of associating states with state vectors, and the latter are related by the unitary operator.

We can use translations to illustrate the idea. The change

$$\mathbf{r} \to \mathbf{r}' = \mathbf{r} + \mathbf{a}$$ (3.136)

describes, in the active interpretation, a movement of an object in which every point of the object is displaced through the vector $\mathbf{a}$ (Fig. 3.3(a)). In the passive interpretation it is a change in the way that points of space are associated with vectors: the point that was called $\mathbf{r}$ is now called $\mathbf{r}'$. This is just what happens if the origin is changed by being displaced through $-\mathbf{a}$ (Fig. 3.3(b)). The distinction extends to wave functions: if the system is actually translated through the vector $\mathbf{a}$, its wave function changes from ψ to ψ', where

$$\psi'(\mathbf{r}) = \psi(\mathbf{r} - \mathbf{a}).$$ (3.137)

Fig. 3.3.
Active and passive transformations.

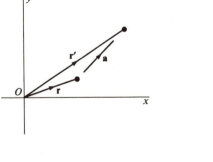
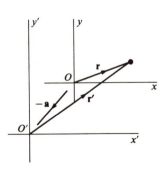

The same change occurs in the passive interpretation if the coordinates are changed by the change of origin shown in Fig. 3.3(*b*). The state which was described in terms of the old coordinates by the wave function ψ will now be described by the wave function ψ'. Since $\psi'(\mathbf{r}') = \psi(\mathbf{r})$, the value of the function at any point of space is unchanged, but its value at particular numerical coordinates will have changed: e.g. $\psi'(1, 0, 0) \neq \psi(1, 0, 0)$.

If we change the way in which (physical) states are described by (mathematical) state vectors, we must also change the way in which (physical) observables A are described by (mathematical) operators $\hat{A}$. Suppose the change is given by a unitary operator U_0 in its passive interpretation, so that the state which was called $|\psi\rangle$ is now called $U_0|\psi\rangle$. A similar argument to that of ●3.9 shows that the operator which now describes the observable A is

$$\hat{A}' = U_0 \hat{A} U_0^{-1}. \tag{3.138}$$

The same applies to the operators representing physical operations Q. The state which was called $U(Q)|\psi\rangle$ is now called $U_0 U(Q)|\psi\rangle$. This must be related to $U_0|\psi\rangle$ by applying the operator $U'(Q)$ which now describes the operation Q:

$$U_0 U(Q)|\psi\rangle = U'(Q) U_0 |\psi\rangle,$$

so that

$$U'(Q) = U_0 U(Q) U_0^{-1}. \tag{3.139}$$

3.3. Groups of operations

Many of the properties of the operations discussed in the previous section are consequences of the fact that the operations have the mathematical structure of a group. Other groups of operations are important in elementary particle theory. The following is a brief summary of some general results from group theory.

A set G of operations forms a **group** if (i) the composition of any two operations in G also belongs to G, (ii) the identity operation (the operation of leaving the system as it is) belongs to G, and (iii) every operation in G has an inverse in G. The composition of operations is usually written as multiplication and called their **product**. (The fourth axiom for a group, that this multiplication should be associative, is automatically satisfied for operations.)

A (unitary) **representation** of a group on a vector space $\mathscr{V}$ is a rule which assigns to each group element Q a (unitary) operator $U(Q)$ on $\mathscr{V}$, in such a way that

$$U(QR) = U(Q)U(R) \quad \text{for all } Q, R \in G. \tag{3.140}$$

A **projective representation** is an assignment of operators $U(Q)$ satisfying

$$U(QR) = \omega(Q, R)U(Q)U(R) \tag{3.141}$$

where $\omega(Q, R)$ is a numerical factor (which has modulus 1 if the operators $U(Q)$ are unitary). We say that the space $\mathscr{V}$ **carries** the representation U. The representation is **irreducible** if it has no non-trivial invariant subspaces, i.e. if there is no subspace of $\mathscr{V}$ (apart from 0 and $\mathscr{V}$) which is taken to itself by every

$U(Q)$. Such a subspace would carry another (smaller) representation of the group, so the original representation could be regarded as being made up of smaller representations: the irreducible representations are the basic building blocks for representations. Two representations U_1 and U_2, carried by vector spaces $\mathscr{V}_1$ and $\mathscr{V}_2$, are **equivalent** if there is an isomorphism $\theta: \mathscr{V}_1 \to \mathscr{V}_2$ such that $\theta U_1(Q) = U_2(Q)\theta$.

It can be seen from (3.75) that the operators $U(T_\mathbf{a})$, defined on the space of wave functions, satisfy

$$U(T_\mathbf{a} T_\mathbf{b}) = U(T_{\mathbf{a}+\mathbf{b}}) = U(T_\mathbf{a})U(T_\mathbf{b}). \tag{3.142}$$

Thus the operators $U(T_\mathbf{a})$ form a representation of the group of translations. However, in the proof of ●3.6 we saw that in general the operators corresponding to a group of unitaristic operations form only a projective representation of the group (see (3.87)).

A **Lie group** (of dimension n) is a group whose elements can be specified by n real parameters, in such a way that if $Q(\xi_1, \ldots, \xi_n)$ denotes the group element with parameters $\xi_1, \ldots, \xi_n$, and

$$Q(\xi_1, \ldots, \xi_n)Q(\eta_1, \ldots, \eta_n) = Q(\zeta_1, \ldots, \zeta_n), \tag{3.143}$$

then the ζ_i are smooth functions of the ξ_i and the η_i. (More generally, a group is a Lie group if it can be divided into subsets which can be parametrised in this way. For fuller details see Gilmore 1974.) A representation U of a Lie group is **differentiable** if $U(Q(\xi_1, \ldots, \xi_n))$ can be differentiated with respect to each ξ_i; we will only consider such representations. We can then take over the notion of hermitian generators from the previous section. A **generator** of a representation of a Lie group G is an operator

$$X = \frac{d}{d\lambda} U(Q_\lambda)\big|_{\lambda=0} \tag{3.144}$$

where Q_λ is a family of elements of G specified by parameters $\xi_1(\lambda), \ldots, \xi_n(\lambda)$ which are smooth functions of λ, such that Q_0 is the identity. If we assume that the identity has all its parameters 0, and take Q_λ to be the element whose parameters are all 0 except for $\xi_i = \lambda$, we get the generator

$$X_i = \frac{\partial}{\partial \xi_i} U(Q(\xi_1, \ldots, \xi_n))\big|_{\xi_i=0}. \tag{3.145}$$

If the representation $U(Q)$ is unitary, then, as in the case of the translation group, X is antihermitian (cf. (3.73)). It follows that $i\hbar X$ is hermitian and represents an observable. In the physics literature the word 'generator' is often used for this observable rather than the antihermitian X; we will continue to distinguish them by calling $i\hbar X$ a 'hermitian generator'.

The basic mathematical facts about the generators of a representation of a Lie group (see Gilmore 1974 for proofs) are:

    LA1   The generators form an n-dimensional real vector space with basis $X_1, \ldots, X_n$.

    LA2   If X and Y are generators, so is $[X, Y]$.

It follows from these that $[X_i, X_j]$ can be expanded as a linear combination of the X_k, i.e.

$$[X_i, X_j] = \sum_k c_{ijk} X_k \tag{3.146}$$

where the c_{ijk} are real numbers. We have

> **LA3** The coefficients c_{ijk} in (3.146) are the same for all representations of G.

The c_{ijk} are called the **structure constants** of G.

An (abstract) **Lie algebra** is a vector space A with a bilinear map from $A \times A$ to A, called the **Lie bracket** and written $[X, Y]$, which satisfies

$$[X, Y] = -[Y, X], \tag{3.147}$$

$$[X, [Y, Z]] + [Y, [Z, X]] + [Z, [X, Y]] = 0. \tag{3.148}$$

A **representation** of a Lie algebra A is a rule which assigns to each $X \in A$ an operator $D(X)$ in such a way that

$$D([X, Y]) = [D(X), D(Y)]. \tag{3.149}$$

Thus **LA1–LA3** say that any Lie group G is associated with a unique Lie algebra, and that the generators of any representation of G form a representation of its Lie algebra.

An example of a Lie group is the three-dimensional **rotation group**. Every element of this group is a rotation about some axis (a straight line through the origin) and through some angle. It is not immediately obvious that these form a group; the fact that they do is a consequence of **Euler's theorem**, which states that every possible motion of a rigid body with one point fixed can be achieved with a single rotation. In particular the product of two rotations is a rotation. Euler's theorem is proved by showing that the rotation group consists of the operations on vectors given by $\mathbf{r} \to R\mathbf{r}$ where R is a 3×3 orthogonal matrix with determinant 1.

Thus the elements of the rotation group can be specified as $R(\mathbf{n}, \theta)$ where $\mathbf{n}$ is a unit vector along the axis of the rotation and θ is the angle of the rotation, with $-\pi < \theta \leqslant \pi$. As parameters for the group we can take the components of the vector $\theta\mathbf{n} = (\xi_1, \xi_2, \xi_3)$; using (3.145) then gives basic generators X_1, X_2, X_3, where

$$X_1 = \frac{\partial}{\partial \xi_1} U(R(\xi_1, \xi_2, \xi_3))\big|_{\xi_i=0} = \frac{\partial}{\partial \theta} U(R(\mathbf{i}, \theta))\big|_{\theta=0}, \quad \text{etc.} \tag{3.150}$$

Using the chain rule we find that the generator of the family of rotations $R(\mathbf{n}, \theta)$ about any axis $\mathbf{n}$ is

$$X = \frac{\partial}{\partial \theta} U(R(\theta n_1, \theta n_2, \theta n_3))\big|_{\theta=0} = n_1 X_1 + n_2 X_2 + n_3 X_3 = \mathbf{n} \cdot \mathbf{X}, \tag{3.151}$$

which illustrates **LA1** (see also problem 3.21).

According to Postulate VII the hermitian generator corresponding to X is the angular momentum operator $\mathbf{n} \cdot \mathbf{J}$, i.e.

$$\mathbf{n} \cdot \mathbf{J} = ihX = ih\mathbf{n} \cdot \mathbf{X} \qquad (3.152)$$

or

$$J_i = ihX_i.$$

Now by taking $V_i = J_i$ in (3.134) we obtain

$$[X_i, X_j] = \varepsilon_{ijk} X_k, \qquad (3.153)$$

which illustrates **LA2** and **LA3**, and shows that the structure constants of the three-dimensional rotation group are $c_{ijk} = \varepsilon_{ijk}$.

We ought to prove (3.153) directly, for it is clearly a property of the rotation group itself and does not depend on any particular application to physics; in other words, we ought to check that Postulate VII is consistent with angular momentum being a vector.

●**3.11**   In any representation of the rotation group, the generators X_i satisfy (3.153).

Proof. We start from the equation

$$SR(\mathbf{n}, \theta) = R(S\mathbf{n}, \theta)S \quad \text{for any rotation } S, \qquad (3.154)$$

a fact about rotations which is depicted in Fig. 3.4. The same equation must hold in any representation U:

$$U(S)U(R(\mathbf{n}, \theta)) = U(R(S\mathbf{n}, \theta))U(S). \qquad (3.155)$$

Differentiating with respect to θ and putting $\theta = 0$,

$$U(S)\mathbf{n} \cdot \mathbf{X} = (S\mathbf{n}) \cdot \mathbf{X} U(S). \qquad (3.156)$$

Now put $S = R(\mathbf{m}, \phi)$. As in (3.129) and Fig. 3.2,

$$\frac{d}{d\phi}[R(\mathbf{m}, \phi)\mathbf{n}]_{\phi=0} = \mathbf{m} \times \mathbf{n}. \qquad (3.157)$$

Hence differentiating (3.155) with respect to ϕ and putting $\phi = 0$ gives

$$(\mathbf{m} \cdot \mathbf{X})(\mathbf{n} \cdot \mathbf{X}) = (\mathbf{n} \cdot \mathbf{X})(\mathbf{m} \cdot \mathbf{X}) + (\mathbf{m} \times \mathbf{n}) \cdot \mathbf{X},$$

i.e.

$$[\mathbf{m} \cdot \mathbf{X}, \mathbf{n} \cdot \mathbf{X}] = (\mathbf{m} \times \mathbf{n}) \cdot \mathbf{X}, \qquad (3.158)$$

which is another way of writing (3.153). ■

Fig. 3.4.
Explanation of (3.154).

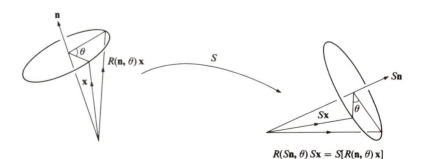

$$R(S\mathbf{n}, \theta)\, S\mathbf{x} = S[R(\mathbf{n}, \theta)\,\mathbf{x}]$$

This can also be expressed by saying that the Lie algebra of the rotation group is isomorphic to the space of three-vectors, with the Lie bracket of two vectors equal to their vector product.

The above proof is not valid if the operators $U(R)$ only form a projective representation of the rotation group, for (3.155) may include a numerical factor. However, the result can be restored by redefining the $U(R)$, as we did in the proof of ●3.6; that is,

> ●**3.12**   If $U(R)$ is a projective representation of the rotation group, it is possible to find a numerical factor $\omega(R)$ so that the projective representation $U'(R) = \omega(R)U(R)$ has generators which satisfy (3.153).

Proof. We can write

$$U(RS) = \eta(R, S)U(R)U(S) \tag{3.159}$$

where $\eta(R, S)$ is a numerical factor; the argument which led from (3.155) to (3.158) will now yield

$$[\mathbf{m} \cdot \mathbf{X}, \mathbf{n} \cdot \mathbf{X}] = (\mathbf{m} \times \mathbf{n}) \cdot \mathbf{X} + \alpha(\mathbf{m}, \mathbf{n})\mathbf{1} \tag{3.160}$$

where

$$\alpha(\mathbf{m}, \mathbf{n}) = \frac{\partial^2}{\partial\theta\,\partial\phi} \eta[R(\mathbf{m}, \theta), R(\mathbf{n}, \phi)]|_{\theta=\phi=0}. \tag{3.161}$$

(3.160) shows that $\alpha(\mathbf{m}, \mathbf{n})$ is pure imaginary if the generators X_i are hermitian, and it is an antisymmetric bilinear function of $\mathbf{m}$ and $\mathbf{n}$. The only such functions for three-vectors are of the form

$$\alpha(\mathbf{m}, \mathbf{n}) = i(\mathbf{m} \times \mathbf{n}) \cdot \mathbf{p} \tag{3.162}$$

where $\mathbf{p}$ is some fixed vector. Define $U'(R)$ by

$$U'(R(\mathbf{n}, \theta)) = e^{i\mathbf{p} \cdot \mathbf{n}\theta}U(R(\mathbf{n}, \theta)); \tag{3.163}$$

then the $U'(R)$ form another projective representation of the rotation group, with generators $\mathbf{X}' = \mathbf{X} + i\mathbf{p}$ which satisfy

$$[\mathbf{m} \cdot \mathbf{X}', \mathbf{n} \cdot \mathbf{X}'] = (\mathbf{m} \times \mathbf{n}) \cdot \mathbf{X}'. \qquad ■ \tag{3.164}$$

Note that the redefinition (3.163) may lead to ambiguities, for the angles θ and $\theta + 2\pi$ define the same rotation $R(n, \theta)$ but may give different values of the factor $e^{i\mathbf{p} \cdot \mathbf{n}\theta}$. The assertion of ●3.12 is only that the factor exists, not that it is unique: it does not imply that any projective representation of the rotation group can be redefined to give a true representation.

The above proof made essential use of some special properties of the rotation group; for a general Lie group it is not true that a projective representation can be redefined so that its generators form a representation of the Lie algebra of the group. An example of physical interest is the Galilean group, which is described in problem 3.24. Classes of Lie groups for which a result like ●3.12 does hold are the semi-simple Lie groups and their irreducible inhomogeneous extensions, which are defined as follows.

A **semi-simple** Lie group is one which has no abelian normal subgroup (i.e.

no subset $H \subset G$ which is itself a Lie group and which satisfies

$$Q_1 Q_2 = Q_2 Q_1, \quad \text{all } Q_1, Q_2 \in H;$$
$$Q R Q^{-1} \in H, \qquad \text{all } Q \in G, R \in H).$$

Let G be any Lie group, $U(Q)$ a representation of G on a vector space V. The **inhomogeneous extension** of G by V (also called the **semi-direct product** of G and V) is the group of operations on V given by

$$v \rightarrow U(Q)v + u \quad \text{for some } Q \in G, u \in V. \tag{3.165}$$

We will denote this group by $G \times) V$. If G is semi-simple and the representation U is irreducible, then both G and $G \times) V$ have the property described in ●3.12.

It is not difficult to see that the rotation group is semi-simple. As an example of an inhomogeneous extension, let G be the rotation group and let V be three-dimensional space, the representation U being given by the usual action of a rotation about the origin. Then the inhomogeneous extension $G \times) V$ consists of the combined operations of rotations and translations. This is the group of all physically possible operations on a rigid body in space (an operation being defined solely by the final position of the body, not by the details of how it was taken there); these are called **rigid motions**. The group consisting of all rigid motions together with all spatial reflections is called the three-dimensional **Euclidean group**.

Some important families of semi-simple Lie groups are the following:

The **orthogonal group** $O(n)$ is the group of all $n \times n$ real orthogonal matrices. The **unitary group** $U(n)$ is the group of all $n \times n$ complex unitary matrices. The **special orthogonal group** $SO(n)$ and the **special unitary group** $SU(n)$ are the subgroups of these consisting of matrices with determinant 1.

Let $R(\lambda)$ be any one-parameter family of orthogonal matrices with $R(0) = 1$; then

$$R(\lambda)^{\mathrm{T}} R(\lambda) = 1. \tag{3.166}$$

Differentiating with respect to λ and putting $\lambda = 0$, we see that the generator $X = dR/d\lambda$ is antisymmetric. It can be shown that any antisymmetric matrix can be obtained in this way, so the Lie algebra of $O(n)$ is the set of all $n \times n$ antisymmetric matrices. This is also the Lie algebra of $SO(n)$. (The relation between these two groups is that $O(n)$ consists of $SO(n)$ together with another separated piece; thus any continuous family of elements of $O(n)$ which contains the identity must lie in $SO(n)$, and so the generators of the two groups are the same.)

A similar argument shows that the Lie algebra of $U(n)$ consists of all $n \times n$ antihermitian (complex) matrices. Finally, the formula†

$$\det [e^{iX}] = e^{i \, \mathrm{tr} \, X}$$

(consider the eigenvalues) shows that the Lie algebra of $SU(n)$ consists of all $n \times n$ antihermitian matrices with trace 0.

† The **trace** tr X of a matrix X is the sum of its diagonal entries, which is equal to the sum of its eigenvalues. It satisfies tr $(XY) = $ tr (YX).

We will see how some of these groups act as physical operations in later chapters (in connection with elementary particles). A group of operations is particularly important if the operations leave the behaviour of the system invariant; in this case the group is called an **invariance group** or **symmetry group** of the system. The importance of the group structure derives from the following simple fact:

> **●3.13** If G is a symmetry group of a system, then every energy eigenspace of the system carries a projective representation of G.

Proof. Let $U(Q)$ be the operator on the state space of the system which represents the operation $Q \in G$. By ●3.4, since Q is an invariance of the system, $U(Q)$ commutes with the Hamiltonian. Let $\mathscr{S}_E$ be an energy eigenspace with energy E; then for each state $|\psi\rangle$ in $\mathscr{S}_E$ we have $H|\psi\rangle = E|\psi\rangle$ and therefore

$$HU(Q)|\psi\rangle = U(Q)H|\psi\rangle = EU(Q)|\psi\rangle, \tag{3.167}$$

i.e. $U(Q)|\psi\rangle$ also belongs to $\mathscr{S}_E$. Thus $U(Q)$ is an operator on $\mathscr{S}_E$, which therefore carries a projective representation of G. ■

The significance of this apparently trivial remark is that there are only a limited number of projective representations of a given group, and so the existence of a symmetry group places restrictions on the energy eigenspaces of a system. In particular, only certain dimensions will be possible for these eigenspaces. Thus the symmetry group controls the degeneracy of the energy levels of the system.

3.4. **The Heisenberg picture**

The description of how things change with time which we have been using up to now is known as the **Schrödinger picture**. The state of a system changes with time; observable quantities are represented by operators which are constant in time, reflecting mathematically the fact that the experimental procedure for determining the value of an observable is always the same. This seems to be the most natural point of view, but it is possible to look from a different angle and see the observables as changing – after all, the position of a moving body does change with time – while the system stays the same. This is the **Heisenberg picture**, in which observables are represented by time-dependent operators while the state is described by a constant state vector. In this picture the 'state' of a system is a concept which encompasses its whole history.

The relationship between the two pictures is given by a sequence of unitary transformations in the passive interpretation. Let $U(t) = e^{-iHt/\hbar}$ where H is the Hamiltonian of the system; $U(t)$ is the operator which describes how states change in time t according to the Schrödinger picture. Now suppose we decide to use $U(t)^{-1}$ to rename all states at time t, so that the state which was called $|\psi(t)\rangle$ will now be called $U(t)^{-1}|\psi(t)\rangle$; in other words, we are specifying each state at time t by saying what state vector it was associated with at time $t = 0$. Then, in this new picture, as the system evolves its state is always represented by the same state vector.

As was discussed on p. 103, changing the way in which states are associated with state vectors entails changing the way in which observables are associated with operators. Consider an observable A, and denote by A_0 the operator describing A in the Schrödinger picture. According to (3.139), A must be described in the Heisenberg picture by the operator

$$A(t) = U(t)^{-1} A_0 U(t), \tag{3.168}$$

which changes with time if A_0 does not commute with $U(t)$, i.e. if it does not commute with the Hamiltonian. Using the equations

$$ih \frac{dU}{dt} = UH, \quad ih \frac{d}{dt}(U^{-1}) = -HU, \tag{3.169}$$

which follow from the exponential form of U, we find that the rate of change of $A(t)$ is given by

$$ih \frac{dA}{dt} = [A, H]. \tag{3.170}$$

This is **Heisenberg's equation of motion**.

Note that the formula (3.26) for the rate of change of an expectation value is an immediate consequence of Heisenberg's equation of motion, since state vectors are independent of time in this picture. More generally, the fact that operators which commute with the Hamiltonian represent conserved quantities is clearly a feature of this equation of motion.

Heisenberg's equation of motion has a very close relationship to the equation of motion of classical mechanics. In classical Hamiltonian mechanics any observable is a function $A(q_1, \ldots, q_n, p_1, \ldots, p_n)$ of the coordinates q_i and momenta p_i of the system. For any two such observables A, B a third one called their **Poisson bracket** $\{A, B\}$ is defined by

$$\{A, B\} = \sum_{i=1}^{n} \left(\frac{\partial A}{\partial q_i} \frac{\partial B}{\partial p_i} - \frac{\partial A}{\partial p_i} \frac{\partial B}{\partial q_i} \right). \tag{3.171}$$

Hamilton's equations (3.5) then give the general equation of motion

$$\frac{dA}{dt} = \{A, H\} \tag{3.172}$$

for any classical observable A. The Poisson bracket has all the algebraic properties (2.79)–(2.81) of the commutator in quantum mechanics; its values for the basic observables q_i and p_i are

$$\left. \begin{aligned} \{q_i, q_j\} &= 0 = \{p_i, p_j\} \\ \{q_i, p_j\} &= \delta_{ij} \end{aligned} \right\}. \tag{3.173}$$

Comparing with the canonical commutation relations (2.116)–(2.117), we see that there is a correspondence between $\{A, B\}$ in classical mechanics and $(ih)^{-1}[A, B]$ in quantum mechanics. Heisenberg's equation of motion (3.170) then corresponds exactly to the classical equation of motion (3.173).

3.5. Time-dependent perturbation theory

In eqs. (3.2) and (3.3) we have in principle a complete description of the time evolution of a quantum system in terms of its Hamiltonian operator H. However, in order to make this description explicit it is necessary to know the eigenvalues and eigenstates of H, so as to expand the solution (3.2) in the form (3.11). We will see some examples of such calculations in Chapter 4 (which can be read before this section if the reader wishes). There are very few Hamiltonians which permit exact calculations; usually it is necessary to use an approximation based on taking the first few terms of an infinite series which is obtained by the methods to be described in this section.

This method is not just a practical technique for doing calculations, for the general form of the results it gives is of considerable theoretical interest. It underlies the description of elementary processes by means of Feynman diagrams, which was sketched in Chapter 1, and it gives rise to an alternative formulation of quantum mechanics which will be described in §3.6.

The idea is to calculate the time development caused by the Hamiltonian H in terms of the eigenstates and eigenvalues of a Hamiltonian H_0 which is used as a standard for reference, either because it can be exactly solved or because its eigenstates are experimentally significant. We write

$$H = H_0 + \varepsilon V \tag{3.174}$$

where ε is a parameter (usually assumed to be small), and we expand the solution in powers of ε.

Write

$$|\tilde{\psi}(t)\rangle = e^{iH_0 t/\hbar}|\psi(t)\rangle, \tag{3.175}$$

$$\tilde{V}(t) = e^{iH_0 t/\hbar} V e^{-iH_0 t/\hbar}. \tag{3.176}$$

These equations define a new picture called the **interaction picture**, which would coincide with the Heisenberg picture if the Hamiltonian was H_0. The effect is to remove the part of the time dependence of states which is due to H_0: using the equation

$$i\hbar \frac{d}{dt}|\psi(t)\rangle = (H_0 + \varepsilon V)|\psi(t)\rangle, \tag{3.177}$$

we find that the equation of motion for $|\tilde{\psi}(t)\rangle$ is

$$i\hbar \frac{d}{dt}|\tilde{\psi}(t)\rangle = \varepsilon \tilde{V}(t)|\tilde{\psi}(t)\rangle. \tag{3.178}$$

The solution of this equation clearly depends on ε. We assume that it can be expanded as a power series in ε:

$$|\tilde{\psi}(t)\rangle = \sum_{n=0}^{\infty} \varepsilon^n |\tilde{\psi}_n(t)\rangle \tag{3.179}$$

with

$$|\tilde{\psi}_n(0)\rangle = 0 \quad \text{if } n > 0$$

since at $t = 0$ the state $|\tilde{\psi}\rangle$ is the initial state $|\psi_0\rangle$, independent of ε. Substituting the expansion (3.179) into the equation of motion (3.178) and equating powers of ε gives a sequence of equations

$$i\hbar \frac{d}{dt}|\tilde{\psi}_0(t)\rangle = 0,$$

$$i\hbar \frac{d}{dt}|\tilde{\psi}_1(t)\rangle = \tilde{V}(t)|\tilde{\psi}_0(t)\rangle.$$

etc.; the nth equation in the sequence is

$$i\hbar \frac{d}{dt}|\tilde{\psi}_n(t)\rangle = \tilde{V}(t)|\tilde{\psi}_{n-1}(t)\rangle. \tag{3.180}$$

These can be solved successively to give

$$|\tilde{\psi}_0(t)\rangle = |\psi_0\rangle, \tag{3.181}$$

$$|\tilde{\psi}_1(t)\rangle = \frac{1}{i\hbar}\int_0^t dt_1\, \tilde{V}(t_1)|\psi_0\rangle, \tag{3.182}$$

$$\vdots$$

$$|\tilde{\psi}_n(t)\rangle = \frac{1}{(i\hbar)^n}\int_0^t dt_n \int_0^{t_n} dt_{n-1}\cdots\int_0^{t_2} dt_1\, \tilde{V}(t_n)\cdots \tilde{V}(t_1)|\psi_0\rangle. \tag{3.183}$$

First-order theory   Suppose $|\psi_0\rangle$ is an eigenstate of H_0, with eigenvalue E_0; at time t it is no longer certain that a measurement of H_0 will give the result E_0. Let E be another eigenvalue of H_0, and let $|\psi_E\rangle$ be the corresponding eigenstate (for simplicity we suppose that E is a non-degenerate eigenvalue). Then the probability that a measurement of H_0 at time t will give the value E is $|a_1(E, t)|^2$ where $a_1(t)$ is given by

$$\begin{aligned}
a_1(E, t) &= \langle \psi_E|\psi(t)\rangle \\
&= \langle \psi_E|\{|\psi_0\rangle + \varepsilon|\psi_1(t)\rangle\} \quad \text{to first order in } \varepsilon \\
&= \varepsilon\langle \psi_E|\psi_1(t)\rangle \quad \text{since } |\psi_0\rangle \text{ and } |\psi_E\rangle \text{ are orthogonal} \\
&= \varepsilon\langle \psi_E|e^{iH_0t/\hbar}|\tilde{\psi}_1(t)\rangle \\
&= \frac{1}{i\hbar}e^{iEt/\hbar}\int_0^t \langle \psi_E|\varepsilon\tilde{V}(t_1)|\psi_0\rangle\, dt_1.
\end{aligned} \tag{3.184}$$

(If E lies in a continuous range of eigenvalues the probability that a measurement of H_0 will give a result lying between E and $E+dE$ is $|a_1(E, t)|^2\rho(E)\,dE$ where $\rho(E)$ is the density of states.) Now a matrix element of $\tilde{V}(t)$ between eigenstates of H_0 is, by (3.176), related to the matrix element of V by

$$\langle \psi_E|\tilde{V}(t)|\psi_0\rangle = e^{-i(E-E_0)t/\hbar}\langle \psi_E|V|\psi_0\rangle. \tag{3.185}$$

If V is independent of time, we can now do the integration in (3.184) to obtain

$$a_1(E, t) = \varepsilon\langle \psi_E|V|\psi_0\rangle \frac{e^{iE_0t/\hbar} - e^{iEt/\hbar}}{E - E_0}. \tag{3.186}$$

Using the formula

$$|e^{i\theta} - e^{i\phi}| = 2\sin\tfrac{1}{2}(\theta - \phi), \tag{3.187}$$

which is easily proved by looking at the Argand diagram, the probability can be written as

$$p(E, t) = |a_1(E, t)|^2 = |\langle \psi_E | \varepsilon V | \psi_0 \rangle|^2 \frac{4 \sin^2 \frac{1}{2} \omega t}{\hbar^2 \omega^2} \qquad (3.188)$$

where

$$\omega = (E - E_0)/\hbar.$$

According to Postulates II and IV, this is the probability that a measurement of H_0 at time t will give the value E, provided that the system is undisturbed by any measurement between times 0 and t. Since such a result at time t would leave the system in the state $|\psi_E\rangle$, this is often described as the probability that the system 'will be found in the state $|\psi_E\rangle$' at time t. It is then applied to situations in which there appears to have been a transition, or 'quantum jump', at some time between 0 and t.

An example of a system which can be described in this way is the hydrogen atom. The full system consists of an electron, a proton and the electromagnetic field (the field has independent degrees of freedom, as is shown by the existence of electromagnetic waves). H_0 is the Hamiltonian describing the electron, the proton and the field by themselves, and also the electrostatic attraction between the electron and the proton (its eigenvalues are determined in Chapter 4); V is the Hamiltonian describing the interaction between the two charged particles and electromagnetic waves. Classically, H_0 gives equations of motion for the electron and the proton whose solutions are orbits like those of the earth and the sun, and equations for the electromagnetic field whose solutions describe electromagnetic waves; V gives equations which describe the radiation by an accelerating electron. The quantum counterpart of the planet-like orbits is a set of 'allowed orbits', i.e. eigenstates of H_0; that of the electromagnetic waves is a set of states of photons; and that of radiation by the accelerating electron is a transition between eigenstates of H_0. This occurs because the full Hamiltonian $H_0 + \varepsilon V$ causes an eigenstate $|\psi_0\rangle$ to evolve into a state $|\psi_0\rangle + \varepsilon |\psi_1(t)\rangle$; if $|\psi_0\rangle$ is an 'excited state', i.e. a state with eigenvalue of H_0 which is higher than the minimum, then $|\psi_E\rangle$ could be a state consisting of an electron–proton state which is an eigenstate of H_0 (with a lower eigenvalue than $|\psi_0\rangle$), together with a state of the electromagnetic field consisting of one photon. A transition from $|\psi_0\rangle$ to $|\psi_E\rangle$ constitutes a quantum jump from the higher electron–proton state to the lower, accompanied by the emission of a photon.

This description can of course be generalised so as to apply to any atom or molecule. A similar description can be given of the radioactive decay of a nucleus, or the decay of an unstable subatomic particle. It is not always clear that the postulates of quantum mechanics, as we have stated them so far, are sufficient to justify deductions about the probability of decay; in particular, Postulate III refers to measurements taking place at a definite time, while an unstable system may be kept under continuous observation to see when it decays. In order to cover this situation, we extend Postulate III as follows:

Postulate III (*continued*). If the observable H_0 is observed continuously, the Hamiltonian being $H = H_0 + \varepsilon V$, then the system will make spontaneous transitions between eigenstates of H_0. The probability that there will be a transition from $|\psi_0\rangle$ to $|\psi_E\rangle$ in time t is

$$p(E, t) = |\langle \psi_E | e^{-iHt/\hbar} | \psi_0 \rangle|^2. \tag{3.189}$$

This and the other parts of Postulate III will be discussed further in Chapter 5.

The first-order expression (3.188) for the transition probability has two parts: the square of the matrix element $\langle \psi_E | \varepsilon V | \psi_0 \rangle$, which measures the extent to which the force described by V links the states $|\psi_0\rangle$ and $|\psi_E\rangle$; and a time-dependent factor which is the same for all processes and depends only on the 'energy difference' $E - E_0$. (This is an inaccurate terminology, since E and E_0 are eigenvalues of H_0 which is not the full energy operator, but it is a common one.) The second factor is plotted as a function of $\omega = (E - E_0)/h$ in Fig. 3.5. Its main features are the central peak, which gets higher and narrower as t increases, and the fall-off at large ω, which goes like ω^{-2} for all t. The total area under the curve is

$$\int_{-\infty}^{\infty} \frac{\sin^2 \frac{1}{2}\omega t}{\omega^2} \, d\omega = \frac{t}{2} \int_{-\infty}^{\infty} \frac{\sin^2 x}{x^2} \, dx = \frac{\pi t}{2}. \tag{3.190}$$

In many circumstances $|\psi_E\rangle$ is one of a set of states with a continuous range of values of E. For example, in the decay of an excited state of the hydrogen atom $|\psi_E\rangle$ consists of a lower-energy state of the atom together with a photon, and the photon can have any energy. (We are again using the term 'energy' loosely and identifying it with 'eigenvalue of H_0'. We will continue to do this without further comment.) In this case $p(E, t)$ must be multiplied by the density of states $\rho(E)$, which depends on the reference Hamiltonian H_0. The physically significant quantity is the total probability that a transition has occurred into a

Fig. 3.5

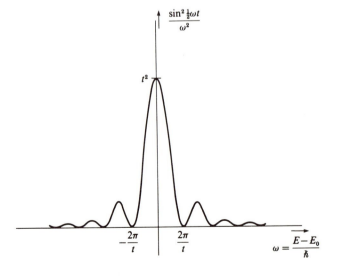

state with energy in some finite range $[E_1, E_2]$, which is

$$P(t) = \int_{E_1}^{E_2} p(E, t)\rho(E)\, dE$$

$$= \frac{4}{\hbar^2} \int_{\omega_1}^{\omega_2} M(\omega) \frac{\sin^2 \frac{1}{2}\omega t}{\omega^2}\, d\omega \tag{3.191}$$

where

$$M(\omega) = |\langle \psi_E | \varepsilon V | \psi_0 \rangle|^2 \rho(E).$$

If E_0 lies outside the interval $[E_1, E_2]$, the range of integration in (3.19) does not include the central peak of Fig. 3.5, which is where the integrand depends significantly on ω; as long as t is large compared with $2\pi/\omega$, it is a good approximation to replace $\sin^2 \frac{1}{2}\omega t$ in (3.191) by its average value of $\frac{1}{2}$, so that

$$P(t) \simeq \frac{2}{\hbar^2} \int_{\omega_1}^{\omega_2} \frac{M(\omega)}{\omega^2}\, d\omega < \frac{\text{const}}{\omega_1}. \tag{3.192}$$

Thus the probability of transition to a state with 'energy' different from the initial value E_0 soon becomes constant, and is small if the 'energy' difference is large.

On the other hand, if $E_1 < E_0 < E_2$ the range of integration in (3.191) will include the central peak of Fig. 3.5, and this will dominate the integral. If $M(\omega)$ varies slowly as a function of ω, we can treat it as constant over the region of the peak, which is narrow if t is large, so that

$$P(t) = \frac{4}{\hbar^2} \int_{\omega_1}^{\omega_2} M(\omega) \frac{\sin^2 \frac{1}{2}\omega t}{\omega^2}\, d\omega \approx \frac{4}{\hbar^2} M(0) \int_{\omega_1}^{\omega_2} \frac{\sin^2 \frac{1}{2}\omega t}{\omega^2}\, d\omega$$

$$= \Gamma(t - K) \quad \text{where } \Gamma = \frac{2\pi}{\hbar^2} M(0) \tag{3.193}$$

and where

$$K = \frac{2}{\pi} \int_{-\infty}^{\omega_1} \frac{\sin^2 \frac{1}{2}\omega t}{\omega^2}\, d\omega + \frac{2}{\pi} \int_{\omega_2}^{\infty} \frac{\sin^2 \frac{1}{2}\omega t}{\omega^2}\, d\omega \simeq \frac{1}{\pi\omega_1} - \frac{1}{\pi\omega_2},$$

which is small and approximately constant.

Eq. (3.193) shows that apart from a small term which soon reaches a constant value, the probability of transition to a state with energy close to the original value E_0 increases steadily with time. Thus there is a constant *transition probability per unit time* which is equal to Γ. This, of course, cannot continue to be true for all t, or the probability would become greater than 1. The reason for this apparent inconsistency is that in calculating the probability in (3.188) we failed to normalise the state $|\psi(t)\rangle$, since this would have involved terms of higher order in ε. Thus (3.193) holds only to lowest order in Γ (which contains ε^2 as a factor). Note that an exponential decay law of the form (1.3) gives the probability of decay in time t as

$$P(t) = 1 - e^{-\Gamma t} \quad \text{where } \Gamma = \tau^{-1}, \tag{3.194}$$

which also gives constant probability per unit time as a first approximation if Γ is small.

If first-order perturbation theory is used to calculate the rate of a decay process like

$$X \rightarrow A + B, \tag{3.195}$$

eq. (3.193) shows that the decay rate is proportional to

$$M(0) = |\langle AB(E_0)|\varepsilon V|X\rangle|^2 \rho(E_0) \tag{3.196}$$

where E_0 is the energy of the X particle, $|AB(E_0)\rangle$ is the state of the A and B particles with energy E_0, and $\rho(E_0)$ is the density of the AB states relative to energy. The appropriate states for decay processes are momentum eigenstates of A and B, for in most theories these states give constant matrix elements $\langle AB(E_0)|V|X\rangle$. Thus the decay rate is proportional to the density of momentum eigenstates relative to energy, which is called the **phase-space factor**.

We will calculate the phase-space factor for the case that the unstable X particle is at rest (so that, by conservation of momentum, the A and B particles have equal and opposite momentum $\pm \mathbf{p}$), and that the eigenvalue of H_0 for an AB state is given by the relativistic formula

$$E = E_A + E_B = \sqrt{(m_A{}^2 c^4 + p^2 c^2)} + \sqrt{(m_B{}^2 c^4 + p^2 c^2)} \tag{3.197}$$

where $p = |\mathbf{p}|$. The AB states are normalised relative to the components of $\mathbf{p}$, for which the element of integration is

$$dp_1 \, dp_2 \, dp_3 = p^2 \, dp \, d\Omega \tag{3.198}$$

where $d\Omega$ ($= \sin \theta \, d\theta \, d\phi$ in spherical polar coordinates) is the element of solid angle for the direction of $\mathbf{p}$. The density of states $\rho(E)$ is defined by

$$dp_1 \, dp_2 \, dp_3 = \rho(E) \, dE \, d\Omega,$$

so

$$\rho(E) = p^2 \frac{dp}{dE}. \tag{3.199}$$

From (3.197) we have

$$\rho(E) = p^2 \left(\frac{dE_A}{dp} + \frac{dE_B}{dp} \right)^{-1} = p^2 \left(\frac{pc^2}{E_A} + \frac{pc^2}{E_B} \right)^{-1} = \frac{p E_A E_B}{E c^2}. \tag{3.200}$$

In the decay (3.195) the initial energy is $E_0 = m_X c^2$. Hence in this case $\rho(E_0)$ can be written in terms of $E_A + E_B = m_X c^2$ and $E_A{}^2 - E_B{}^2 = (m_A{}^2 - m_B{}^2)c^2$ as

$$\rho(E_0) = \frac{p}{4 m_X{}^3} \{ m_X{}^4 - (m_A{}^2 - m_B{}^2)^2 \}. \tag{3.201}$$

The momentum p can be obtained from (3.197) (or using four-vectors) as

$$p = \frac{1}{2 m_X} \left[(m_X + m_A + m_B)(m_X + m_A - m_B)(m_X - m_A + m_B)(m_X - m_A - m_B) \right]^{\frac{1}{2}}.$$

This phase-space factor is larger for smaller masses m_A, m_B, since then the momentum p is greater; thus we have as a rule of thumb

> The rate of decay is greater for decays which release more kinetic energy.

This also holds for decays into more than two particles.

Second-order theory   The second-order term obtained from (3.183) is

$$|\psi_2(t)\rangle = e^{iH_0t/\hbar}|\tilde{\psi}_2(t)\rangle$$

$$= \frac{1}{(i\hbar)^2} e^{iH_0t/\hbar} \int_0^t dt_2 \int_0^{t_2} dt_1 \, \tilde{V}(t_2)\tilde{V}(t_1)|\psi_0\rangle. \tag{3.202}$$

If we include this in the calculation of the probability of transition to a state $|\psi_E\rangle$, we will have

$$p(E, t) = |a(E, t)|^2 = |a_1(E, t) + a_2(E, t)|^2 \tag{3.203}$$

where

$$a_n(E, t) = \varepsilon^n \langle \psi_E|\psi_n(t)\rangle. \tag{3.204}$$

Suppose H_0 has a complete set of states $|\psi_k\rangle$ with eigenvalues E_k (we will write our equations as if these were discrete, for the sake of simplicity, but they can easily be adapted to cover the continuous case). Then we can use the resolution of the identity (2.68) to insert the complete set $|\psi_k\rangle$ in (3.202); this gives

$$a_2(E, t) = \frac{e^{iEt/\hbar}}{(i\hbar)^2} \int_0^t dt_2 \int_0^{t_2} dt_1 \sum_k \langle \psi_E|\varepsilon\tilde{V}(t_2)|\psi_k\rangle\langle \psi_k|\varepsilon\tilde{V}(t_1)|\psi_0\rangle. \tag{3.205}$$

Using (3.185), we can extract the time dependence of the integrand in this expression and do the integrations to get

$$a_2(E, t) = \sum_k \langle \psi_E|\varepsilon V|\psi_k\rangle\langle \psi_k|\varepsilon V|\psi_0\rangle \frac{1}{E_k - E_0}$$

$$\times \left\{ \frac{e^{iE_0t/\hbar} - e^{iEt/\hbar}}{E - E_0} - \frac{e^{iE_kt/\hbar} - e^{iEt/\hbar}}{E_k - E_0} \right\}. \tag{3.206}$$

The terms in the curly brackets are quotients of the same type as that appearing in (3.186), which gave rise to the function shown in Fig. 3.5 when the squared modulus was calculated. Such a quotient is only significant when its denominator is small. Thus the second term in the curly brackets only contributes when $E_k \approx E_0$, i.e. only for a few values of k. The first term, however, contributes for all k as long as $E \approx E_0$. This means that to a good approximation we can ignore the second term; $a_2(E, t)$ then has the same time dependence as $a_1(E, t)$, as given by (3.186), and the factor $\langle \psi_E|\varepsilon V|\psi_0\rangle$ in a_1 is replaced by

$$\sum_k \frac{\langle \psi_E|\varepsilon V|\psi_k\rangle\langle \psi_k|\varepsilon V|\psi_0\rangle}{E_k - E_0}. \tag{3.207}$$

Interpretation: transition   amplitudes   The full result of the perturbation theory calculation is that the probability of transition from $|\psi_0\rangle$ to $|\psi_E\rangle$ by time t is

$$p(t) = |a(t)|^2 = |a_1(t) + a_2(t) + \cdots|^2 \tag{3.208}$$

where

$$a_1(t) = \frac{1}{i\hbar} \int_0^t \langle \psi_E | \varepsilon \tilde{V}(t_1) | \psi_0 \rangle \, dt_1, \tag{3.209}$$

$$a_2(t) = \frac{1}{(i\hbar)^2} \int_0^t dt_2 \int_0^{t_2} dt_1 \sum_k \langle \psi_E | \varepsilon \tilde{V}(t_2) | \psi_k \rangle \langle \psi_k | \varepsilon \tilde{V}(t_1) | \psi_0 \rangle, \tag{3.210}$$

$$\vdots$$

$$a_n(t) = \frac{1}{(i\hbar)^n} \int_0^t dt_n \int_0^{t_n} dt_{n-1} \cdots \int_0^{t_2} dt_1 \sum_{k_1 \cdots k_{n-1}} \langle \psi_E | \varepsilon \tilde{V}(t_n) | \psi_{k_{n-1}} \rangle$$

$$\cdots \langle \psi_{k_1} | \varepsilon \tilde{V}(t_1) | \psi_0 \rangle \tag{3.211}$$

(these formulae differ slightly from (3.184) and (3.198) in that we have removed the common phase factor $e^{iEt/\hbar}$, which does not affect the probability). If $a(t)$ was the total probability, these formulae would have the natural interpretation that at time s the probability of transition from $|\psi\rangle$ to $|\phi\rangle$ in a time interval dt is $(i\hbar)^{-1} \langle \phi | \varepsilon \tilde{V}(s) | \psi \rangle \, dt$. Then $a_1(t)$ in (3.209) would represent the total probability that there is a transition direct from $|\psi_0\rangle$ to $|\psi_E\rangle$ at some time t_1 between 0 and t; $a_2(t)$ in (3.210) would represent the total probability that there is a transition from $|\psi_0\rangle$ to some other state $|\psi_k\rangle$ at some time t_1, followed by a transition from $|\psi_k\rangle$ to $|\psi_E\rangle$ at a later time t_2; and so on, $a_n(t)$ in (3.211) representing the total probability of an indirect transition in n stages.

In fact, of course, each quantity a_k mentioned in the last paragraph is not a probability but a complex number whose squared modulus is a probability. It is characteristic of quantum mechanics that it deals with such complex numbers, which are added and multiplied as if they were probabilities. They are called **probability amplitudes**. The probability amplitude for a process consisting of two *successive* stages is the *product* of the probability amplitudes for the individual stages; the probability amplitude for a process which can happen in a number of *alternative* ways is the *sum* of the probability amplitudes for the individual alternatives. When a probability is calculated by taking the squared modulus of such a sum of probability amplitudes, the cross-terms give rise to the interference effects which typify quantum physics.

We have seen that in the cases of $a_1(t)$ and $a_2(t)$ the interference effects guarantee that the probability of transition from $|\psi_0\rangle$ to $|\psi_E\rangle$ is small unless their 'energies' (eigenvalues of H_0) are approximately equal; this continues to be true for the higher-order terms $a_n(t)$. Thus the intermediate states $|\psi_k\rangle$ in (3.210) and (3.211) are states which are unlikely to be the result of an actual transition from $|\psi_0\rangle$. Nevertheless the probability amplitudes for transitions to these states make important contributions to the total probability amplitude for the actual transition to $|\psi_E\rangle$. These transitions are called **virtual** transitions.

Dyson's form of the
perturbation series

Our series solution of the equation

$$ih\frac{d}{dt}|\tilde\psi(t)\rangle = \varepsilon\tilde V(t)|\tilde\psi(t)\rangle \tag{3.178}$$

is

$$|\tilde\psi(t)\rangle = \sum_{n=0}\left(\frac{\varepsilon}{ih}\right)^n\int_0^t dt_n\int_0^{t_n}dt_{n-1}\cdots\int_0^{t_2}dt_1\,\tilde V(t_n)\cdots\tilde V(t_1)|\psi_0\rangle. \tag{3.212}$$

The n-fold integral is the integral of the product $\tilde V(t_n)\cdots\tilde V(t_1)$ over the region of $\mathbb{R}^n$ consisting of the points $(t_1,\ldots,t_n)$ which satisfy

$$0\leqslant t_1\leqslant t_2\leqslant\cdots\leqslant t_n\leqslant t. \tag{3.213}$$

There are $n!$ regions like this one, corresponding to the $n!$ possible orderings of $(t_1,\ldots,t_n)$, which together make up the region $0\leqslant t_i\leqslant t$. Geometrically, this is a dissection of the n-dimensional cube into $n!$ n-simplices (an n-**simplex** is the n-dimensional analogue of a tetrahedron). We could obtain the integral by integrating the product of the $\tilde V(t_i)$ over any of these simplices, provided we put the factors in the right order. To express this in symbols, we define the **time-ordered product** of n time-dependent operators $X(t_i)$ to be

$$\mathrm{T}[X(t_1)\cdots X(t_n)] = X(t_{i_1})\cdots X(t_{i_n})$$

where

$$t_{i_1}\leqslant t_{i_2}\leqslant\cdots\leqslant t_{i_n}. \tag{3.214}$$

Then the integral in (3.212) can be written as

$$\int_R dt_1\cdots dt_n\,\mathrm{T}[\tilde V(t_n)\cdots\tilde V(t_1)]|\psi_0\rangle \tag{3.215}$$

where R is any one of the simplicial regions mentioned above. Since these all fit together to make up the cubic region $0\leqslant t_i\leqslant t$, the sum of all the integrals (3.215) is the integral over this cube:

$$\sum_R\int_R dt_1\cdots dt_n\,\mathrm{T}[\tilde V(t_n)\cdots\tilde V(t_1)]$$

$$= \int_0^t dt_1\cdots\int_0^t dt_n\,\mathrm{T}[\tilde V(t_n)\cdots\tilde V(t_1)]$$

$$= \mathrm{T}\left[\left(\int_0^t\tilde V(t')\,dt'\right)^n\right], \tag{3.216}$$

which defines an extension of the time-ordered product symbol T. Since the integrals (3.215) are all equal, any one of them is equal to $1/n!$ of their sum (3.216); hence we can put this into (3.212) to obtain

$$|\tilde\psi(t)\rangle = \sum_{n=0}^\infty\frac{1}{n!}\left(\frac{1}{ih}\right)^n\mathrm{T}\left[\left(\int_0^t\tilde V(t')\,dt'\right)^n\right]|\psi_0\rangle$$

$$= \mathrm{T}\left[\exp\left(\frac{1}{ih}\int_0^t\tilde V(t')\,dt'\right)\right]|\psi_0\rangle. \tag{3.217}$$

As well as being pretty, this formula is useful in quantum field theory. It shows the relation between the perturbation solution of (3.162) and the exponential solution of the similar equation (3.4) in which the operator on the right-hand side is constant.

Exponential decay   We have seen that to first order, perturbation theory gives a probability for survival of an unstable state which is of the form

$$P^*(t) = 1 - P(t) = 1 - \Gamma t. \tag{3.218}$$

To first order in the decay constant Γ, this agrees with the exponential form

$$P^*(t) = e^{-\Gamma t}, \tag{3.219}$$

which would be obtained if quantum transitions occurred as a classical Poisson process, as in the simple discussion of radioactivity on p. 5. We will now derive this exponential law from quantum mechanics by an argument which, though not exact, is not restricted to the first order of perturbation theory. We will also see that the approximation in this derivation is unavoidable, because the decay law in quantum mechanics cannot be exactly exponential.

As before, we assume that the unstable state $|\psi_0\rangle$ is an eigenstate of a reference Hamiltonian H_0 and that the decay is caused by a small term εV in the full Hamiltonian $H = H_0 + \varepsilon V$. Write

$$|\psi_1\rangle = V|\psi_0\rangle; \tag{3.220}$$

by redefining ε, if necessary, we can assume that $|\psi_1\rangle$ is normalised. We think of $|\psi_1\rangle$ as the state immediately after the decay. It is not necessarily an eigenstate of H_0 (for example, if an unstable particle decays to form two different particles, we would expect the two final particles to separate after the decay); thus we must suppose that $|\psi_1\rangle$ belongs to a subspace $\mathscr{S}'$ consisting of states of the decay products, and that H_0 acts as an operator inside $\mathscr{S}'$. The states of the decay products must be recognisably different from $|\psi_0\rangle$, so $|\psi_0\rangle$ must be orthogonal to the subspace $\mathscr{S}'$. Thus we can take the full state space to be $\mathscr{S}_0 \oplus \mathscr{S}'$ where $\mathscr{S}_0$ is the one-dimensional subspace containing $|\psi_0\rangle$.

Since V is hermitian,

$$\langle \psi_0 | V | \psi_1 \rangle = \overline{\langle \psi_1 | V | \psi_0 \rangle} = 1. \tag{3.221}$$

Thus $V|\psi_1\rangle$ necessarily contains a component in $\mathscr{S}_0$. We will assume that this is the only component of $V|\psi_1\rangle$ (we can always arrange this by changing the way we split the total Hamiltonian into H_0 and V). Now V is completely defined by (3.220) and the equation

$$V|\psi'\rangle = \langle \psi_1 | \psi' \rangle |\psi_0\rangle \quad \text{for } |\psi'\rangle \in \mathscr{S}'. \tag{3.222}$$

We can write the state at time t as

$$|\psi(t)\rangle = f(t)|\psi_0\rangle + |\psi'(t)\rangle \tag{3.223}$$

with $|\psi'(t)\rangle \in \mathscr{S}'$; then the equation of motion gives

$$ i\hbar \frac{df}{dt}|\psi_0\rangle + i\hbar \frac{d}{dt}|\psi'(t)\rangle $$

$$ = (H_0 + \varepsilon V)\{f(t)|\psi_0\rangle + |\psi'(t)\rangle\} $$

$$ = E_0 f(t)|\psi_0\rangle + \varepsilon f(t)|\psi_1\rangle + H_0|\psi'(t)\rangle + \varepsilon\langle\psi_1|\psi'(t)\rangle|\psi_0\rangle. \qquad (3.224) $$

Equating components in $\mathscr{S}_0$ and $\mathscr{S}'$,

$$ i\hbar \frac{df}{dt} = E_0 f(t) + \varepsilon\langle\psi_1|\psi'(t)\rangle, \qquad (3.225) $$

$$ i\hbar \frac{d}{dt}|\psi'(t)\rangle = \varepsilon f(t)|\psi_1\rangle + H_0|\psi'(t)\rangle. \qquad (3.226) $$

After multiplying by the integrating factor $e^{iH_0 t/\hbar}$, the second equation can be integrated to give

$$ |\psi'(t)\rangle = \frac{\varepsilon}{i\hbar}\int_0^t dt' f(t') e^{-iH_0(t-t')/\hbar}|\psi_1\rangle. \qquad (3.227) $$

This expresses the state of the decay products as a superposition of states, one for each time t' in the interval $(0, t)$: the state corresponding to t' describes the possibility that $|\psi_0\rangle$ decayed to $|\psi_1\rangle$ between times t' and $t' + dt'$, and the decay state then evolved for a time $t - t'$ according to the Hamiltonian H_0. The coefficient of this state is the product of $f(t')$, the probability amplitude that $|\psi_0\rangle$ survived undecayed at t', and the decay amplitude $(i\hbar)^{-1}\varepsilon \, dt'$.

Putting (3.227) into (3.225) gives an integro-differential equation for the non-decay amplitude $f(t)$:

$$ \frac{dF}{dt} = -\frac{\varepsilon^2}{\hbar^2}\int_0^t q(t-t')F(t')\,dt' \qquad (3.228) $$

where

$$ F(t) = e^{iH_0 t/\hbar}f(t) \qquad (3.229) $$

and

$$ q(t) = \langle\psi_1|e^{-iH_0 t/\hbar}|\psi_1\rangle. \qquad (3.230) $$

Now $q(t)$ is the amplitude for finding the decay products in their initial state $|\psi_1\rangle$ after a time t. If the decay products disperse quickly, $q(t)$ will vanish after some short time τ. Let us see if (3.228) has a solution $F(t)$ which varies only slightly over time intervals of the order of τ. If F satisfies this condition, the integral can be replaced by $AF(t)$ where $A = \int_0^\tau q(t')\,dt'$, so that the equation becomes

$$ \frac{dF}{dt} = -\frac{\varepsilon^2 A}{\hbar^2}F(t). \qquad (3.231) $$

Thus the condition on $F(t)$ is satisfied if $\varepsilon^2 A/\hbar^2$ is small compared with τ^{-1} or, since A is of order τ, if $\varepsilon/\hbar$ is small compared with τ^{-1}. If this is so there is a solution for $F(t)$ of exponential form, and the probability that the unstable state has not decayed in time t is

$$|f(t)|^2 = |F(t)|^2 = e^{-\lambda t} \quad \text{with } \lambda = 2 \operatorname{Re} [\varepsilon^2 A/\hbar^2].$$ (3.232)

This result is necessarily approximate. In our model it follows immediately from (3.228) that the exponential decay law must break down at small times, for this equation shows that the derivative $F'(t)$ must vanish at $t=0$. In other calculations it has been found that the exponential law also breaks down at large times (see Fonda, Ghirardi & Rimini 1978, Peres 1980).

There is a general argument to show that the time development of a state cannot be exactly exponential. Suppose the state $|\psi_0\rangle$ develops to $|\psi(t)\rangle$, and that

$$\langle \psi_0 | \psi(t) \rangle = e^{-\gamma t}$$ (3.233)

where $\gamma = (\tfrac{1}{2}\Gamma + iE_0)/\hbar$ is a complex number. For $t<0$ we have

$$\langle \psi_0 | \psi(t) \rangle = \langle \psi_0 | e^{-iHt/\hbar} | \psi_0 \rangle = \overline{\langle \psi_0 | e^{iHt/\hbar} | \psi_0 \rangle}$$
$$= \overline{\langle \psi_0 | \psi(-t) \rangle} = e^{\bar{\gamma} t}.$$

so

$$\langle \psi_0 | \psi(t) \rangle = \exp \left[-(iE_0 t + \tfrac{1}{2}\Gamma |t|)/\hbar \right].$$ (3.234)

This holds for all t. Now expand $|\psi_0\rangle$ in eigenstates of the exact Hamiltonian:

$$|\psi_0\rangle = \int \rho(E) |\psi(E)\rangle \, dE.$$ (3.235)

Then

$$\langle \psi_0 | \psi(t) \rangle = \langle \psi_0 | e^{-iHt/\hbar} | \psi_0 \rangle = \int |\rho(E)|^2 e^{-iEt/\hbar} \, dE.$$ (3.236)

Hence, by the Fourier inversion formula (2.93),

$$|\rho(E)|^2 = \frac{1}{2\pi\hbar} \int_{-\infty}^{\infty} \langle \psi_0 | \psi(t) \rangle e^{iEt/\hbar} \, dt$$
$$= \frac{\Gamma/2\pi}{\tfrac{1}{2}\Gamma^2 + (E - E_0)^2}.$$ (3.237)

But the energy of a system always has some minimum value $E_{\min}$, so $\rho(E)$ must vanish for $E < E_{\min}$; whereas (3.237) is strictly positive for all E.

The function (3.237) is called the **Breit–Wigner** energy distribution. It is shown in Fig. 3.6. Although it must be cut off at some point on the left, an approximation to this distribution is found in a wide range of quantum

Fig. 3.6.
The Breit–Wigner distribution.

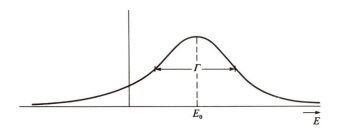

systems, corresponding to the exponential decay law which is usually found to be approximately true for unstable systems. The parameter Γ is called the **width** of the distribution. The inverse relation between Γ and the lifetime $\tau = \hbar/\Gamma$ is often described as an example of the 'time–energy uncertainty relation', but it should be noted that Γ is not the same as the uncertainty ΔE. For the distribution (3.237), indeed, ΔE is infinite.

3.6. Feynman's formulation of quantum mechanics

In the previous section we saw that the probability of an event in quantum mechanics has a strange structure. If the event can be achieved by a number of alternative processes, each process consisting of a succession of intermediate events, then the total probability of it is calculated, not by the classical formula in which the probabilities of alternative processes are added and the probabilities of successive processes are multiplied, but by a similar-looking formula in which probabilities are replaced by probability *amplitudes* – complex numbers whose squared moduli give the probabilities.

In Feynman's formulation of quantum mechanics these probability amplitudes are taken to be fundamental: the theory starts from the assignment of probability amplitudes to basic physical processes. For a single simple particle, a 'basic process' is described in the same way as in classical mechanics: it is the motion of a particle from a point $\mathbf{r}_1$ at time t_1 to another point $\mathbf{r}_2$ at time t_2 along a definite trajectory $\mathbf{r}(t)$. Thus the probability amplitude is to be a function of the whole trajectory $\{\mathbf{r}(t): t_1 \leqslant t \leqslant t_2\}$, which we will denote by $[\mathbf{r}(t)]$.

In classical mechanics an important function of the trajectory of a particle is the **action** S. This is an integral of the form

$$S([\mathbf{r}(t)]) = \int_{t_1}^{t_2} L(\mathbf{r}(t), \dot{\mathbf{r}}(t))\, dt \tag{3.238}$$

where L is a function of the position and velocity of the particle called the **Lagrangian**. L depends on the forces acting on the particle; for a particle of mass m moving in a potential $V(\mathbf{r})$ it is

$$L = \tfrac{1}{2}m\dot{\mathbf{r}}^2 - V(\mathbf{r}). \tag{3.239}$$

A Lagrangian L, and consequently an action S, exists for any mechanical system; if the configuration of the system is specified by coordinates $q_1, \ldots, q_n$, L is a function of $q_1, \ldots, q_n$ and $\dot{q}_1, \ldots, \dot{q}_n$ which is normally the difference between the kinetic energy and the potential energy of the system, as in (3.239); S is the time integral of L, as in (3.238). The significance of S is that the trajectory followed by the system when it follows the equations of motion is that which makes S minimum (or, more generally, stationary); this is the **principle of least action**.

Feynman's postulate is that the probability amplitudes for all trajectories are equal in modulus, but their phases are given by the action (in units of $\hbar$, which has the same dimensions as action). Thus the probability amplitude for a trajectory with action S is proportional to $e^{iS/\hbar}$. The total probability amplitude for a particle to move from $\mathbf{r}_A$ at time t_A to $\mathbf{r}_B$ at time t_B must be the sum of the probability amplitudes for all the ways this could have happened, i.e. all the trajectories $\mathbf{r}(t)$ with $\mathbf{r}(t_A) = \mathbf{r}_A$ and $\mathbf{r}(t_B) = \mathbf{r}_B$. Since these trajectories form a continuously infinite set, the sum must be some sort of integral; the total amplitude must be of the form

$$I(\mathbf{r}_A, t_A; \mathbf{r}_B, t_B) = \int e^{iS([\mathbf{r}(t)])/\hbar} \, d[\mathbf{r}(t)] \tag{3.240}$$

where, in some sense, the integral is over all trajectories with the given endpoints at t_A and t_B.

The integral (3.240) is called a **path integral**. It can be defined as a limit of integrals over polygonal paths, in the following sense. Divide the interval $[t_A, t_B]$ into N subintervals by means of intermediate times $t_A = t_0, \ldots, t_N = t_B$. Let us call this dissection D. Given D, and given $N + 1$ points $\mathbf{r}_A = \mathbf{r}_0, \mathbf{r}_1, \ldots, \mathbf{r}_N = \mathbf{r}_B$, we can define a polygonal trajectory $\mathbf{r}(t)$ which passes through $\mathbf{r}_n$ at time t_n and moves with constant velocity between these points:

$$\mathbf{r}(t) = \mathbf{r}_n + \frac{t - t_n}{t_{n+1} - t_n} (\mathbf{r}_{n+1} - \mathbf{r}_n) \quad \text{if } t_n \leqslant t \leqslant t_{n+1}. \tag{3.241}$$

Let $S_D(\mathbf{r}_0, \ldots, \mathbf{r}_N)$ be the action of this polygonal trajectory, and let

$$I_D(\mathbf{r}_A, t_A; \mathbf{r}_B, t_B) = \int \exp\left[\frac{i}{\hbar} S_D(\mathbf{r}_0, \ldots, \mathbf{r}_N) \right] \frac{d^3\mathbf{r}_1}{K_1} \cdots \frac{d^3\mathbf{r}_{N-1}}{K_{N-1}} \tag{3.242}$$

where

$$K_n = \left\{ \frac{2i\pi\hbar}{m} (t_n - t_{n-1}) \right\}^{\frac{3}{2}} \tag{3.243}$$

(m is the mass of the particle). The path integral $I(\mathbf{r}_A, t_A; \mathbf{r}_B, t_B)$ is defined to be the limit of I_D as the dissection D becomes infinitely fine, i.e. as $n \to \infty$ and $\max (t_{n+1} - t_n) \to 0$. (The factors K_i, as we will see in the proof of ●3.12, are necessary if the path integral I is to be a continuous function of $\mathbf{r}_B$.)

Clearly there are formidable mathematical problems surrounding the existence of this limit. We will ignore these and proceed purely formally, making no attempt to be rigorous. In this spirit we can state the basic assumption, generalised to an arbitrary mechanical system, as

> **Feynman's postulate.** Consider a mechanical system with coordinates denoted collectively by q, and with dynamics determined by an action functional $S[q(t)]$. The probability amplitude for the system to move through the sequence of configurations $q(t)$ is
>
> $$\exp\left[\frac{i}{\hbar} S[q(t)] \right] d[q(t)]. \tag{3.244}$$

This refers only to systems with counterparts in classical mechanics, and not to the purely quantum mechanical systems which we will encounter in later chapters, such as particles with spin or other internal properties. Feynman's postulate can be extended to such cases through the medium of quantum field theory (see §7.3).

Let us now see how the equation of motion of conventional quantum mechanics can be derived from Feynman's postulate.

● **3.14** Let $S([\mathbf{r}(t)])$ be the action for a simple particle moving in a potential $V(\mathbf{r})$ (see (3.238)–(3.239)), let $I(\mathbf{r}_A, t_A; \mathbf{r}_B, t_B)$ be the probability amplitude given by (3.240), and let

$$\psi(\mathbf{r}, t) = I(\mathbf{r}_0, t_0; \mathbf{r}, t) \tag{3.245}$$

for some fixed $\mathbf{r}_0, t_0$. If ψ fulfils the conditions **W1–W3** (p. 39), then it satisfies the Schrödinger equation (3.7).

Proof. Look at the construction of $\psi(\mathbf{r}, t + \delta t)$ as a path integral. It is the limit of a set of integrals defined with reference to a dissection D of the interval $[t_0, t + \delta t]$; if δt is small enough we can consider the last subinterval in the dissection to be $[t, t + \delta t]$. Then the amplitude $I_D(\mathbf{r}_0, t_0; \mathbf{r}, t + \delta t)$ is given by (3.242), which is an integral over a set of polygonal trajectories each consisting of a polygonal trajectory from $(\mathbf{r}_0, t_0)$ to $(\mathbf{r}', t)$ for some $\mathbf{r}' = \mathbf{r}_{N-1}$, followed by a straight line from $(\mathbf{r}', t)$ to $(\mathbf{r}, t + \delta t)$. Now if a trajectory is divided into two contiguous parts, its action can be written as the sum of the actions for the two parts, as can be seen from (3.238); hence (3.242) gives

$$I_D(\mathbf{r}_0, t_0; \mathbf{r}, t + \delta t)$$
$$= \int \exp\left[\frac{i}{\hbar} S_D(\mathbf{r}_0, \ldots, \mathbf{r}_{N-2}, \mathbf{r}')\right]$$
$$\times \exp\left[\frac{i}{\hbar} S(\mathbf{r}', \mathbf{r})\right] \frac{d^3\mathbf{r}_1}{K_1} \cdots \frac{d^3\mathbf{r}_{N-2}}{K_{N-2}} \frac{d^3\mathbf{r}'}{K_{N-1}} \tag{3.246}$$

with

$$K_{N-1} = \left(\frac{2i\pi\hbar\,\delta t}{m}\right)^{\frac{3}{2}}. \tag{3.247}$$

Here $S(\mathbf{r}', \mathbf{r})$ is the action of the straight-line trajectory from $(\mathbf{r}', t)$ to $(\mathbf{r}, t + \delta t)$; this can be written as

$$S(\mathbf{r}', \mathbf{r}) = \frac{U(\mathbf{r}', \mathbf{r})}{\delta t} - W(\mathbf{r}', \mathbf{r})\,\delta t \tag{3.248}$$

where

$$\frac{U}{\delta t} = \int_t^{t+\delta t} \tfrac{1}{2} m\dot{\mathbf{r}}^2\, dt' = \int_t^{t+\delta t} \tfrac{1}{2} m\left(\frac{\mathbf{r} - \mathbf{r}'}{\delta t}\right)^2 dt' = \frac{m(\mathbf{r} - \mathbf{r}')^2}{2\,\delta t} \tag{3.249}$$

and

$$W\,\delta t = \int_t^{t+\delta t} V\left(\mathbf{r}' + \frac{t' - t}{\delta t}(\mathbf{r}' - \mathbf{r})\right) dt' = \delta t \int_0^1 V(\mathbf{r}' + s(\mathbf{r} - \mathbf{r}'))\, ds. \tag{3.250}$$

Thus both U and W depend only on $\mathbf{r}$ and $\mathbf{r}'$ and not on δt.

In the limit as the dissection of $[t_0, t]$ becomes infinitely fine but the last subinterval $(t, t + \delta t)$ is kept fixed, (3.246) becomes

$$\psi(\mathbf{r}, t + \delta t) = \left(\frac{m}{2i\pi\hbar\,\delta t}\right)^{\frac{3}{2}} \int \psi(\mathbf{r}', t) \exp\left[\frac{im}{2\hbar\,\delta t}(\mathbf{r} - \mathbf{r}')^2 - \frac{i\,\delta t}{\hbar} W(\mathbf{r}', \mathbf{r})\right] d^3\mathbf{r}'. \tag{3.251}$$

If ψ is continuous, this should give $\psi(\mathbf{r}, t)$ as $\delta t \to 0$. Now the function e^{ix^2} is integrable, with integral

$$\int_{-\infty}^{\infty} e^{ix^2} \, dx = (i\pi)^{\frac{1}{2}} \tag{3.252}$$

(see problem 3.35(i)); it follows that

$$\lim_{a \to 0} \frac{1}{a} e^{ix^2/a^2} = (i\pi)^{\frac{1}{2}} \delta(x) \tag{3.253}$$

for if f is any bounded continuous function we have

$$\lim_{a \to 0} \frac{1}{a} \int_{-\infty}^{\infty} f(x) e^{ix^2/a^2} \, dx = \lim_{a \to 0} \int_{-\infty}^{\infty} f(ay) e^{iy^2} \, dy = (i\pi)^{\frac{1}{2}} f(0) \tag{3.254}$$

(a more careful version of this argument is outlined in problem 3.35(ii)). Putting $a^2 = 2\hbar \, \delta t/m$, and considering all three components of the vector $\mathbf{r} - \mathbf{r}'$, we obtain from (3.253)

$$\lim_{\delta t \to 0} \left(\frac{m}{2\hbar \, \delta t}\right)^{\frac{3}{2}} \exp\left[\frac{im}{2\hbar \, \delta t} (\mathbf{r} - \mathbf{r}')^2\right] = (i\pi)^{\frac{3}{2}} \delta(\mathbf{r} - \mathbf{r}'). \tag{3.255}$$

Thus the right-hand side of (3.251) does indeed give $\psi(\mathbf{r}, t)$ as $\delta t \to 0$. (This is the reason for the factors K_n in the definition (3.242)–(3.243) of the path integral.)

Now we expand $\psi(\mathbf{r}, t + \delta t)$ as

$$\psi(\mathbf{r}, t + \delta t) = \psi(\mathbf{r}, t) + \frac{\partial \psi}{\partial t} \, \delta t + O(\delta t^2) \tag{3.256}$$

and find $\partial \psi/\partial t$ by differentiating (3.251) with respect to δt and letting $\delta t \to 0$. The differentiation gives

$$\left(\frac{m}{2i\pi\hbar \, \delta t}\right)^{\frac{3}{2}} \int \int \left\{-\frac{3}{2 \, \delta t} - \frac{im}{2\hbar \, \delta t^2} (\mathbf{r} - \mathbf{r}')^2 - \frac{i}{\hbar} W(\mathbf{r}', \mathbf{r})\right\} \Psi \Phi \, d^3\mathbf{r}' \tag{3.257}$$

where

$$\Psi = \psi(\mathbf{r}', t) \exp\left[-\frac{i \, \delta t}{\hbar} W(\mathbf{r}', \mathbf{r})\right],$$

$$\Phi = \exp\left[\frac{im}{2\hbar \, \delta t} (\mathbf{r} - \mathbf{r}')^2\right].$$

Using ∇' to denote differentiation with respect to $\mathbf{r}'$, we have

$$\nabla'^2 \Phi = \frac{im}{\hbar \, \delta t} \left\{3 + \frac{im}{\hbar \, \delta t} (\mathbf{r} - \mathbf{r}')^2\right\} \Phi, \tag{3.258}$$

so that (3.257) can be written as

$$\left(\frac{m}{2i\pi\hbar \, \delta t}\right)^{\frac{3}{2}} \int \int \left\{-\frac{\hbar}{2im} \Psi \nabla'^2 \Phi - \frac{i}{\hbar} W \Psi \Phi\right\} d^3\mathbf{r}'. \tag{3.259}$$

Now if ψ satisfies the conditions to be a wave function, Ψ decreases fast enough at large $\mathbf{r}'$ to make it possible to apply Green's theorem and convert the integral to

$$\left(\frac{m}{2\pi i\hbar\,\delta t}\right)^{\frac{3}{2}} \int \Phi \left\{-\frac{\hbar}{2im}\nabla'^2\Psi - \frac{i}{\hbar}W\Psi\right\} d^3\mathbf{r}'. \tag{3.260}$$

Taking the limit as $\delta t \to 0$ and using (3.255), we get the part of the integrand in curly brackets, evaluated at $\mathbf{r}'=\mathbf{r}$. Since $W(\mathbf{r},\mathbf{r})=V(\mathbf{r})$, this gives

$$\frac{\partial\psi}{\partial t} = -\frac{\hbar}{2im}\nabla^2\psi - \frac{i}{\hbar}V\psi,$$

which is the Schrödinger equation. ■

Bones of Chapter 3

Postulate III (*continued*). Transition probabilities 114
Postulate VI. Undisturbed time development 79
Postulate VII. Translations and rotations 94
Feynman's postulate 124

Further reading

The general references on quantum mechanics given in Chapter 2 are equally relevant to this chapter. There is a great number of textbooks (e.g. Schiff 1968) which can be consulted for details of calculations with specific Hamiltonians. The relevance of group theory to quantum mechanics was emphasised in one of the first quantum mechanics textbooks, Weyl 1928. Other textbooks on group theory and quantum mechanics are Gilmore 1974 and Cornwell 1984. For Feynman's formulation see Feynman & Hibbs 1965.

Problems on Chapter 3

1. A quantum system can exist in two states $|a_0\rangle$ and $|a_1\rangle$, which are normalised eigenstates of an observable A with eigenvalues 0 and 1 respectively. The Hamiltonian operator is defined by

$$H|a_0\rangle = \alpha|a_0\rangle + \beta|a_1\rangle, \quad H|a_1\rangle = \beta|a_0\rangle + \alpha|a_0\rangle$$

where α and β are real. If the system is in the state $|a_0\rangle$ at time $t=0$, show that its state at time t is $e^{-i\alpha t}(\cos \beta t|a_0\rangle - i \sin \beta t|a_1\rangle)$.

The observable A is measured at $t=T$, but the value is lost. It is measured again at $t=2T$. Find the probability that the second measurement gives the result 0.

2. It is possible that the three neutrinos mentioned in Chapter 1 are different states of a single system, being eigenstates of an experiment whose results are called ν_e, ν_μ and ν_τ, and that they are not stationary states. It is also possible that the masses (energy eigenvalues) of this system are not exactly 0. Let $|\psi_i\rangle$ ($i = 1, 2, 3$) be the normalised eigenstates of energy, with eigenvalues $m_i c^2$, and suppose the eigenstates of the 'neutrino type' experiment are

$$|\nu_e\rangle = \tfrac{1}{2}|\psi_1\rangle + \sqrt{3/2}|\psi_2\rangle, \quad |\nu_\mu\rangle = \tfrac{3}{4}|\psi_1\rangle - \sqrt{3/4}|\psi_2\rangle - \tfrac{1}{2}|\psi_3\rangle.$$

If the system starts at time $t=0$ in the electron-neutrino state, find the probabilities that at time t it will be found to be in (a) the μ-neutrino state, (b) the τ-neutrino state.

3. A quantum system can exist in two states $|\psi_1\rangle$ and $|\psi_2\rangle$, which are eigenstates of the Hamiltonian with eigenvalues E_1 and E_2. An observable A has eigenvalues ± 1 and eigenstates $|\psi_\pm\rangle = 2^{-\frac{1}{2}}(|\psi_1\rangle \pm |\psi_2\rangle)$. This observable is measured at the times $t = 0, \tau, 2\tau, \ldots$. The (normalised) state of the system at $t=0$, just before the first measurement, is $c_1|\psi_1\rangle + c_2|\psi_2\rangle$. If p_n denotes the probability that the measurement at $t=n\tau$ gives the result $A=1$, show that

$$p_{n+1} = \tfrac{1}{2}(1 - \cos \alpha) + p_n \cos \alpha, \quad \text{where } \alpha = (E_1 - E_2)/\hbar,$$

and deduce that

$$p_n = \tfrac{1}{2}(1 - \cos^n \alpha) + \tfrac{1}{2}|c_1 + c_2|^2 \cos^n \alpha.$$

What happens in the limit as $n \to \infty$ with $n\tau = t$ fixed? [See ●5.7.]

4. Show that if the time-dependent Schrödinger equation (3.8) has a solution of the form $\Psi(\mathbf{r}_1, \ldots, \mathbf{r}_n)f(t)$, then Ψ must be an eigenfunction of the Hamiltonian and $f(t) = e^{-iEt/\hbar}$ where E is the eigenvalue associated with Ψ.

5. For any observable A, show that $\tau_A \Delta H \geqslant \tfrac{1}{2}\hbar$, where H is the Hamiltonian and $\tau_A = |d\langle A\rangle/dt|^{-1} \Delta A$.

6. If $[H, A] = -i\lambda\hbar A$, show that $\Delta A = Ce^{\lambda t}$ where C is constant.

7. Find the energy levels of a particle moving in space and confined to a cubical box of side a, but subject to no other forces.

8. At time $t=0$ the wave function of a free particle of mass m moving in one dimension is $\psi_0(x) = G(x, a_0) = (a_0\sqrt{\pi})^{-\frac{1}{2}} \exp(-x^2/2a_0^2)$. By writing ψ_0 as a Fourier transform (see problem 2.15), find the wave function at time t and show that $|\psi(x, t)| = G(x, a)$ where $a^2 = a_0^2 + \hbar^2 t^2/m^2 a_0^2$.

9. Find the probability current density for the wave function $\psi(x) = Ae^{ikx} + Be^{-ikx}$ and show that it is the same as the current density of two beams of classical particles with densities $|A|^2$ and $|B|^2$ moving in opposite directions. Is this true if k is replaced by k' in the second term?

10. A particle is moving along the x-axis in a potential which vanishes for $x < a$ and $x > b$. If the wave function in the regions of zero potential is $\psi(x) =$

$Ae^{-ikx} + Be^{ikx}$ for $x < a$, $\psi(x) = Ce^{-ikx}$ for $x < b$, show that $|A|^2 = |B|^2 + |C|^2$. What is the physical significance of this? (See problem 9.)

11. Let $V(x)$ be a function which is infinitely differentiable everywhere except at a finite number of points x_i. Show that if ψ is a superposition of solutions of the Schrödinger equation (3.45) and satisfies the continuity conditions (3.53) at x_i, then ψ is infinitely differentiable at x_i.

12. Suppose the function $\psi(x)$ is an eigenbra of the Hamiltonian $H = p^2/2m + V(x)$, so that it satisfies (3.55) for all ϕ satisfying the continuity conditions (3.53). Show that ψ satisfies the Schrödinger equation (3.45) if and only if it too satisfies (3.53).

13. A particle of mass m is moving in one dimension in a potential in the form of a barrier with value V_0 in the region $0 \leqslant x \leqslant a$ and 0 everywhere else. Find the eigenfunction of the Hamiltonian which has the form $e^{-ikx} + Be^{ikx}$ for $x < 0$ and Ce^{-ikx} for $x > a$. [This can be regarded as the wave function of a particle which is incident on the barrier from the left; there is probability $|B|^2$ that it will be reflected from the barrier and probability $|C|^2$ that it will pass through it. The fact that $|C|^2 \neq 0$ even if $E < V_0$, when a classical particle would be unable to penetrate the barrier, is the **quantum tunnelling effect**.]

In the case $E > V_0$, find the relative probability that the particle will be found in the interval $[0, a]$, compared with the probability that it will be found in an interval of the same length in the region $x > a$. Find the limit as $\hbar \to 0$, and compare with the ratio of the times spent in these intervals by a classical particle with energy E.

14. A particle is moving in space in a potential which vanishes in a certain region D. It has a stationary state in which its wave function in D is $\psi(r) = f(r)e^{ikr}$ where f is a real function of a scalar variable and k is a real constant. Find the function f and the energy of the state. [Use $\nabla^2 f(r) = f''(r) + 2f'(r)/r$.]

Calculate the probability current in D. Is it possible for D to consist of all space except for a neighbourhood of the origin?

15. Find $\langle x \rangle$ and Δx for the nth stationary state of a free particle in one dimension restricted to the interval $[0, a]$. Show that as $n \to \infty$ these become the classical values.

16. Show that if a system is only invariant under translations in the direction $\mathbf{n}$, then the momentum component $\mathbf{n} \cdot \mathbf{p}$ is conserved.

17. Show that the system of a free particle moving in space is invariant under translations.

18. If a particle moves in a potential $V(\mathbf{r})$, find $U(T_{\mathbf{a}})VU(T_{\mathbf{a}})^{-1}$. Deduce that the system is invariant under translations only if there is no force on the particle. Is the same true classically?

19. Describe the translation operators for a system of two simple particles moving in space. If they are subject to forces deriving from a potential $V(\mathbf{r}_1, \mathbf{r}_2)$, find conditions on V for the system to be invariant under translations.

20. For photons travelling in a certain medium the states $|\phi_x\rangle \pm i|\phi_y\rangle$ are eigenstates of the Hamiltonian with energies E_+ and E_-. Show that this

system is invariant under rotations about the z-axis, and describe the propagation in the z-direction of plane polarised light in this medium.

21. Show that the rotation $R(\mathbf{n}, \theta)$ is given by

    $$R(\mathbf{n}, \theta)\mathbf{x} = \mathbf{x} \cos \theta + (\mathbf{x} \cdot \mathbf{n})\mathbf{n}(1 - \cos \theta) + \mathbf{n} \times \mathbf{x} \sin \theta$$

    and verify that the generator of rotations about the axis $\mathbf{n}$, acting on the space of wave functions, is $\mathbf{n} \cdot \mathbf{J}$ where $\mathbf{J} = -i\hbar \mathbf{r} \times \mathbf{V}$.

22. An operator V has even parity if $PV = VP$, odd parity if $PV = -VP$. Show that position and momentum both have odd parity, but angular momentum has even parity. Show that the expectation value of V in a state $|\psi\rangle$ vanishes if V has odd parity and $|\psi\rangle$ is an eigenstate of P.

23. **Wigner's theorem.** Let $|\psi\rangle \to |T\psi\rangle$ be a map of state space which satisfies $|\langle T\phi | T\psi\rangle| = |\langle \phi | \psi\rangle|$ for all $|\phi\rangle, |\psi\rangle$. Fill in the details of the following proof that phase factors $\omega(\psi)$ can be found so that the map $|\psi\rangle \to U|\psi\rangle = \omega(\psi)|T\psi\rangle$ is either linear or antilinear.

    Choose an orthonormal basis $|\psi_i\rangle$, and let $|\phi_{ij}\rangle = |\psi_i\rangle + |\psi_j\rangle$. Show that U can be defined so that $U|\phi_{ij}\rangle = |T\psi_i\rangle + y_{ij}|\psi_j\rangle$ with $|y_{ij}| = 1$ and $y_{1j} = 1$. Let $|\xi\rangle = \sum c_i |\psi_i\rangle$ be any vector; then $U|\xi\rangle$ can be defined so that $U|\xi\rangle = \sum c_i' |T\psi_i\rangle$ with $c_1 = 1$ and $|c_i'| = |c_i|$. Show that either $c_i' = c_i$ (say that c_i is a 'linear coordinate') or $c_i' = \omega \bar{c}_i$ where $\omega = c_1/\bar{c}_1$ (say that c_i is an 'antilinear coordinate'); and that if c_i is a linear coordinate but c_j is an antilinear one, then either $c_i = \bar{y}_{ij}\omega \bar{c}_i$ or $c_j = \bar{y}_{ij}\omega \bar{c}_j$. Deduce that there is a vector whose ith and jth coordinates are either both linear or both antilinear. By considering $|\xi\rangle$ together with $c_1|\psi_1\rangle + d_i|\psi_i\rangle$, show that all vectors have ith coordinates of the same kind (either linear or antilinear), and conclude that U is either linear or antilinear.

24. Let $U(\lambda)$ be a set of non-singular operators depending on a real parameter λ. Show that $(d/d\lambda)[U(\lambda)^{-1}] = -U^{-1}(dU/d\lambda)U^{-1}$.

25. Let Q_λ be a group of operations labelled by a real parameter λ and satisfying $Q_{\lambda+\mu} = Q_\lambda Q_\mu$. Let U be a representation of the group on a vector space V, with generator X; let A_0 be any operator on V and let $A(\lambda) = U(Q_\lambda)A_0 U(Q_\lambda)^{-1}$. Show that $dA/d\lambda = [X, A(\lambda)]$ for all λ.

26. Let $U(\theta) = U(R(\mathbf{k}, \theta))$ and let A be any operator. By setting up a differential equation for $U(\theta)AU(\theta)^{-1}$, show that (taking $\hbar = 1$)

    $$U(\theta)AU(\theta)^{-1} = A + \sum_{n=1}^{\infty} \frac{(-i\theta)^n}{n!} [J_z, [J_z, \ldots, [J_z, A] \cdots]]$$

    where the nth term in the sum contains an n-fold commutator.

27. For a system of n spinless particles an operator $U(B_\mathbf{v})$ is defined by $U(B_\mathbf{v})\psi(\mathbf{r}_1, \ldots, \mathbf{r}_1) = \exp\{i(m_1\mathbf{r}_1 + \cdots + m_n\mathbf{r}_n) \cdot \mathbf{v}\}\psi(\mathbf{r}_1, \ldots, \mathbf{r}_n)$ where $m_1, \ldots, m_n$ are the masses of the particles. Show that $U(B_\mathbf{v})^{-1}\mathbf{p}_i U(B_\mathbf{v}) = \mathbf{p}_i + m_i\mathbf{v}$, and deduce that $U(B_\mathbf{v})$ represents the operation of giving the whole system a velocity $\mathbf{v}$.

    The **Galilean group** consists of translations $T_\mathbf{a}$, rotations R, time translations T_τ and boosts $B_\mathbf{v}$, which are transformations of space-time $\mathbb{R}^4$ acting as follows: $T_\mathbf{a}: (t, \mathbf{r}) \to (t, \mathbf{r} + \mathbf{a})$, $R: (t, \mathbf{r}) \to (t, R\mathbf{r})$, $T_\tau: (t, \mathbf{r}) \to (t + \tau, \mathbf{r})$, $B_\mathbf{v}: (t, \mathbf{r}) \to (t, \mathbf{r} + \mathbf{v}t)$. Show that $T_\mathbf{a}$ and $B_\mathbf{v}$ commute but $U(T_\mathbf{a})U(B_\mathbf{v}) =$

$\omega U(B_v)U(T_a)$ where $|\omega|=1$. Find the hermitian generators of the boost operators $U(B_v)$.

28. Show that the factors $\omega(Q, R)$ in the definition (3.141) of a projective representation satisfy $\omega(Q, RS)\omega(R, S)=\omega(Q, R)\omega(QR, S)$, and that if $\omega(Q, R)=\theta(Q)\theta(R)/\theta(QR)$ for some function θ on the group, there is a true representation associated with the projective one.

29. Let U be a representation of a group G on a complex vector space V. For any $Q \in G$ and $v \in V$ define an operator $T(Q, v)$ on $V \oplus \mathbb{C}$ by $T(Q, v)(w, c)=(U(Q)w+cv, c)$. Show that T is a representation of the inhomogeneous extension $G \times)V$. Deduce that the Lie algebra of $G \times)V$ is isomorphic to $L \oplus V$ where L is the Lie algebra of G, and determine the Lie brackets.

30. Let U be a unitary representation of a group G on a vector space V. Show that if W is an invariant subspace of the representation, so is its orthogonal complement $W^\perp$. Deduce that if V is finite-dimensional it can be written as $V=V_1 \oplus \cdots \oplus V_n$ where each V_i carries an irreducible representation of G. [The representation U is said to be **completely reducible**.]

By considering the representation T of problem 29, show that a non-unitary representation need not be completely reducible.

31. Show that the space of three-vectors forms a Lie algebra with Lie bracket given by the cross product.

Let $\Omega(\mathbf{a})$ be the 3×3 matrix defined by $\Omega(\mathbf{a})\mathbf{x}=\mathbf{a} \times \mathbf{x}$ where $\mathbf{a}$ and $\mathbf{x}$ are three-vectors. Show that $\Omega(\mathbf{a})$ is antisymmetric and that $[\Omega(\mathbf{a}), \Omega(\mathbf{b})]=\Omega(\mathbf{a} \times \mathbf{b})$. Deduce that the Lie algebra of SO(3) is isomorphic to the Lie algebra defined in the first sentence of this problem.

32. A particle of mass m and electric charge e, moving in one dimension, is confined to an interval of length a and is subject to an electric field E. Initially it is in the eigenstate of kinetic energy with eigenvalue $E_k=k^2\pi^2\hbar^2/2ma^2$ where k is an integer. Find, to first order in e^2, the probability that after a time t its kinetic energy will be found to be E_l where $k \neq l$.

33. A system has Hamiltonian $H_0+\varepsilon V$ and makes transitions between eigenstates of H_0. There are three such eigenstates $|\psi_1\rangle, |\psi_2\rangle$ and $|\psi_3\rangle$ with eigenvalues E_i such that $E_3-E_2=E_2-E_1=E \neq 0$. If V is independent of time, with $\langle\psi_2|V|\psi_1\rangle=0$, and if the system is in state $|\psi_1\rangle$ at time $t=0$, find the probability of finding it in the state $|\psi_2\rangle$ at time t, to lowest non-vanishing order.

34. A system makes transitions between eigenstates of H_0 under the action of the time-dependent Hamiltonian $H_0+\varepsilon V_0 \cos \omega t$. Find an expression for the probability of transition from $|\psi_1\rangle$ to $|\psi_2\rangle$ in time t, where $|\psi_1\rangle$ and $|\psi_2\rangle$ are eigenstates of H_0 with eigenvalues E_1 and E_2. Show that this probability is small unless $E_2-E_1 \simeq \omega\hbar$.

[This shows that a charged particle in an oscillating electric field with frequency v will exchange energy with the field only in multiples of $E=hv$.]

35. Fill in the following details in the proof of ●3.14.

(i) By considering an integral over a triangular contour with corners at 0, R and $R+iR$, show that $\int_0^\infty \exp(ix^2)\,dx=\frac{1}{2}(i\pi)^{\frac{1}{2}}$.

(ii) Let ϕ be an integrable function of a real variable with $\int_{-\infty}^{\infty} \phi(x)\,dx = k < \infty$. Let f be any continuous function of a real variable, and let $I(s; a, b) = \int_{a}^{b} f(x) s\phi(sx)\,dx$. Show that

(a)   If $0 < a < b$, $I(s; a, b) \to 0$ as $s \to \infty$;

(b)   If $-a < 0 < b$, $I(s; a, b) \to kf(0)$ as $s \to \infty$.

Deduce that

$$\underset{s \to \infty}{\mathrm{Lim}}\,[s\phi(sx)] = k\,\delta(x).$$

4

Some quantum systems

In this chapter we will investigate mathematically some of the operators whose physical significance was explained in Chapters 2 and 3. In particular, we will find the eigenvalues of the operators and the number of independent eigenvectors associated with each, so obtaining a complete description of the corresponding physical observable.

4.1. Angular momentum   In §3.3 we saw that the components of angular momentum (J_x, J_y, J_z), which are the hermitian generators of rotations about the origin, satisfy the commutation relations

$$[J_x, J_y] = i\hbar J_z, \quad [J_y, J_z] = i\hbar J_x, \quad [J_z, J_x] = i\hbar J_y. \tag{4.1}$$

Since no two of these operators commute, they will not in general have simultaneous eigenvalues. However, the operator

$$\mathbf{J}^2 = J_x{}^2 + J_y{}^2 + J_z{}^2 \tag{4.2}$$

commutes with all three of J_x, J_y and J_z, as can easily be verified by using the identity (2.80) for a commutator containing a product. This means that we can look for simultaneous eigenvalues of $\mathbf{J}^2$ and any one of the individual components, say J_z.

Consider a system for which $\mathbf{J}^2$ and J_z are a complete set of commuting observables, so that there is a complete set of states $|\lambda, \mu\rangle$ consisting of simultaneous eigenstates with eigenvalues λ for $\mathbf{J}^2$ and μ for J_z:

$$\mathbf{J}^2|\lambda, \mu\rangle = \lambda|\lambda, \mu\rangle,$$
$$J_z|\lambda, \mu\rangle = \mu|\lambda, \mu\rangle. \tag{4.3}$$

We define the operators

$$J_\pm = J_x \pm iJ_y. \tag{4.4}$$

Then J_+ and J_- are hermitian conjugates of each other:

$$J_+ = J_-{}^\dagger, \tag{4.5}$$

and their products can be expressed in terms of $\mathbf{J}^2$ and J_z:

$$J_+ J_- = \mathbf{J}^2 - J_z^2 + \hbar J_z, \tag{4.6}$$

$$J_- J_+ = \mathbf{J}^2 - J_z^2 - \hbar J_z. \tag{4.7}$$

From (4.1) we obtain the commutators of J_+ and J_- with J_z as

$$[J_z, J_+] = \hbar J_+, \tag{4.8}$$

$$[J_z, J_-] = -\hbar J_-. \tag{4.9}$$

The crucial properties of $J_\pm$ are the following:

1. $J_\pm |\lambda, \mu\rangle$ are also eigenstates of $\mathbf{J}^2$ with eigenvalue λ. (4.10)

For $\mathbf{J}^2$ commutes with J_+, and so

$$\mathbf{J}^2 J_\pm |\lambda, \mu\rangle = J_\pm \mathbf{J}^2 |\lambda, \mu\rangle = \lambda J_\pm |\lambda, \mu\rangle.$$

2. Either $J_+ |\lambda, \mu\rangle = 0$ or $J_+ |\lambda, \mu\rangle$ is an eigenstate of J_z with eigenvalue $\mu + \hbar$. (4.11)

For from (4.8) we have

$$J_z J_+ |\lambda, \mu\rangle = (J_+ J_z + \hbar J_+)|\lambda, \mu\rangle = (\mu + \hbar) J_+ |\lambda, \mu\rangle.$$

3. Either $J_- |\lambda, \mu\rangle = 0$ or $J_- |\lambda, \mu\rangle$ is an eigenstate of J_z with eigenvalue $\mu - \hbar$. (4.12)

This is proved in the same way as (4.11), using (4.9).

Because of (4.11) and (4.12), J_+ and J_- are known as **raising and lowering operators** for J_z. In symbols, we have

$$J_\pm |\lambda, \mu\rangle = c_\pm |\lambda, \mu \pm \hbar\rangle. \tag{4.13}$$

To determine the factors $c_\pm$, note that since the states $|\lambda, \mu\rangle$ are normalised we have, writing $|\phi_\pm\rangle = J_\pm |\lambda, \mu\rangle$,

$$|c_+|^2 = \langle \phi_+ | \phi_+ \rangle = \langle \lambda, \mu | J_+^\dagger J_+ | \lambda, \mu \rangle$$
$$= \langle \lambda, \mu | (\mathbf{J}^2 - J_z^2 - \hbar J_z) | \lambda, \mu \rangle, \quad \text{using (4.5) and (4.7)},$$
$$= \lambda - \mu^2 - \mu\hbar. \tag{4.14}$$

Similarly,

$$|c_-|^2 = \langle \phi_- | \phi_- \rangle = \lambda - \mu^2 + \mu\hbar. \tag{4.15}$$

(Only the modulus of $c_\pm$ can be determined since any state vector can be multiplied by an arbitrary phase factor.)

From (4.14), using the positive-definite property (2.29) of the inner product, we have

$$\lambda - \mu^2 - \mu\hbar \geqslant 0, \tag{4.16}$$

$$\lambda - \mu^2 - \mu\hbar = 0 \Leftrightarrow |\phi_+\rangle = J_+ |\lambda, \mu\rangle = 0. \tag{4.17}$$

Similarly, (4.15) gives

$$\lambda - \mu^2 + \mu\hbar \geqslant 0, \tag{4.18}$$

$$\lambda - \mu^2 + \mu\hbar = 0 \Leftrightarrow |\phi_-\rangle = J_- |\lambda, \mu\rangle = 0. \tag{4.19}$$

The statements (4.11)–(4.12) and (4.16)–(4.19) are sufficient to determine the possible values of λ and μ. From (4.11) we see that, unless $J_+|\lambda,\mu\rangle = 0$, an eigenvalue μ of J_z is accompanied by a higher eigenvalue $\mu + \hbar$, which in turn is accompanied by an eigenvalue $\mu + 2\hbar$, and so on; the eigenvalues rise in steps of $\hbar$, like a ladder, and will only stop if they reach a value μ_{max} for which

$$J_+|\lambda,\mu_{max}\rangle = 0. \tag{4.20}$$

There must be such a maximum value, for if μ continued indefinitely it would reach a value which violated the inequality (4.16). According to (4.17), this maximum value is given in terms of λ by

$$\lambda = \mu_{max}(\mu_{max} + \hbar) \tag{4.21}$$

(the value of λ is the same for every state in the ladder of eigenstates $|\lambda,\mu\rangle$, $J_+|\lambda,\mu\rangle,\ldots$, because of (4.10)).

Similarly, the lowering operator produces a sequence of eigenvalues going down from μ in steps of $\hbar$, which must reach a minimum value μ_{min} or else (4.18) would be violated, and for this value we must have

$$J_-|\lambda,\mu_{min}\rangle = 0 \tag{4.22}$$

and therefore, according to (4.19),

$$\lambda = \mu_{min}(\mu_{min} - \hbar). \tag{4.23}$$

From (4.21) and (4.23) we find

$$(\mu_{max} + \mu_{min})(\mu_{max} - \mu_{min} + \hbar) = 0. \tag{4.24}$$

Since $\mu_{max} \geqslant \mu_{min}$, it follows that $\mu_{min} = -\mu_{max}$. Write $\mu_{max} = j\hbar$; then the difference between μ_{max} and μ_{min} is $2j\hbar$. But we can get from μ_{min} to μ_{max} by going up in steps of $\hbar$; hence $2j$ is an integer. From (4.21) we find the value of λ as $j(j+1)\hbar^2$. The values of μ that are contained in our ladder of eigenvalues of J_z are $-j\hbar, (-j+1)\hbar,\ldots,(j-1)\hbar, j\hbar$; no other values are possible for states with the same eigenvalue of $\mathbf{J}^2$, because we started with a general eigenvalue of J_z and showed that it belonged to this set.

Finally, let us note that because of (4.6) and (4.7), if we start with a state $|\lambda,\mu\rangle$ and apply first the raising operator J_+ and then the lowering operator J_-, or vice versa, we arrive back at a multiple of the state we started with. Thus the ladder of states with a given value of λ span a space $\mathscr{S}$ such that the operators J_x, J_y, J_z act entirely inside $\mathscr{S}$.

From now on we will replace the labels λ and μ by j and m, where $\lambda = j(j+1)\hbar^2$ and $\mu = m\hbar$:

we write $|j\,m\rangle$ instead of $|j(j+1)\hbar^2, m\hbar\rangle$.

Then our results can be summarised as

●**4.1** The possible eigenvalues of $\mathbf{J}^2$ are $j(j+1)\hbar^2$ where $2j$ is an integer. For each of these there are $2j+1$ states

$|j\,m\rangle, \quad m = -j, -j+1,\ldots,j-1, j$

on which the components of the angular momentum vector $\mathbf{J}$ act according to

$$J_z|j\,m\rangle = m\hbar|j\,m\rangle, \tag{4.25}$$

$$J_\pm|j\,m\rangle = \sqrt{(j(j+1)-m(m\pm1))}\hbar|j\,m\pm1\rangle \tag{4.26}$$

where

$$J_\pm = J_x \pm iJ_y. \quad\blacksquare$$

For the remainder of this section we will assume that the units are such that $\hbar = 1$.

Experimental confirmation of these results is provided by the **Stern–Gerlach experiment**, in which a beam of atoms or other particles is directed past a pole of a magnet onto a screen where they make a mark (Fig. 4.1). It is found that the beam splits into a number of pieces, so that the marks made by the particles on the screen form several distinct small patches. These are equally spaced along a line parallel to the axis of the magnet, and are placed symmetrically on either side of the point where the original line of the beam meets the screen.

In classical physics, a spinning object made of electrically charged material acts like a magnet, with a magnetic moment (= pole strength × length) proportional to its internal angular momentum (i.e. angular momentum about its centre of mass). On moving past a pole of a magnet it will be deflected by an amount proportional to the component of angular momentum parallel to the axis of the magnet, say J_z. Thus the Stern–Gerlach experiment, in which the beam is split into $2j+1$ parts separated along the z-direction, so that there are only $2j+1$ possible values for the deflection of an individual particle, gives evidence that there are only $2j+1$ possible values for J_z for each particle.

The number j is characteristic of the type of particle in the beam; it is called the **spin** of the particle.

Fig. 4.1.
The Stern–Gerlach
experiment.

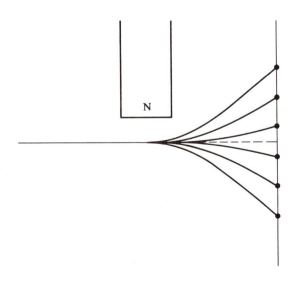

Orbital angular momentum

For a simple particle, which has no internal properties and in particular no spin, the angular momentum about the origin is given by the classical expression $\mathbf{J} = \mathbf{r} \times \mathbf{p}$. Such angular momentum is called **orbital** angular momentum, and is often denoted by $\mathbf{L}$. We have seen (p. 95) that for this system $\mathbf{L}$ coincides with the angular momentum required by Postulate VII. The corresponding operator (taking $\hbar = 1$) is

$$\mathbf{L} = -i\mathbf{r} \times \mathbf{V}. \tag{4.27}$$

In spherical polar coordinates (r, θ, ϕ) the operators L_z, $L_\pm$ and $\mathbf{L}^2$ are

$$L_z = -i\frac{\partial}{\partial\phi}, \tag{4.28}$$

$$L_\pm = e^{\pm i\phi}\left(\pm\frac{\partial}{\partial\theta} + i\cot\theta\frac{\partial}{\partial\phi}\right) \tag{4.29}$$

and

$$\mathbf{L}^2 = -\frac{1}{\sin^2\theta}\left\{\left(\sin\theta\frac{\partial}{\partial\theta}\right)^2 + \frac{\partial^2}{\partial\theta^2}\right\}. \tag{4.30}$$

The eigenfunctions of the operator L_z given by (4.28) are of the form $\psi(\mathbf{r}) = f(r, \theta)e^{im\phi}$. Since $(r, \theta, \phi + 2\pi)$ are the coordinates of the same point of space as (r, θ, ϕ), this function must be unchanged if ϕ is increased by 2π, which means that m must be an *integer* (not half an odd integer). A similar point applies to the operator $\mathbf{L}^2$ given by (4.30). This is the angular part of the Laplacian operator ∇^2, and in the study of that operator it is found that $\mathbf{L}^2$ has eigenvalues $l(l+1)$ where l is an integer (again, not half an odd integer). The simultaneous eigenfunctions of $\mathbf{L}^2$ and L_z are

$$Y_{lm}(\theta, \phi) = P_l{}^m(\cos\theta)e^{im\phi}, \quad m = -l, -l+1, \ldots, l \tag{4.31}$$

where $P_l{}^m$ is an associated Legendre polynomial. The functions Y_{lm} are called **spherical harmonics**; it is also sometimes useful to consider the **solid harmonics**

$$S_{lm}(r, \theta, \phi) = r^l Y_{lm}(\theta, \phi). \tag{4.32}$$

The only special property of these functions that we will need is the following:

●**4.2** Let ψ be an eigenfunction of $\mathbf{L}^2$ with eigenfunction $l(l+1)$. Then ψ has parity $(-1)^l$, i.e.

$$P\psi(\mathbf{r}) = \psi(-\mathbf{r}) = (-1)^l\psi(\mathbf{r}). \tag{4.33}$$

Proof. Consider the function

$$\Phi(\mathbf{r}) = (x + iy)^l = r^l\sin^l\theta\, e^{il\phi}.$$

Using (4.28) and (4.30), you can readily verify that

$$\mathbf{L}^2\Phi = l(l+1)\Phi, \quad L_z\Phi = l\Phi. \tag{4.34}$$

It follows that

$$\Phi(\mathbf{r}) = r^l Y_{ll}(\theta, \phi)$$

so that

$$L_-{}^{l-m}\Phi(\mathbf{r}) = r^l Y_{lm}(\theta, \phi) = S_{lm}(\mathbf{r}). \tag{4.35}$$

Since Φ is a homogeneous polynomial of degree l,

$$\Phi(-\mathbf{r}) = (-1)^l \Phi(\mathbf{r}).$$

The parity operator P commutes with rotation operators and therefore with the angular momentum operators, and in particular with the lowering operator L_-; from (4.35) it follows that all the solid harmonics $S_{lm}(\mathbf{r})$ satisfy

$$S_{lm}(-\mathbf{r}) = (-1)^l S_{lm}(\mathbf{r}). \tag{4.36}$$

Now any eigenfunction of $\mathbf{L}^2$ with eigenvalue $l(l+1)$ must have the form

$$\psi(\mathbf{r}) = \sum_m f_m(r) Y_{lm}(\theta, \phi) = \sum r^{-l} f_m(r) S_{lm}(\mathbf{r}) \tag{4.37}$$

for some functions $f_m(r)$. Hence (4.39) gives

$$\psi(-\mathbf{r}) = \sum (-1)^l r^{-l} f_m(r) S_{lm}(\mathbf{r}) = (-1)^l \psi(\mathbf{r}). \ \blacksquare$$

Since the eigenvalues of $\mathbf{L}^2$ are of the form $l(l+1)$ where l is an integer, eigenstates $|jm\rangle$ of $\mathbf{J}^2$ and J_z in which j (and therefore m) has an odd half cannot be realised as wave functions. Nevertheless, operators satisfying the commutation relations for angular momentum and having such half-odd-integral eigenvalues certainly exist mathematically – they are exhibited in (4.25)–(4.26) – and physical states with these eigenvalues occur in nature, since there are particles for which the Stern–Gerlach experiment gives a splitting into an even number of beams (as in Fig. 4.1, where the number of beams (equal to $2j+1$) is six, so that $j = \frac{5}{2}$).

Spin   The 'internal' angular momentum of a particle which is revealed by the Stern–Gerlach experiment is called the **spin** of the particle. (The same word is used for the number j.) This is angular momentum which the particle has even when it is at rest, i.e. when it has zero eigenvalues for $\mathbf{p}$ and therefore for $\mathbf{r} \times \mathbf{p}$. In describing the Stern–Gerlach experiment we implied that this was like the internal angular momentum of a classical spinning body, which has angular momentum about its centre of mass obtained by adding up terms like $\mathbf{r} \times \mathbf{p}$ for all the parts of the body ($\mathbf{r}$ being the position vector relative to the centre of mass). For a composite particle like an atom, it makes sense in quantum mechanics to talk about the angular momentum about the centre of mass, and it may be possible to identify this with the spin. However, this cannot be the case if the spin j is half an odd integer, for the angular momentum about the centre of mass is obtained by adding together orbital angular momenta, with integer eigenvalues, and, as we will see in the next section, this can only give rise to integer eigenvalues. Since particles with half-odd-integer spin do exist, we must conclude that it is possible for a particle to have a spin which does not arise from the motion of its parts about its centre of mass. In particular, it is possible for a truly elementary particle (which has no constituents) to have an intrinsic spin.†

† This should not really be surprising, in view of the relation between mechanics and geometry expounded in §3.2. Geometrically, rotations cannot be reduced to translations; correspondingly, one should not expect angular momentum to be reducible to linear momentum. This view can be held in classical mechanics also; it dates back to Euler.

There is a general rule relating the spin of a particle to its statistics (i.e. whether it is a fermion or a boson). This can be proved as a theorem in relativistic quantum field theory, but we will assume it as a basic law of nature:

The spin-statistics law. All bosons have integer spin; all fermions have half-odd-integer spin.

Of the particles described in Chapter 1, all the fermions (the baryons, leptons and quarks) have spin $\frac{1}{2}$; the gauge bosons (the photon, $W^\pm$, Z^0 and gluons) have spin 1; and the mesons (π, K and η) have spin 0. There are, however, other baryons and mesons with different spins.

Spin-$\frac{1}{2}$ particles With $j = \frac{1}{2}$ there are two possible values of m, namely $m = \pm\frac{1}{2}$; we will simplify our notation still further by denoting the corresponding eigenstates of J_z by $|+\rangle$ and $|-\rangle$ (these are often called **spin up** and **spin down** states). From (4.25)–(4.26) we find the action of the components of $\mathbf{J}$ on these states as

$$\left.\begin{aligned}
J_x|+\rangle &= \tfrac{1}{2}|+\rangle, & J_x|-\rangle &= \tfrac{1}{2}|+\rangle \\
J_y|+\rangle &= \tfrac{1}{2}i|-\rangle, & J_y|-\rangle &= -\tfrac{1}{2}i|+\rangle \\
J_z|+\rangle &= \tfrac{1}{2}|+\rangle, & J_y|-\rangle &= -\tfrac{1}{2}|-\rangle.
\end{aligned}\right\} \tag{4.38}$$

The matrices representing these (as in §2.4) are the components of $\frac{1}{2}\boldsymbol{\sigma}$ where

$$\sigma_x = \begin{pmatrix} 0 & 1 \\ 1 & 0 \end{pmatrix}, \quad \sigma_y = \begin{pmatrix} 0 & -i \\ i & 0 \end{pmatrix}, \quad \sigma_z = \begin{pmatrix} 1 & 0 \\ 0 & -1 \end{pmatrix}. \tag{4.39}$$

These are known as the **Pauli matrices**. Their matrix products are given by

$$\sigma_x{}^2 = \sigma_y{}^2 = \sigma_z{}^2 = 1, \tag{4.40}$$

$$\sigma_x\sigma_y = -\sigma_y\sigma_x = i\sigma_z, \tag{4.41}$$

and cyclic permutations of x, y, z, which can be summarised as

$$\sigma_i\sigma_j = \delta_{ij} + i\varepsilon_{ijk}\sigma_k \tag{4.42}$$

or

$$(\mathbf{a}\cdot\boldsymbol{\sigma})(\mathbf{b}\cdot\boldsymbol{\sigma}) = \mathbf{a}\cdot\mathbf{b} + i(\mathbf{a}\times\mathbf{b})\cdot\boldsymbol{\sigma}. \tag{4.43}$$

The general state of a spin-$\frac{1}{2}$ particle (as far as its spin is concerned) is a superposition $c_1|+\rangle + c_2|-\rangle$. This state vector is called a **spinor**. An actual particle has not only spin properties but also properties due to its motion in space, of the sort which are described by a wave function $\psi(\mathbf{r})$. As in the case of a photon (p. 70), these two aspects can be combined by making the coefficients c_1 and c_2 into wave functions: thus the full state space of the particle consists of **spinor wave functions** $\psi_1(\mathbf{r})|+\rangle + \psi_2(\mathbf{r})|-\rangle$, which can also be written as

$$\begin{pmatrix} \psi_1(\mathbf{r}) \\ \psi_2(\mathbf{r}) \end{pmatrix}.$$

Rotation operators If $\mathbf{n}$ is any unit vector, $\mathbf{n}\cdot\mathbf{J}$ is the hermitian generator of rotations about the axis $\mathbf{n}$; hence, according to ●3.6, the operator representing the rotation through the angle θ about $\mathbf{n}$ is

$$U(R(\mathbf{n}, \theta)) = e^{-i\theta\mathbf{n}\cdot\mathbf{J}}. \tag{4.44}$$

In the case of a spin-$\frac{1}{2}$ particle we can evaluate the exponential by using

$$(\mathbf{n}\cdot\boldsymbol{\sigma})^2 = 1, \tag{4.45}$$

which follows from (4.43); this gives the matrix

$$U(R(\mathbf{n}, \boldsymbol{\theta})) = e^{-\frac{1}{2}i\theta\mathbf{n}\cdot\boldsymbol{\sigma}} = \cos\tfrac{1}{2}\theta + i\mathbf{n}\cdot\boldsymbol{\sigma}\sin\tfrac{1}{2}\theta. \tag{4.46}$$

Notice that when $\theta = 2\pi$ this is -1. At first sight this looks wrong, since a rotation through 2π is the identity operation and one might expect it to be represented by the identity operator. However, since a state vector $-|\psi\rangle$ describes the same physical state as $|\psi\rangle$, the operator -1 can validly represent the identity operation. The operators (4.46) form a projective representation of the rotation group, as discussed in §3.3, which cannot be redefined to give a true representation. This happens whenever j is half an odd integer, as can be see by considering the effect of a rotation about the z-axis on a state $|j\,m\rangle$:

$$U(R(\mathbf{k}, \theta))|j\,m\rangle = e^{-i\theta J_z}|j\,m\rangle = e^{-im\theta}|j\,m\rangle$$
$$= -|j\,m\rangle \quad \text{when } \theta = 2\pi \tag{4.47}$$

since m is also half an odd integer.

In the terminology introduced in §3.3, these facts about rotation operators can be expressed as follows:

> ●**4.3**   For each eigenvalue $j(j+1)$ of $\mathbf{J}^2$ there is a $(2j+1)$-dimensional projective representation of the rotation group, in which $\mathbf{J}^2$ acts as a multiple of the identity. This representation is unique (up to equivalence). It is a true representation if and only if j is an integer. ∎

We will denote this representation (or the vector space on which it acts) by $\mathscr{D}_j$. It defines a set of $(2j+1) \times (2j+1)$ matrices $D^j(R)$ with matrix elements $d^j_{mn}(R)$ $(m, n = -j, \ldots, j)$, where R is any rotation and

$$U(R)|j\,m\rangle = \sum_{n=-j}^{j} d^j_{nm}(R)|j\,n\rangle; \tag{4.48}$$

and also a set of three matrices representing the generators, forming a vector of matrices $\mathbf{t}^j = (t^j_x, t^j_y, t^j_z)$ with matrix elements t^j_{nm}, where

$$\mathbf{J}|j\,m\rangle = \sum_{n=-j}^{j} \mathbf{t}^j_{nm}|j\,n\rangle \tag{4.49}$$

(in the case $j = \frac{1}{2}$, $\mathbf{t}^j = \frac{1}{2}\boldsymbol{\sigma}$). From (4.25)–(4.26) we find the explicit formulae

$$(t^j_{\pm})_{nm} = (t^j_x \pm it^j_y)_{nm} = \sqrt{(j(j+1) - m(m\pm1))}\,\delta_{n,m\pm1}, \tag{4.50}$$

$$(t^j_z)_{nm} = m\delta_{nm}, \tag{4.51}$$

There is one obvious representation of the rotation group which at first sight does not seem to be included in ●4.3, namely the **vector representation** in which the vector space consists of three-dimensional geometrical vectors and the rotation operator acts by rotating these vectors in the geometrical sense.

Thus there is a complete set of three states $|x\rangle, |y\rangle, |z\rangle$ corresponding to the unit vectors $\mathbf{i}, \mathbf{j}, \mathbf{k}$, and a rotation about the z-axis, for example, is represented by the unitary operator $U(R(\mathbf{k}, \theta))$ defined by

$$
\begin{aligned}
U(R(\mathbf{k}, \theta))|x\rangle &= \cos \theta |x\rangle + \sin \theta |y\rangle \\
U(R(\mathbf{k}, \theta))|y\rangle &= -\sin \theta |x\rangle + \cos \theta |y\rangle \\
U(R(\mathbf{k}, \theta))|z\rangle &= |z\rangle.
\end{aligned}
\tag{4.52}
$$

Differentiating with respect to θ and putting $\theta = 0$ gives J_z as

$$
J_z|x\rangle = i|y\rangle, \quad J_z|y\rangle = -i|x\rangle, \quad J_z|z\rangle = 0.
\tag{4.53}
$$

There are similar expressions for J_x and J_y.

We can now show that this representation is the same as that given by ●4.1 with $j = 1$. Define the states $|1\,m\rangle$ by

$$
|1 \pm 1\rangle = \frac{1}{\sqrt{2}} (|x\rangle \pm i|y\rangle), \quad |1\,0\rangle = |z\rangle;
\tag{4.54}
$$

then (4.53) gives

$$
J_z|1 \pm 1\rangle = \pm|1 \pm 1\rangle, \quad J_z|1\,0\rangle = 0
\tag{4.55}
$$

which is the same as (4.25) with $j = 1$. The expressions for J_x and J_y similar to (4.53) likewise yield (4.26) with $j = 1$. Thus the vector representation is isomorphic to the representation $\mathcal{D}_1$.

The rotation group and SU(2)

In the case $j = \frac{1}{2}$ the unitary operator $U(R)$ representing a rotation can be identified with a 2×2 matrix, as in (4.46). Naturally, this matrix is unitary. Furthermore, its determinant is 1, as follows from the identity

$$
\det (a_0 + \mathbf{a} \cdot \boldsymbol{\sigma}) = \begin{vmatrix} a_0 + a_3 & a_1 - ia_2 \\ a + ia_2 & a_0 - a_3 \end{vmatrix} = a_0^2 - \mathbf{a}^2,
\tag{4.56}
$$

which holds for any (complex) scalar a_0 and vector $\mathbf{a}$. Conversely, any unitary 2×2 matrix with determinant 1 (i.e. any element of SU(2)) must be of the form (4.46) and therefore could represent some rotation. However, if we specify that the angle θ should satisfy $0 \leqslant \theta < 2\pi$, we only get half the elements of SU(2); the other matrices are given by values of θ between 2π and 4π. Thus for every rotation there are two possible 2×2 matrices: if $R(\mathbf{n}, \theta)$ can be represented by the matrix $U(\mathbf{n}, \theta)$ it can also be represented by

$$
U(\mathbf{n}, \theta + 2\pi) = -U(\mathbf{n}, \theta).
\tag{4.57}
$$

Each matrix in SU(2), however, corresponds to just one rotation.

This two-to-one correspondence between elements of SU(2) and rotations is described mathematically by a map ϕ from SU(2) to the rotation group, associating each matrix U in SU(2) with the rotation $\phi(U)$ that it represents; then (4.57) implies that $\phi(-U) = \phi(U)$. The rotation $\phi(U)$ can be simply defined in terms of the set V of hermitian 2×2 matrices with zero trace; any such matrix is of the form

$$\begin{pmatrix} a_3 & a_1 - ia_2 \\ a_1 + ia_2 & -a_3 \end{pmatrix} = \mathbf{a} \cdot \boldsymbol{\sigma} \qquad (4.58)$$

where (a_1, a_2, a_3) are real, so V is a three-dimensional real vector space which can be identified with the space of physical vectors $\mathbf{a}$. From (4.43) we have

$$\text{tr}\,(AB) = 2\mathbf{a} \cdot \mathbf{b} \qquad \text{if } A = \mathbf{a} \cdot \boldsymbol{\sigma} \text{ and } B = \mathbf{b} \cdot \boldsymbol{\sigma}, \qquad (4.59)$$

so the scalar product between vectors is given in V by the inner product

$$\langle A, B \rangle = \tfrac{1}{2} \text{tr}\,(AB). \qquad (4.60)$$

Now, given any U in SU(2) and any other 2×2 matrix X, we define a matrix $\phi(U)X$ by

$$\phi(U)X = UXU^\dagger. \qquad (4.61)$$

Then $\phi(U)X$ is hermitian if X is, because of the identity $(XY)^\dagger = Y^\dagger X^\dagger$, and it has the same trace as X, for

$$\text{tr}\,(UXU^\dagger) = \text{tr}\,(U^\dagger UX) = \text{tr}\,X \qquad (4.62)$$

since U is unitary. Hence $\phi(U)$ maps the space V to itself. Moreover, from (4.60) we have

$$\langle \phi(U)A, \phi(U)B \rangle = \langle A, B \rangle, \qquad (4.63)$$

so $\phi(U)$ is an orthogonal operator on V. It is clear from (4.61) that $\phi(U_1 U_2) = \phi(U_1)\phi(U_2)$, so ϕ is a homomorphism of SU(2) into the group of orthogonal operators on V. Our previous remarks based on (4.46) can be put in group-theoretical terms by saying that the image of ϕ is the group of all rotations of V, and its kernel contains just the two elements ± 1. (See problem 4.5.)

Intrinsic parity   The state space of a particle with spin is $\mathscr{S} \otimes \mathscr{W}$, where $\mathscr{S}$ is its 'internal' (spin) space and $\mathscr{W}$ is the space of wave functions. The rotation operators act in $\mathscr{W}$, taking $\psi(\mathbf{r})$ to $\psi(R^{-1}\mathbf{r})$, and in $\mathscr{S}$ as shown by (4.44). Similarly, the parity operator P, which acts in $\mathscr{W}$ by taking $\psi(\mathbf{r})$ to $\psi(-\mathbf{r})$, may also act in $\mathscr{S}$. Now P commutes with all rotations, and therefore with the angular momentum operators; hence, like $\mathbf{J}^2$, it has just one eigenvalue in $\mathscr{S}$. Since $P^2 = 1$, this eigenvalue must be ± 1. Like the spin, this is characteristic of the type of particle; it is called its **intrinsic parity**. This applies even if the particle has spin 0, when $\mathscr{S}$ is one-dimensional and so $\mathscr{S} \otimes \mathscr{W}$ is isomorphic to $\mathscr{W}$. Thus if we regard the state vector of a spin-j particle as a $(2j+1)$-component wave function, like the spinor wave function we introduced for spin-$\tfrac{1}{2}$ particles, the effect of the parity operator is given by

$$(P\psi)(\mathbf{r}) = \varepsilon\psi(-\mathbf{r}) \qquad (4.64)$$

where ε is the intrinsic parity of the particle.

Of the particles mentioned in Chapter 1, the massive fermions have positive parity, their antiparticles have negative parity, and both the gauge bosons and the light mesons have negative parity. The neutrinos are a special case, as will be discussed in the next subsection.

Massless particles The spin space of a particle can be defined as the space of momentum eigenstates with eigenvalue **0**. This space is taken into itself by rotations; hence rotation operators, and therefore angular momentum operators, are defined on it, and so it must be of the form described in ●4.1.

This argument breaks down in the case of massless particles, which must travel at the speed of light and therefore have no zero-momentum eigenstates. For such particles it is impossible to isolate a spin space on which rotations act without affecting the momentum. All one can do is consider the space $\mathcal{S}_{\mathbf{p}}$ of eigenstates of momentum with a particular non-zero momentum **p**. This space is invariant under rotations about the direction of **p**, which means that it carries an operator $\hat{\mathbf{p}} \cdot \mathbf{J}$ where $\hat{\mathbf{p}}$ is a unit vector in the direction of **p**, but no other components of angular momentum. (The other components exist as operators on the full state space of the particle, but they do not leave the space $\mathcal{S}_{\mathbf{p}}$ invariant.) The observable $\hat{\mathbf{p}} \cdot \mathbf{J}$ is called the **helicity** of the particle. Since it is a component of angular momentum, its eigenvalues must be integers or half-integers, but they need not make up a full range $-j, \ldots, j$. A complete explanation of the situation requires the theory of relativistic transformations (the Poincaré group); see Cornwell 1984, Chapter 17.

Of the massless particles mentioned in Chapter 1, the graviton has helicity ± 2, the photon ± 1, the neutrinos $-\frac{1}{2}$ and the antineutrinos $+\frac{1}{2}$. The neutrinos spin about their direction of motion like a left-handed screw; they are said to be **left-handed**, and antineutrinos **right-handed**.

The operation of reflection in a plane parallel to **p**, applied to a particle with momentum **p**, would leave its momentum as **p** but would reverse its helicity; it would make a left-handed particle right-handed, and vice versa (see Fig. 4.2). Thus such an operation cannot be applied to a neutrino, since there are no right-handed neutrinos. Now a reflection in a plane can be obtained by combining the parity operation with a rotation through π in the plane. Since rotations can be applied to any state, we conclude that *there is no parity operator on the state space of a neutrino*.

Fig. 4.2.
The effect of reflection on helicity.

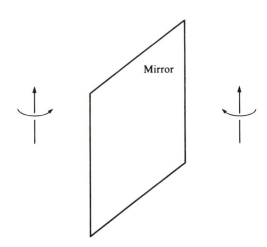

Mirror

The spins and parities of particles are summarised in Table 4.1. Some of these parity values are purely a matter of convention, since the intrinsic parity of a particle does not always have an absolute meaning (see Gibson & Pollard 1976).

4.2. Addition of angular momentum

Consider two systems (which we will think of as 'particles') which have individual angular momentum vectors $\mathbf{J}_1$ and $\mathbf{J}_2$. As was explained in §3.2, these can be regarded as commuting observables for the combined system, for which the angular momentum is $\mathbf{J} = \mathbf{J}_1 + \mathbf{J}_2$. The theory of angular momentum developed above applies both to the individual angular momenta $\mathbf{J}_1$ and $\mathbf{J}_2$, and to the total angular momentum $\mathbf{J}$. We will now look at the relationship between these.

Since we are only concerned with angular momentum, we will continue to assume that there are no other observables, i.e. that $\mathbf{J}_1^2$ and J_{1z} form a complete set of observables for particle 1, and $\mathbf{J}_2^2$ and J_{2z} likewise for particle 2. Suppose further that $\mathbf{J}_1^2$ takes the single value $j_1(j_1 + 1)$ (i.e. particle 1 has spin j_1), so that the state space $\mathscr{S}_1$ of particle 1 is $(2j_1 + 1)$-dimensional, with a complete set of states $|j_1 m_1\rangle$ where m_1 (taking values $-j_1, -j_1 + 1, \ldots, j_1$) is the eigenvalue of J_{1z}. Similarly, suppose particle 2 has spin j_2, so that it has a $(2j_2 + 1)$-dimensional state space $\mathscr{S}_2$ with complete set of states $|j_2 m_2\rangle$ ($m_2 = -j_2, \ldots, j_2$). Then the state space of the two-particle system is $\mathscr{S}_1 \otimes \mathscr{S}_2$; this has dimension $(2j_1 + 1)(2j_2 + 1)$ with a complete set of states $|j_1 m_1\rangle |j_2 m_2\rangle$, which we will write as $|j_1 m_1, j_2 m_2\rangle$. Since j_1 and j_2 are fixed, these states can be identified by the eigenvalues m_1 and m_2; thus J_{1z} and J_{2z} form a complete set of commuting observables ($\mathbf{J}_1^2$ and $\mathbf{J}_2^2$ being multiples of the identity).

Now consider the total angular momentum operators $\mathbf{J}^2$ and J_z acting on

Table 4.1. *Spins and parities*

	Spin	Helicity	Parity
Quarks u, d, s, c, b, t	$\frac{1}{2}$	$\pm\frac{1}{2}$	$+$
Octet baryons n, p, Λ, Σ, Ξ	$\frac{1}{2}$	$\pm\frac{1}{2}$	$+$
Decuplet baryons Δ, Σ^*, Ξ^*, Ω^-	$\frac{3}{2}$	$\pm\frac{1}{2}$, $\pm\frac{3}{2}$	$+$
Charged leptons e^-, μ^-, τ^-	$\frac{1}{2}$	$\pm\frac{1}{2}$	$+$

The antiparticle of a fermion always has the same spin as the fermion and the opposite parity.

	Spin	Helicity	Parity
Neutrinos ν_e, ν_μ, ν_τ		$-\frac{1}{2}$	
Antineutrinos $\bar{\nu}_e$, $\bar{\nu}_\mu$, $\bar{\nu}_\tau$		$+\frac{1}{2}$	
Graviton		± 2	$+$
Photon		± 1	$-$
$W^\pm$, Z^0	1	0, ± 1	$-$
Gluons	1	± 1	$-$
Octet mesons π, K, $\bar{K}$, η	0	0	$-$

the state space $\mathscr{S}_1 \otimes \mathscr{S}_2$. Since $J_z = J_{1z} + J_{2z}$, each state $|j_1 m_1, j_2 m_2\rangle$ is an eigenstate of J_z with eigenvalue $m_1 + m_2$. For a given eigenvalue M of J_z, there will be a number of eigenstates, since there are a number of pairs (m_1, m_2) with $m_1 + m_2 = M$; the possibilities are shown in Fig. 4.3. From this diagram it can be seen that among the states $|j_1 m_1, j_2 m_2\rangle$ are one with $M = j_1 + j_2$, two with $M = j_1 + j_2 - 1$, three with $M = j_1 + j_2 - 2$, and so on until we reach $m = j_1 - j_2$ (if $j_2 < j_1$) or $M = j_2 - j_1$ (if $j_1 < j_2$). If we continue to reduce M in steps of 1, the number of states for each value of M remains the same until we reach $M = -|j_1 - j_2|$; thereafter the number of states decreases as we decrease M, until finally there is just one state with $M = -(j_1 + j_2)$.

Let $\mathscr{Z}_M$ be the subspace of $\mathscr{S}_1 \otimes \mathscr{S}_2$ spanned by the states $|j_1 m_1, j_2 m_2\rangle$ with $m_1 + m_2 = M$; then $\mathscr{Z}_M$ contains all the eigenstates of J_z with eigenvalue M. Thus the raising operator J_+ maps $\mathscr{Z}_M$ into $\mathscr{Z}_{M+1}$. We have just seen that for $|j_2 - j_2| \leqslant M \leqslant j_1 + j_2$, the dimension of $\mathscr{Z}_M$ is greater (by 1) than that of $\mathscr{Z}_{M+1}$; hence J_+ must have a null vector in $\mathscr{Z}_M$, i.e. a state $|\Psi\rangle$ such that $J_+ |\Psi\rangle = 0$. Using $J_- J_+ = \mathbf{J}^2 - J_z^2 - J_z$, it follows that this null vector is an eigenstate of $\mathbf{J}^2$, the eigenvalue being $J(J+1)$ with $J = M$. According to ●4.1, this state stands at the head of a ladder of simultaneous eigenstates of $\mathbf{J}^2$ and J_z, all with the same eigenvalue $J(J+1)$ for $\mathbf{J}^2$ and with the eigenvalues for J_z ranging from $-J$ to J. There is one of these ladders for each value of J between $|j_1 - j_2|$ and $j_1 + j_2$; the states in them are shown in Fig. 4.4, in which each blob stands for a simultaneous eigenstate of $\mathbf{J}^2$ and J_z, the blob with coordinates (J, M) having eigenvalues $J(J+1)$ and M.

In Fig. 4.4 each horizontal row represents a complete set of states for the subspace $\mathscr{Z}_M$; each vertical column is one of the ladders of states described in

Fig. 4.3.
Simultaneous eigenstates of J_{1z} and J_{2z}.

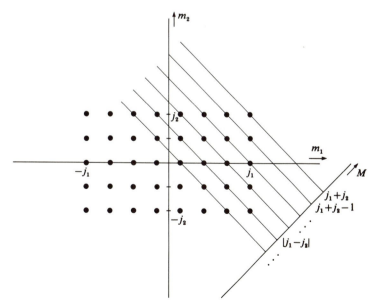

●4.1. We will denote the subspace spanned by this ladder of states by $\mathscr{D}_J$; as in
●4.3, the rotation operators act in this subspace according to the $(2J+1)$-dimensional representation of the rotation group.

The results of this discussion can be summarised as follows:

> ●**4.4**   Let S_1 and S_2 be two systems for which the z-component of angular momentum constitutes a complete set of observables, the eigenvalue of $\mathbf{J}^2$ being $j_1(j_1+1)$ for S_1 and $j_2(j_2+1)$ for S_2. Then the eigenvalues of $\mathbf{J}^2$ for the combined system $S_1 S_2$ are $J(J+1)$ with
>
> $$J = |j_1 - j_2|, |j_1 - j_2| + 1, \ldots, j_1 + j_2.$$
>
> In terms of representations of the rotation group, this can be written as
>
> $$\mathscr{D}_{j_1} \otimes \mathscr{D}_{j_2} \cong \mathscr{D}_{|j_1-j_2|} \oplus \cdots \oplus \mathscr{D}_{j_1+j_2}. \quad \blacksquare \tag{4.65}$$

This is easily extended to cover the case where the systems S_1 and S_2 have more than one eigenvalue for $\mathbf{J}_1{}^2$ and $\mathbf{J}_2{}^2$. If, for example, S_1 has eigenvalues $j_1(j_1+1)$ and $j_1'(j_1'+1)$ for $\mathbf{J}_1{}^2$, then we write its state space as $\mathscr{S}_1 = \mathscr{D}_{j_1} \oplus \mathscr{D}_{j_1'}$ and we have

$$
\begin{aligned}
\mathscr{S}_1 \otimes \mathscr{S}_2 &= (\mathscr{D}_{j_1} \oplus \mathscr{D}_{j_1'}) \otimes \mathscr{D}_{j_2} = \mathscr{D}_{j_1} \otimes \mathscr{D}_{j_2} \oplus \mathscr{D}_{j_1'} \otimes \mathscr{D}_{j_2} \\
&= \mathscr{D}_{|j_1-j_2|} \oplus \cdots \oplus \mathscr{D}_{j_1+j_2} \oplus \mathscr{D}_{|j_1'-j_2|} \oplus \cdots \oplus \mathscr{D}_{j_1'+j_2}. \tag{4.66}
\end{aligned}
$$

This arises in combining three or more systems with angular momentum: one first combines two of the systems, say S_1 and S_2, and then combines each

Fig. 4.4.
Simultaneous eigenstates of $\mathbf{J}^2$ and J_z.

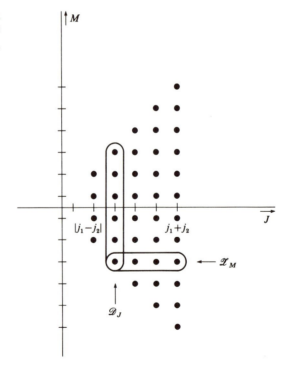

component of $S_1 S_2$ with the third system S_3. For example, if $j_1 = j_2 = j_3 = 1$ this procedure gives

$$\mathscr{D}_1 \otimes \mathscr{D}_1 \otimes \mathscr{D}_1 = (\mathscr{D}_0 \oplus \mathscr{D}_1 \oplus \mathscr{D}_2) \otimes \mathscr{D}_1$$
$$= \mathscr{D}_1 \oplus (\mathscr{D}_0 \oplus \mathscr{D}_1 \oplus \mathscr{D}_2) \oplus (\mathscr{D}_1 \oplus \mathscr{D}_2 \oplus \mathscr{D}_3). \qquad (4.67)$$

Note that since orbital angular momentum is characterised by integer values of j, the combination of any number of systems with orbital angular momentum can only give rise to integer values of j. This is the justification for the remark on p. 138 that half-odd-integer spin cannot be explained in terms of orbital motion.

Figs. 4.3 and 4.4 show two different complete sets of states for the combined system $S_1 S_2$, corresponding to the two sides of (4.65). The states of Fig. 4.3 are the original states $|j_1 m_1, j_2 m_2\rangle$; those of Fig. 4.4 will be written as $|JM\rangle$. All of these state vectors are taken to be normalised. In each of the complete sets the states have different eigenvalues for a pair of hermitian operators, and so each is an orthonormal set. Now any state can be expanded in terms of either of these complete sets; in particular, each state $|j_1 m_1, j_2 m_2\rangle$ can be written as

$$|j_1 m_1, j_2 m_2\rangle = \sum_{J,M} c_{JM} |JM\rangle. \qquad (4.68)$$

The coefficient c_{JM} can be obtained by taking the inner product with the state $|JM\rangle$:

$$c_{JM} = \langle JM | j_1 m_1, j_2 m_2 \rangle. \qquad (4.69)$$

These coefficients are called **Clebsch–Gordan coefficients**. A way of calculating them is outlined in problem 4.6. The calculation involves an arbitrary choice of phase at one point; this reflects the fact that the state vectors $|JM\rangle$ are not uniquely defined, but can be multiplied by an arbitrary phase factor. The phases can be chosen so that the coefficients are all real. There is a table of Clebsch–Gordan coefficients at the back of this book.

Since the Clebsch–Gordan coefficients relate two orthonormal complete sets of states, they form the elements of a unitary matrix (the rows being labelled by J and M, the columns by m_1 and m_2) – in fact, since the coefficients are real, the matrix is orthogonal. Thus

$$\sum_{m_1, m_2} \langle JM | j_1 m_1, j_2 m_2 \rangle \langle J'M' | j_1 m_1, j_2 m_2 \rangle = \delta_{JJ'} \delta_{MM'}, \qquad (4.70)$$

$$\sum_{J,M} \langle JM | j_1 m_1, j_2 m_2 \rangle \langle JM | j_1 m'_1, j_2 m'_2 \rangle = \delta_{m_1 m'_1} \delta_{m_2 m'_2}. \qquad (4.71)$$

The Wigner–Eckart theorem Not only states, but also observables may be subject to transformation by rotations. For example, in §3.2 we considered a set of observables V_i which form the components of a vector. We saw that the appropriate transformation for an observable A was $A \rightarrow U(R)AU(R)^{-1}$ with infinitesimal version $A \rightarrow [J_i, A]$ (see ●3.8); and that if V_i are the components of a vector the three-dimensional space of operators consisting of linear combinations of the V_i (i.e.

the set of operators $\mathbf{a} \cdot \mathbf{V}$) is invariant under these transformations. This can be generalised. The components of a vector are equivalent to the basic states of a system with angular momentum $j = 1$, so the space of operators $\mathbf{a} \cdot \mathbf{V}$ behaves like the space $\mathscr{D}_1$. In the generalisation we consider a space of operators corresponding to any $\mathscr{D}_j$.

An **irreducible set of operators** (under the rotation group) of **spin type** j is a set of $2j + 1$ operators T^j_m $(m = -j, \ldots, j)$ satisfying

$$[J_z, T^j_m] = mT^j_m,$$
$$[J_\pm, T^j_m] = \sqrt{(j(j+1) - m(m \pm 1))} T^j_{m \pm 1} \tag{4.72}$$

where $\mathbf{J}$ is the total angular momentum of the system. These equations generalise (3.131), which can be put in this form by taking $T^1_0 = V_z$, $T^1_{\pm 1} = V_x \pm iV_y$ (cf. (4.54)). They can be written as

$$[\mathbf{J}, T^j_m] = \sum \mathbf{t}^j_{nm} T^j_n \tag{4.73}$$

and are equivalent to the statement that

$$U(R)T^j_m U(R)^{-1} = \sum_n d^j_{nm}(R) T^j_n \tag{4.74}$$

where $d^j_{nm}(R)$ is the matrix representing R in the representation $\mathscr{D}_j$, given by (4.48), and $\mathbf{t}^j_{nm}$, given by (4.49), is the vector of matrices corresponding to the angular momentum operators in this representation.

The following theorem says that applying such an irreducible set of operators to a system with angular momentum j' is similar to combining it with another system with angular momentum j.

●**4.5   The Wigner–Eckart theorem.** Let T^j_m be an irreducible set of operators of spin type j, and let $|jm\alpha\rangle$ be eigenstates of a set of commuting operators which include $\mathbf{J}^2$ and J_z, where $j(j+1)$ is the eigenvalue of $\mathbf{J}^2$, m is that of J_z, and α stands for the eigenvalues of the other operators. Then the matrix elements of T^j_m are proportional to Clebsch–Gordan coefficients:

$$\langle j''m''\alpha'' | T^j_m | j'm'\alpha' \rangle = \langle j''m'' | jm, j'm' \rangle \langle j''\alpha'' \| T^j \| j'\alpha' \rangle \tag{4.75}$$

where $\langle j''\alpha'' \| T^j \| j'\alpha' \rangle$ is independent of m, m', m''.

Proof. Consider the $(2j+1)(2j'+1)$ states $T^j_m | j'm'\alpha' \rangle$. On the space spanned by these states we can define operators $\mathbf{J}_1$ which act like angular momentum operators on the index m, and similar operators $\mathbf{J}_2$ to act on the index m'. The action of the actual angular momentum operators is

$$\mathbf{J}T^j_m | j'm'\alpha' \rangle = [\mathbf{J}, T^j_m] | j'm'\alpha' \rangle + T^j_m \mathbf{J} | j'm'\alpha' \rangle$$
$$= (\mathbf{J}_1 + \mathbf{J}_2) T^j_m | j'm'\alpha' \rangle, \tag{4.76}$$

i.e. the states $T^j_m | j'm'\alpha' \rangle$ behave like product states describing particles with spins j and j'. Hence, in analogy with (4.68)–(4.69), they can be written as

$$T^j_m | j'm'\alpha' \rangle = \sum_{J,M} \langle JM | jm, j'm' \rangle | JM\alpha' \rangle_1 \tag{4.77}$$

where the states $|JM\alpha'\rangle_1$ behave like the two-particle states with definite total angular momentum, i.e. they are eigenstates of $\mathbf{J}^2$ and J_z. Since $\mathbf{J}^2$ and J_z are hermitian, we have

$$\langle j''m''\alpha''|JM\alpha'\rangle_1 = 0 \quad \text{unless } J = j'' \text{ and } M = m''. \tag{4.78}$$

Moreover

$$\langle j''\, m'' + 1\,\alpha''|j''\, m''+1\,\alpha'\rangle_1 = \frac{\langle j''m''\alpha''|J_+^{\,\dagger}J_+|j''m''\alpha'\rangle_1}{j''(j''+1) - m''(m''+1)}$$

$$= \frac{\langle j''m''\alpha''|(\mathbf{J}^2 - J_z^{\,2} - J_z)|j''m''\alpha'\rangle_1}{j''(j''+1) - m''(m''+1)}$$

$$= \langle j''m''\alpha''|j''m''\alpha'\rangle_1.$$

Thus this quantity is independent of m''. Calling it $\langle j''\alpha''\|T^j\|j'\alpha'\rangle$, we find that (4.75) follows from (4.77) and (4.78). ∎

$\langle j''\alpha''\|T^j\|j'\alpha'\rangle$ is called a **reduced matrix element**.

4.3. Two-particle systems

The full state space of a system of two particles is

$$(\mathscr{S}_1 \otimes \mathscr{W}_1) \otimes (\mathscr{S}_2 \otimes \mathscr{W}_2) \cong (\mathscr{S}_1 \otimes \mathscr{S}_2) \otimes (\mathscr{W}_1 \otimes \mathscr{W}_2) \tag{4.79}$$

where $\mathscr{S}_1$ and $\mathscr{S}_2$ are the spin spaces of the two particles and $\mathscr{W}_1$ and $\mathscr{W}_2$ (both isomorphic to $\mathscr{W}$) are the spaces of wave functions. We will call a state in $\mathscr{S}_1 \otimes \mathscr{S}_2$ a **spin state** and a state in $\mathscr{W}_1 \otimes \mathscr{W}_2$ an **orbital state**; thus the basic state of the two-particle system is a product of a spin state and an orbital state (the general state being a superposition of such product states).

Consider the orbital states first. $\mathscr{W}_1 \otimes \mathscr{W}_2$ is the space of wave functions $\psi(\mathbf{r}_1, \mathbf{r}_2)$, where $\mathbf{r}_1$ and $\mathbf{r}_2$ are the positions of the two particles. In the classical mechanics of two particles it is a good idea to change variables from $\mathbf{r}_1$ and $\mathbf{r}_2$ to the **centre-of-mass position R** and the **relative position r**, which are defined by

$$\mathbf{R} = \frac{1}{M}(m_1\mathbf{r}_1 + m_2\mathbf{r}_2), \tag{4.80}$$

$$\mathbf{r} = \mathbf{r}_1 - \mathbf{r}_2, \tag{4.81}$$

where m_1 and m_2 are the masses of the particles and $M = m_1 + m_2$. In quantum mechanics this change of position variables goes with a change of the momentum variables associated with the partial derivatives; using the chain rule for partial differentiation, we find that the momenta associated with $\mathbf{R}$ and $\mathbf{r}$ are

$$\mathbf{P} = -i\hbar \frac{\partial}{\partial \mathbf{R}} = -i\hbar\left(\frac{\partial}{\partial \mathbf{r}_1} + \frac{\partial}{\partial \mathbf{r}_2}\right) = \mathbf{p}_1 + \mathbf{p}_2, \tag{4.82}$$

$$\mathbf{p} = -i\hbar \frac{\partial}{\partial \mathbf{r}} = -i\hbar\left(\frac{m_2}{M}\frac{\partial}{\partial \mathbf{r}_1} - \frac{m_1}{M}\frac{\partial}{\partial \mathbf{r}_2}\right) = \frac{m_2\mathbf{p}_1 - m_1\mathbf{p}_2}{M}. \tag{4.83}$$

If the forces on the particles are derived from a potential function $V(\mathbf{r}_1, \mathbf{r}_2)$, i.e. if they are

$$\mathbf{F}_1 = -\frac{\partial V}{\partial \mathbf{r}_1} \quad \text{and} \quad \mathbf{F}_2 = -\frac{\partial V}{\partial \mathbf{r}_2}, \tag{4.84}$$

then the Hamiltonian is

$$H = \frac{\mathbf{p}_1^2}{2m_1} + \frac{\mathbf{p}_2^2}{2m_2} + V(\mathbf{r}_1, \mathbf{r}_2), \tag{4.85}$$

which can be expressed in the new variables, using (4.82)–(4.83), as

$$H = \frac{\mathbf{P}^2}{2M} + \frac{\mathbf{p}^2}{2\mu} + V(\mathbf{R}, \mathbf{r}) \tag{4.86}$$

where

$$\frac{1}{\mu} = \frac{1}{m_1} + \frac{1}{m_2}. \tag{4.87}$$

μ is called the **reduced mass** of the system.

Now suppose the two particles form an isolated system, i.e. each particle is subject only to forces exerted by the other. Then by Newton's third law of motion the forces on the two particles are equal and opposite and so (4.83) gives

$$\frac{\partial V}{\partial \mathbf{r}_1} = -\frac{\partial V}{\partial \mathbf{r}_2}. \tag{4.88}$$

According to (4.81), this means that $\partial V/\partial \mathbf{R} = 0$, i.e. V is a function of $\mathbf{r}$ only. Thus the Hamiltonian (4.85) looks like the Hamiltonian of two (fictitious) particles, one of mass M situated at the centre of mass $\mathbf{R}$ and moving as a free particle, and one of mass μ which moves in a field of force derived from the potential $V(\mathbf{r})$.

The total orbital angular momentum of the two-particle system is

$$\begin{aligned}\mathbf{L} &= \mathbf{r}_1 \times \mathbf{p}_1 + \mathbf{r}_2 \times \mathbf{p}_2 \\ &= \mathbf{R} \times \mathbf{p} + \mathbf{r} \times \mathbf{p}\end{aligned} \tag{4.89}$$

from (4.80)–(4.83). This is the sum of the orbital angular momentum associated with $\mathbf{R}$, i.e. with the motion of the centre of mass, and that associated with $\mathbf{r}$, which is an internal quantity of the two-particle system. The latter term, $\mathbf{r} \times \mathbf{p}$, is called the **internal angular momentum**.

This change of variables shows that the space $\mathcal{W}_1 \otimes \mathcal{W}_2$ can also be written as $\mathcal{W}_{int} \otimes \mathcal{W}_{c.m.}$, where $\mathcal{W}_{int}$ and $\mathcal{W}_{c.m.}$ are the spaces of wave functions with arguments $\mathbf{r}$ and $\mathbf{R}$ respectively. The full state space of the two-particle system can now be put in a form analogous to that of a single particle; it is

$$\mathcal{S} \otimes \mathcal{W}_{c.m.} \quad \text{where } \mathcal{S} = \mathcal{S}_1 \otimes \mathcal{S}_2 \otimes \mathcal{W}_{int}. \tag{4.90}$$

We will focus attention on the internal space $\mathcal{S}$ by assuming a fixed wave function $\psi_0(\mathbf{R})$ in $\mathcal{W}_{c.m.}$ (usually we take ψ_0 to be an eigenfunction of the total momentum with eigenvalue $\mathbf{0}$).

The total internal angular momentum of the two-particle system is thus obtained by adding three angular momenta, namely the two spins and the relative orbital angular momentum $\mathbf{r} \times \mathbf{p}$. Since the last-named has eigenvalues $j(j+1)$ with j an integer, a composite particle with two constituents will have integer spin if its constituents both have integer spin or both have half-odd-integer spin, and it will have half-odd-integer spin if one of its constituents has integer spin and the other has half-odd-integer spin. This is similar to the rule for combining fermions and bosons (●2.7), and shows the consistency of the spin-statistics law with this rule. Generalised in the obvious way to a system of n particles, it also provides the justification for the assertion on p. 6 that the spin of nuclei is not consistent with the assertion that they are composed of protons and electrons; for example, if the nucleus of ^{3}He consisted of three protons and an electron its spin would be an integer whereas in fact it has spin $\frac{1}{2}$.

If the two particles in our system are identical, the possible states are restricted by the requirements of statistics (i.e. that the state should be symmetric or antisymmetric). If $\mathbf{r}_1$ and $\mathbf{r}_2$ are interchanged, the centre-of-mass position $\mathbf{R}$ is unchanged (since $m_1 = m_2$ for identical particles) and so the state space $\mathscr{W}_{\text{c.m.}}$ is unaffected; thus we can continue to restrict attention to $\mathscr{S}_1 \otimes \mathscr{S}_2 \otimes \mathscr{W}_{\text{int}}$, i.e. to products of spin states and relative orbital states. If the particles are fermions, so that the product state is antisymmetric, then either the spin state is symmetric and the orbital state is antisymmetric, or vice versa; if the particles are bosons, the spin state and the orbital state are either both symmetric or both antisymmetric. We will now look separately at spin states and orbital states to determine the consequences of symmetry and antisymmetry.

For the *spin state*, suppose the particles have spin j, so that $\mathscr{S}_1 \cong \mathscr{S}_2 \cong \mathscr{D}_j$. Then we want to know which are the symmetric and antisymmetric states in $\mathscr{D}_j \otimes \mathscr{D}_j$. The answer is

> ●4.6 In $\mathscr{D}_j \otimes \mathscr{D}_j$ the symmetric states are those with $J = 2j, 2j-2$, ..., the antisymmetric ones those with $J = 2j-1, 2j-3, \ldots$.

Proof. Since $\mathbf{J}_1$ and $\mathbf{J}_2$ act identically on the individual spin spaces, the exchange operator X satisfies

$$X\mathbf{J}_1 = \mathbf{J}_2 X \quad \text{and} \quad X\mathbf{J}_2 = \mathbf{J}_1 X. \tag{4.91}$$

Hence $X\mathbf{J} = \mathbf{J}X$; in particular X commutes with the raising and lowering operators $J_\pm$. It follows that if, in the ladder of states with a given value of J, there is a symmetric state, then all the states in that ladder are symmetric; and likewise for antisymmetric states.

As in Fig. 4.4, let $\mathscr{Z}_M$ be the subspace spanned by the states $|j_1 m_1, j_2 m_2\rangle$ with a given value of $M = m_1 + m_2$. Then X takes $\mathscr{Z}_M$ into itself. Since X commutes with J_+, the unique state $|\psi_M\rangle$ in $\mathscr{Z}_M$ which satisfies $J_+|\psi_M\rangle = 0$ is an eigenstate of X, i.e. either symmetric or antisymmetric. Since this state stands at the head of the ladder $\mathscr{D}_J$ with $J = M$, all the states in the ladder are either symmetric or antisymmetric.

Now if $M = 2j - 2k + 1 > 0$, the subspace $\mathscr{Z}_M$ has k symmetric states $|jm_1, jm_2\rangle + |jm_2, jm_1\rangle$ and k antisymmetric states $|jm_1, jm_2\rangle - |jm_2, jm_1\rangle$, all linearly independent; if $M = 2j - 2k$, $\mathscr{Z}_M$ again has k symmetric and k

antisymmetric states, together with an additional symmetric state $|jm, jm\rangle$ where $m = j - k$. In order to make up these numbers of symmetric and antisymmetric states, the ladders $\mathscr{D}_J$ must consist of alternately symmetric and antisymmetric states, starting with $\mathscr{D}_{2j}$ which is symmetric since its top state $|jj, jj\rangle$ is symmetric. This gives the pattern stated. ∎

In particular, for $j = \frac{1}{2}$ the state with $J = 0$ (in which the spins are said to be 'antiparallel') is antisymmetric; the states with $J = 1$ ('parallel spins') are symmetric. For $j = 1$ the states with $J = 0$ and $J = 2$ are symmetric and those with $J = 1$ are antisymmetric. This last fact is related, via the isomorphism between $\mathscr{D}_1$ and the vector representation, to the properties of the products of two vectors: the scalar product (corresponding to $\mathscr{D}_0$) is symmetric (i.e. commutative), while the vector product (corresponding to $\mathscr{D}_1$) is antisymmetric (anticommutative).

For the *orbital state* we have

●**4.7** In a state with relative orbital angular momentum l, the orbital state is symmetric if l is even, antisymmetric if l is odd.

Proof. If $\mathbf{r}_1$ and $\mathbf{r}_2$ are interchanged, $\mathbf{r}$ becomes $-\mathbf{r}$. Hence on $\mathscr{W}_{\text{int}}$ the exchange operator X is the same as the parity operator P:

$$X\psi(\mathbf{r}) = P\psi(\mathbf{r}) = \psi(-\mathbf{r}). \tag{4.92}$$

The result now follows from ●4.2. ∎

Example If a particle decays into two π^0 mesons, its spin must be an even integer. For by conservation of angular momentum, the spin of the decaying particle must be the same as that of the internal angular momentum of the $2\pi^0$ state. Since pions have spin 0, there is only an orbital state to consider; since they are bosons, this must be symmetric; so by ●4.7 the internal angular momentum l must be even.

For example, there is a spin-1 meson ρ^0 which decays into $\pi^+ + \pi^-$ but not into $2\pi^0$.

●4.7 makes it possible to determine the intrinsic parity of a two-particle system. We have seen that for a single particle, with state space $\mathscr{S} \otimes \mathscr{W}$, the parity operator P acts on the spin space $\mathscr{S}$ as well as on the wave function space $\mathscr{W}$. A two-particle system can be treated in a similar way, with $\mathscr{S} = \mathscr{S}_1 \otimes \mathscr{S}_2 \otimes \mathscr{W}_{\text{int}}$. In this space P no longer acts as a multiple of the identity, for in the internal wave-function space $\mathscr{W}_{\text{int}}$ it takes $\psi(\mathbf{r})$ to $\psi(-\mathbf{r})$. Now ●4.2 shows that the eigenstates of parity are the eigenstates of relative orbital angular momentum l. These are usually the states of interest, since there may be only one value of l for which the particles bind together to form a composite particle; more generally, the energy (and therefore, in a relativistic context, the mass) of the composite particle may depend on l. For these eigenstates, since parity is a multiplicative quantum number, ●4.2 gives the rule

●**4.8** The intrinsic parity of a two-particle state is $\varepsilon_1\varepsilon_2(-1)^l$, where

ε_1 and ε_2 are the intrinsic parities of the two particles and l is their relative orbital angular momentum. ∎

Examples

1. The relative orbital angular momentum of the quark and antiquark in a π-meson must be even. For the quark and the antiquark have opposite intrinsic parities, so the intrinsic parity of the pion is $-(-1)^l$. But pions have negative intrinsic parity, so l must be even.

As we will see in the next two sections, the lowest energy in a two-particle system usually occurs for $l=0$, so it is assumed that this is the case for the mesons π, K and η, which are the lightest quark–antiquark composites.

2. The deuteron d, the nucleus of deuterium (heavy hydrogen), has spin 1. It forms an 'atom' with a π^- meson in which the relative orbital angular momentum is 0. This decays into two neutrons:

$$\pi^- + d \rightarrow n + n. \tag{4.93}$$

If the Hamiltonian governing this process is invariant under reflections, the deuteron must have intrinsic parity $+$.

For if the Hamiltonian is invariant under reflections, parity is conserved. By ●4.8, the intrinsic parity of the 2n state is $(-1)^l$ where l is the relative orbital angular momentum; and the intrinsic parity of the initial state is $\varepsilon_\pi \varepsilon_d (-1)^0 = -\varepsilon_d$. Hence $\varepsilon_d = (-1)^{l+1}$. Now we use conservation of angular momentum. Since the spin of the pion and the relative orbital angular momentum in the initial state are both 0, the total angular momentum initially is the spin of the deuteron, which is 1. The final total angular momentum is the sum of the total spin, which can be 0 or 1, and the relative orbital angular momentum l. Since neutrons are fermions, the final state must be antisymmetric. If the total spin was 0, the spin state would be antisymmetric (by ●4.6), so the orbital state would have to be symmetric and by ●4.7 l would be even; then the total angular momentum would not be 1. Hence the total spin must be 1, the spin state is symmetric, so the orbital state is antisymmetric, l is odd and therefore the intrinsic parity of the deuteron is $(-1)^{l+1} = +1$.

4.4. The hydrogen atom

In this section we will determine the energy levels (i.e. the eigenvalues of the Hamiltonian) for a system of two particles attracted to each other by a force which is inversely proportional to the square of the distance between them. This is a simplified model of the hydrogen atom, the inverse-square force being the electrostatic attraction between the electron and the proton (it omits the magnetic force between the particles and the emission of radiation by them).

Since the force is a function only of the relative position $\mathbf{r} = \mathbf{r}_1 - \mathbf{r}_2$, the two-body problem can be reduced to a one-body problem. We look for eigenstates in $\mathcal{W}_{\text{int}}$ (i.e. eigenfunctions $\psi(\mathbf{r})$) of the Hamiltonian

$$H = \frac{\mathbf{p}^2}{2\mu} - \frac{\gamma}{r} \tag{4.94}$$

since $V(\mathbf{r}) = -\gamma/r$ is a potential function for the force

$$\mathbf{F} = -\frac{\gamma}{r^2}\frac{\mathbf{r}}{r} = -\boldsymbol{\nabla}V. \tag{4.95}$$

Here γ is the constant in the inverse-square law (for the hydrogen atom, $\gamma = e^2/4\pi\varepsilon_0$ where e is the magnitude of the charge on the electron), μ is the reduced mass given by (4.87) and $\mathbf{p}$ is the relative momentum (4.83). The variables $\mathbf{r}$ and $\mathbf{p}$ can be treated just like the position and momentum vectors of a single particle; in particular, they satisfy the basic commutation relations (2.116)–(2.117).

In this one-particle system the particle is attracted to a fixed point, the origin, by the force (4.95). Classically, it can move in a closed orbit consisting of a plane ellipse with one focus at the origin. The other possible motions are unbounded orbits in which the particle gets indefinitely far from the origin. These two types of orbit are distinguished by the sign of the energy $E = H(\mathbf{p}, \mathbf{r})$, which is a constant of the motion; if $E < 0$, (4.94) shows that r cannot become indefinitely large and so the particle moves in a closed orbit. The shape and orientation of the orbit are described by two vector constants of the motion, the angular momentum $\mathbf{L} = \mathbf{r} \times \mathbf{p}$, which is perpendicular to the plane of the orbit, and the **Laplace–Runge–Lenz vector** (or 'Lenz vector' for short)

$$\mathbf{M} = \mathbf{p} \times \mathbf{L} - \frac{\mu\gamma}{r}\mathbf{r}, \tag{4.96}$$

which points along the major axis of the ellipse. The size and shape of the ellipse are determined by the energy E and the magnitude of $\mathbf{M}$; the major axis has length $\gamma/|E|$ and the eccentricity is $|\mathbf{M}|/\gamma\mu$.

The vector $\mathbf{M}$ can be used to study this system in quantum mechanics also, but because $\mathbf{p}$ and $\mathbf{L}$ do not commute the product in (4.96) must be defined carefully. In order to obtain a hermitian result we take each product AB in the classical expression to be the *symmetrised* product $\frac{1}{2}(AB + BA)$, for which it is convenient to use the notation

$$\{A, B\} = AB + BA \tag{4.97}$$

(this is called the **anticommutator** of A and B). Thus for the quantum Lenz vector we take

$$M_i = \frac{1}{2}\varepsilon_{ijk}\{p_j, L_k\} - \frac{\mu\gamma}{r}x_i. \tag{4.98}$$

Then we have

⬤**4.9** The Hamiltonian H of (4.94), the orbital angular momentum $\mathbf{L}$, and the Lenz vector $\mathbf{M}$ of (4.98) satisfy the following equations:

$$[H, L_i] = [H, M_i] = 0; \tag{4.99}$$

$$[L_i, L_j] = i\hbar\varepsilon_{ijk}L_k; \tag{4.100}$$

$$[L_i, M_j] = i\hbar\varepsilon_{ijk}M_k; \tag{4.101}$$

$$[M_i, M_j] = -2\mu i\hbar\varepsilon_{ijk}L_k H; \tag{4.102}$$

$$\mathbf{L}\cdot\mathbf{M} = \mathbf{M}\cdot\mathbf{L} = 0; \tag{4.103}$$

$$\mathbf{M}^2 - 2\mu H(\mathbf{L}^2 + \hbar^2) = \gamma^2\mu^2. \tag{4.104}$$

Proof. Since H is a scalar observable, it commutes with the total angular momentum by ●3.9. It clearly commutes with the spin and the centre-of-mass angular momentum, so this means that it commutes with the three components of $\mathbf{L}$. Hence, using the rule (2.80) for the commutator with a product,

$$[H, M_i] = \tfrac{1}{2}\varepsilon_{ijk}\{[H, p_j], L_k\} - \tfrac{1}{2}\gamma\{p_j, [p_j, x_i/r]\}$$

$$= \frac{1}{2}i\hbar\gamma\varepsilon_{ijk}\left\{\frac{x_j}{r^3}, L_k\right\} + \frac{1}{2}i\hbar\gamma\left\{p_j, \frac{\delta_{ij}}{r} - \frac{x_ix_j}{r^3}\right\}, \tag{4.105}$$

using (3.133) for the commutators with p_j. Now write

$$L_k = \varepsilon_{klm}x_l p_m = \varepsilon_{klm}p_m x_l: \tag{4.106}$$

the first term in (4.105) becomes

$$\frac{1}{2}i\hbar\gamma\varepsilon_{ijk}\varepsilon_{klm}\left(\frac{x_j}{r^3}x_l p_m + p_m x_l\frac{x_j}{r^3}\right) = \frac{1}{2}i\hbar\gamma(\delta_{il}\delta_{jm} - \delta_{im}\delta_{jl})\left\{p_m, \frac{x_jx_l}{r^3}\right\},$$

which cancels the second term.

This proves (4.99). The next two equations are examples of the basic commutation relation (3.134), since $\mathbf{L}$ is the only relevant part of the angular momentum. (4.102) is an exercise in commutator and tensor algebra, for which we will trace a path through the manipulations. First calculate the commutators

$$[p_i, M_j] = i\hbar\left(p_i p_j - \mathbf{p}^2\delta_{ij} + \frac{\mu\gamma}{r}\delta_{ij} - \frac{\mu\gamma}{r^3}x_ix_j\right), \tag{4.107}$$

$$\left[\frac{x_i}{r}, M_j\right] = i\hbar\left(\frac{1}{r}\varepsilon_{ijk}L_k - \frac{1}{2r^3}\varepsilon_{jmn}\{x_ix_m, L_n\}\right.$$

$$\left.+\frac{1}{2}\left\{p_i, \frac{x_j}{r}\right\} - \frac{1}{2}\delta_{ij}\left\{p_m, \frac{x_j}{r}\right\}\right). \tag{4.108}$$

The latter yields (remembering that $\mathbf{r}\cdot\mathbf{L} = \mathbf{p}\cdot\mathbf{L} = 0$)

$$\varepsilon_{ijk}\left[\frac{x_i}{r}, M_j\right] = -2i\hbar\frac{L_k}{r}. \tag{4.109}$$

Substituting from (4.98) for M_i, and using (4.107) and (4.109), we find

$$\varepsilon_{ijk}[M_i, M_j] = i\hbar(-\mathbf{p}^2 + 3\mu\gamma/r)L_k + \tfrac{1}{2}i\hbar\varepsilon_{ijk}\{p_i, M_j\}. \tag{4.110}$$

Now substituting for M_j gives

$$\varepsilon_{ijk}\{p_i, M_j\} = (-2\mathbf{p}^2 + 2\mu\gamma/r)L_k, \tag{4.111}$$

so that

$$\varepsilon_{ijk}[M_i, M_j] = 2i\hbar(-\mathbf{p}^2 + 2\mu\gamma/r)L_k = 2i\hbar(-2\mu H)L_k. \tag{4.112}$$

Eq. (4.102) follows by multiplying by ε_{ijk}.

To prove (4.103) and (4.104), write M_i in the form

$$M_i = \varepsilon_{ijk} p_j L_k - i\hbar p_i - \mu\gamma \frac{x_i}{r}. \tag{4.113}$$

Since $\mathbf{p} \cdot \mathbf{L} = \mathbf{r} \cdot \mathbf{L} = 0$, this gives

$$\mathbf{M} \cdot \mathbf{L} = \varepsilon_{ijk} p_j L_k L_i = \tfrac{1}{2}\varepsilon_{ijk} p_j [L_k, L_i] = \tfrac{1}{2}i\hbar\mathbf{p} \cdot \mathbf{L} = 0.$$

Finally, calculating $M_i M_i$ from (4.113) and using the basic commutation relations gives (4.104). ∎

A **bound state** is a stationary state with a negative eigenvalue for H; the bound states are the quantum states corresponding to the closed orbits of the classical theory. We will now use ●4.9 to find the bound states.

Let $\mathscr{B}_E$ be the space of eigenstates of H with eigenvalue E. If these are bound states we can write $-2\mu E = \kappa^2$ where κ is real. Since $\mathbf{L}$ and $\mathbf{M}$ commute with H, they leave $\mathscr{B}_E$ invariant; as operators on $\mathscr{B}_E$ their commutation relations are (4.100)–(4.101) and

$$[M_i, M_j] = i\hbar\kappa^2 \varepsilon_{ijk} L_k. \tag{4.114}$$

Let

$$\mathbf{P} = \tfrac{1}{2}(\mathbf{L} + \kappa^{-1}\mathbf{M}), \quad \mathbf{Q} = \tfrac{1}{2}(\mathbf{L} - \kappa^{-1}\mathbf{M}); \tag{4.115}$$

then (4.100)–(4.101) and (4.114) give

$$[P_i, P_j] = i\hbar\varepsilon_{ijk} P_k, \tag{4.116}$$

$$[Q_i, Q_j] = i\hbar\varepsilon_{ijk} Q_k, \tag{4.117}$$

$$[P_i, Q_j] = 0,$$

while (4.103)–(4.104) become

$$\mathbf{P}^2 - \mathbf{Q}^2 = 0, \tag{4.118}$$

$$\mathbf{P}^2 + \mathbf{Q}^2 = \frac{\gamma^2\mu^2}{2\kappa^2} - \frac{1}{2}\hbar^2. \tag{4.119}$$

Eqs. (4.116)–(4.117) show that $\mathbf{P}$ and $\mathbf{Q}$ both satisfy the commutation relations of angular momentum. The arguments of ●4.1 therefore show that $\mathbf{P}^2$ and $\mathbf{Q}^2$ have eigenvalues of the form $j(j+1)\hbar^2$ where $2j$ is an integer. (4.118)–(4.119) show that these eigenvalues are equal and are given by

$$j(j+1) = \frac{\gamma^2\mu^2}{4\kappa^2\hbar^2} - \frac{1}{4}. \tag{4.120}$$

Hence

$$E = -\frac{\kappa^2}{2\mu} = -\frac{\mu\gamma^2}{2(2j+1)^2\hbar^2}. \tag{4.121}$$

Since each component of $\mathbf{P}$ commutes with each component of $\mathbf{Q}$, every eigenspace of P_3 is invariant under $\mathbf{Q}$ and therefore contains the full ladder of $2j+1$ eigenstates of Q_3. Thus the space $\mathscr{B}_E$ is of the form $\mathscr{D}_j \otimes \mathscr{D}_j$, i.e. it is like the spin space of two particles of spin j, their individual angular momenta being $\mathbf{P}$ and $\mathbf{Q}$. The total spin would then be $\mathbf{P} + \mathbf{Q} = \mathbf{L}$; according to ●4.4, this has eigenvalues $l(l+1)$ with $l = 0, 1, \ldots, 2j$.

Writing $2j = n$, and using the value $\gamma = e^2/4\pi\varepsilon_0$, we can summarise these results as follows.

●**4.10** The energy eigenvalues of the bound states of the hydrogen atom are

$$E_n = -\frac{1}{(4\pi\varepsilon_0)^2} \frac{\mu e^4}{2n^2\hbar^2} \tag{4.122}$$

where n is a positive integer. With this energy there are states with relative orbital angular momentum $l = 0, 1, \ldots, n-1$. ■

We now see how quantum mechanics explains the spectrum of hydrogen, which was mentioned in Chapter 1. The frequencies v_{mn} of (1.1) are related to the differences between the energies E_n of (4.122) by Planck's relation:

$$E_n - E_m = hv_{mn} \quad \text{if} \quad R = \frac{\mu e^4}{8\varepsilon_0^2 \hbar^3}. \tag{4.123}$$

This is indeed the measured value of the constant R. Thus the energy of each photon in the radiation with frequency v_{mn} is the energy lost by an atom of hydrogen in changing from one bound state to another. (As explained in §3.5, the bound states we have determined are not exactly stationary states, since we have omitted the part of the Hamiltonian describing the coupling to the electromagnetic field, which causes transitions between bound states.) The state of lowest energy is called the **ground state**; it is a truly stationary state.

The periodic table   There is a close relation between the set of bound states of the hydrogen atom, as described in ●4.10, and the set of chemical elements, with the structure of the periodic table. It is as if the Z electrons in the ground state of an atom with atomic number Z were all in states $|n\,l\,m\rangle|s\rangle$ where $|n\,l\,m\rangle$ is one of the bound states of the hydrogen atom ($m = -l, \ldots, l$ is the eigenvalue of L_z) and $|s\rangle$ is a spin state. Because of the Pauli exclusion principle, the electrons must all be in different states; since there are two spin states, there can be at most two electrons in any orbital state $|n\,l\,m\rangle$. As Z increases, the states are filled in the order shown in Fig. 4.5, which also shows the resulting elements and should be compared with the periodic table (Fig. 1.1).

This description of the states of atoms is called the **aufbau** ('building-up') principle. It would be justified if all the electrons independently were moving in the fixed field of the nucleus, so that the Hamiltonian would be a sum of terms like (4.94), one for each electron; then the stationary states would be products of one-electron states which would be independent of the number of electrons in the atom. But in fact the electrons all repel each other, so that each electron moves in a changing field due to the other electrons: quantum-mechanically, the Hamiltonian is not just a sum of one-electron terms. The success of the aufbau principle can be explained by assuming that the effect of the electronic repulsion is the same as that of a constant average field, so that the electrons can be treated as a cloud of negative charge which shields (or 'screens') the field of the nucleus. This produces an effective spherically symmetric potential which, although not the same as the inverse-square potential of (4.94), will have similar eigenvalues. Thus the states available to the electrons can be labelled as $|n\,l\,m\rangle$ like those of the hydrogen atom, but their energy eigenvalues will be changed; in particular, states with the same value of n will no longer have the same energy. The tendency is for the energy to increase with l, so that the order of the energy eigenvalues becomes as shown in Fig. 4.5.

4.5. **The harmonic oscillator**

A **harmonic oscillator** is a particle which is attracted to a fixed point O by a force which is proportional to its distance from O. We will first consider the one-dimensional problem, so that classically the particle is described by a

Fig. 4.5.
The Aufbau principle: for each l there are $2(2l+1)$ states to be filled.

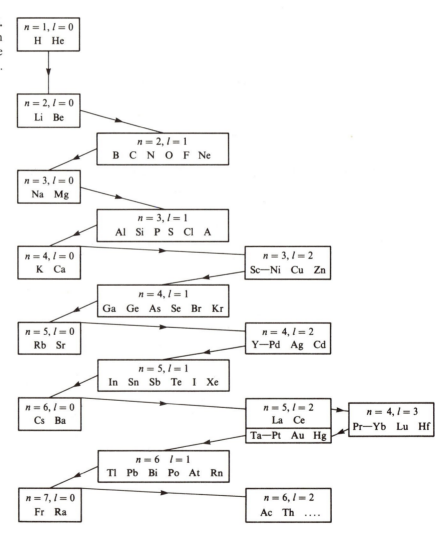

single coordinate x which satisfies an equation of motion of the form

$$\frac{d^2x}{dt^2} = -\omega^2 x \tag{4.124}$$

in which ω is the angular frequency of the oscillator. The Hamiltonian for this system is

$$H = \frac{p^2}{2m} + \frac{1}{2}m\omega^2 x^2. \tag{4.125}$$

The quantum-mechanical eigenvalues of this Hamiltonian can be found by a similar technique to that which was used for angular momentum in §4.1. Let

$$a = \frac{1}{\sqrt{(2\hbar m\omega)}}(p - im\omega x), \quad a^\dagger = \frac{1}{\sqrt{(2\hbar m\omega)}}(p + im\omega x); \tag{4.126}$$

then the standard commutator $[x, p] = i\hbar$ gives

$$[a, a^\dagger] = 1, \tag{4.127}$$

and H can be written in terms of a and $a^\dagger$ as

$$H = \hbar\omega(a^\dagger a + \tfrac{1}{2}). \tag{4.128}$$

From (4.127)–(4.128) we obtain

$$[H, a] = \hbar\omega[a^\dagger, a]a = -\hbar\omega a \tag{4.129}$$

and similarly

$$[H, a^\dagger] = \hbar\omega a^\dagger. \tag{4.130}$$

Now let $|\psi\rangle$ be an eigenstate of H with eigenvalue E. Then (4.130) implies that $a^\dagger|\psi\rangle$ is an eigenstate of H with eigenvalue $E + \hbar\omega$ (provided $a^\dagger|\psi\rangle \neq 0$); for

$$Ha^\dagger|\psi\rangle = (a^\dagger H + \hbar\omega a^\dagger)|\psi\rangle = (E + \hbar\omega)a^\dagger|\psi\rangle. \tag{4.131}$$

Similarly, (4.129) implies that $a|\psi\rangle$ is an eigenstate of H with eigenvalue $E - \hbar\omega$, unless $a|\psi\rangle = 0$. Let $|\phi\rangle = a|\psi\rangle$; then

$$\langle\phi|\phi\rangle = \langle\psi|a^\dagger a|\psi\rangle = \langle\psi|\left(\frac{H}{\hbar\omega} - \frac{1}{2}\right)|\psi\rangle = \frac{E}{\hbar\omega} - \frac{1}{2} \tag{4.132}$$

if $|\psi\rangle$ is normalised. Hence, by the positive-definiteness of the inner product,

$$E \geqslant \tfrac{1}{2}\hbar\omega \tag{4.133}$$

and

$$E = \tfrac{1}{2}\hbar\omega \Leftrightarrow a|\psi\rangle = 0. \tag{4.134}$$

We can write $H = \hbar\omega(aa^\dagger - \tfrac{1}{2})$ and apply the same argument to $a^\dagger|\psi\rangle$; this yields

$$E \geqslant -\tfrac{1}{2}\hbar\omega \quad \text{and} \quad E = -\tfrac{1}{2}\hbar\omega \Leftrightarrow a^\dagger|\psi\rangle = 0. \tag{4.135}$$

Thus starting with an eigenvalue E, we can find a descending chain of eigenvalues $E - \hbar\omega, E - 2\hbar\omega, \ldots$ by repeatedly applying the lowering operator a, and an ascending chain $E + \hbar\omega, E + 2\hbar\omega, \ldots$ by applying $a^\dagger$. The descending chain must terminate, otherwise (4.133) would be violated; it can only do so by reaching an eigenstate $|\psi\rangle$ for which $a|\psi\rangle = 0$, and then (4.134) shows that the corresponding eigenvalue is $\tfrac{1}{2}\hbar\omega$. On the other hand, the ascending chain can never terminate, for if it did there would be an eigenstate satisfying $a^\dagger|\psi\rangle=0$, and (4.135) shows that this would have an eigenvalue which violates (4.133). Thus the eigenvalues of H are $\frac{1}{2}\hbar\omega$, $\frac{3}{2}\hbar\omega$, ..., $(n+\frac{1}{2})\hbar\omega$, ...

Because of (4.128), applying the lowering operator a and then the raising operator $a^\dagger$ to an eigenstate of H brings you back to (a multiple of) the state you started with. Thus for a single spinless particle, for which x (or p, or $a^\dagger$) constitutes a complete set of commuting observables, there is just one state for each eigenvalue of H. We denote the normalised eigenstate with eigenvalue $(n+\frac{1}{2})\hbar\omega$ by $|n\rangle$; then the operators a and $a^\dagger$ act by

$$a|n\rangle=c_n|n-1\rangle, \quad a^\dagger|n\rangle=d_n|n+1\rangle \tag{4.136}$$

for some coefficients c_n and d_n. To find these, note that

$$|c_n|^2=\langle n|a^\dagger a|n\rangle=\langle n|\left(\frac{H}{\hbar\omega}-\frac{1}{2}\right)|n\rangle=n$$

$$=d_{n-1}c_n\langle n\,|\,n\rangle=d_{n-1}c_n. \tag{4.137}$$

We can choose the phases of the states so that c_n is real and positive; then $c_n=\sqrt{n}$ and $d_n=\sqrt{(n+1)}$.

To summarise,

> ●**4.11** The harmonic-oscillator Hamiltonian has eigenvalues $(n+\frac{1}{2})\hbar\omega$ where n can be any non-negative integer. It can be written as
>
> $$H=(a^\dagger a+\tfrac{1}{2})\hbar\omega$$
>
> where the operators $a,a^\dagger$ satisfy
>
> $$[a,a^\dagger]=1$$
>
> and act on the eigenstates $|n\rangle$ of H by
>
> $$\left.\begin{array}{l}a|n\rangle=\sqrt{n}|n-1\rangle\\a^\dagger|n\rangle=\sqrt{(n+1)}|n+1\rangle\end{array}\right\}. \quad ■ \tag{4.138}$$

The three-dimensional harmonic oscillator

The classical equation of motion for a harmonic oscillator in three dimensions is

$$\frac{d^2\mathbf{r}}{dt^2}=-\omega^2\mathbf{r}. \tag{4.139}$$

The ith component of this equation involves the ith coordinate x_i only; thus each coordinate separately satisfies the simple harmonic equation (4.124). The Hamiltonian is

$$H=\frac{\mathbf{p}^2}{2m}+\frac{1}{2}m\omega^2r^2, \tag{4.140}$$

which is the sum of three terms, each having the form of the one-dimensional Hamiltonian (4.125). There are three raising and lowering operators

$$a_i^\dagger=\frac{1}{\sqrt{(2\hbar m\omega)}}(p_i-im\omega x_i), \quad a_i=\frac{1}{\sqrt{(2\hbar m\omega)}}(p_i+im\omega x_i) \tag{4.141}$$

whose commutation relations are

$$[a_i, a_j^\dagger] = \delta_{ij} \\ [a_i, a_j] = 0 = [a_i^\dagger, a_j^\dagger] \Bigg\}, \tag{4.142}$$

and in terms of which the Hamiltonian is

$$H = (a_i^\dagger a_i + \tfrac{3}{2})\hbar\omega. \tag{4.143}$$

Each of the raising operators then creates a series of energy eigenstates

$$|n\rangle_i = \frac{1}{\sqrt{n!}} (a_i^\dagger)^n |0\rangle \tag{4.144}$$

where $|0\rangle$ is the unique ground state, which satisfies

$$a_1|0\rangle = a_2|0\rangle = a_3|0\rangle = 0. \tag{4.145}$$

The general eigenstate of H is

$$|l\,m\,n\rangle = \frac{1}{\sqrt{(l!\,m!\,n!)}} (a_1^\dagger)^l (a_2^\dagger)^m (a_3^\dagger)^n |0\rangle, \tag{4.146}$$

which has energy $E = (l + m + n + \tfrac{3}{2})\hbar\omega$. Thus the energy eigenvalues of the three-dimensional harmonic oscillator are $(N + \tfrac{3}{2})\hbar\omega$, where N can be any non-negative integer, and the number of independent eigenstates for a given N is the number of ways N can be written as $N = l + m + n$, which is $\tfrac{1}{2}(N + 1)(N + 2)$. But this can be made clearer by looking at the angular momentum of the states.

The components of (orbital) angular momentum are

$$L_i = \varepsilon_{ijk} x_j p_k = i\hbar\varepsilon_{ijk} a_j^\dagger a_k, \tag{4.147}$$

using (4.141) and (4.142). This gives

$$\mathbf{L}^2 = -\hbar^2 (a_j^\dagger a_k a_j^\dagger a_k - a_j^\dagger a_k a_k^\dagger a_j),$$

which can be put in the form

$$\mathbf{L}^2 = -\hbar^2 [A^\dagger A - N(N + 1)] \tag{4.148}$$

where

$$A = a_i a_i, \tag{4.149}$$

$$N = a_i^\dagger a_i = \frac{H}{\hbar\omega} - \frac{3}{2}. \tag{4.150}$$

Now since H is a scalar observable, it commutes with the angular momentum operators and so each eigenspace of H can be split into spaces $\mathscr{D}_l$ which are eigenspaces of $\mathbf{L}^2$ with eigenvalue $l(l + 1)$. Being a scalar product, A is also a scalar operator (though not an observable, since it is not hermitian) and commutes with $\mathbf{L}^2$, so it preserves the eigenvalue of $\mathbf{L}^2$; also, since A contains two lowering operators, it reduces the value of N by 2. Thus if there is a state $|n\,l\rangle$ with eigenvalues n and $l(l + 1)$ for N and $\mathbf{L}^2$, there is also a state $A|n\,l\rangle$ with eigenvalues $n - 2$ and $l(l + 1)$, unless $A|n\,l\rangle = 0$. From (4.148),

$$\langle n\,l|A^\dagger A|n\,l\rangle = n(n + 1) - l(l + 1), \tag{4.151}$$

so, as in (4.133)–(4.134),

$$n \geqslant l \quad \text{and} \quad n = l \Leftrightarrow A|n\,l\rangle = 0. \tag{4.152}$$

Since the number of states with eigenvalue n for N increases with n, the operator A (mapping a larger space into a smaller one) must have a null vector $|n\,l\rangle$ among the eigenvectors of N; (4.152) then shows that this has $l=n$. Applying $A^\dagger$ repeatedly to this state gives states $|n\,l\rangle$ with $n=l, l+2, l+4, \ldots$. Thus for each n there are states $|n\,l\rangle$ with $l=n, n-2, \ldots$. But there is no state with $l=n-1$, for then $A|n\,l\rangle$ would be non-zero (by the second part of (4.152)) but it would have $N=n-2=l-1$, violating the first part of (4.152).

The structure of the set of stationary states of the three-dimensional harmonic oscillator is shown in Fig. 4.6(*b*). The states of the hydrogen atom are shown for comparison in Fig. 4.6(*a*).

4.6. Annihilation and creation operators

All the quantum systems considered so far have consisted of a fixed number of particles with given forces between them (i.e. a given Hamiltonian). In classical mechanics this is practically the definition of a physical system. But if we are to describe processes like those mentioned in Chapter 1, e.g. the decay of a neutron into a proton, an electron and an antineutrino, we will have to consider systems in which the number of particles may change. In this section we will develop a formalism to do this, which is mathematically very similar to the formalism developed in the previous section to describe the harmonic oscillator.

Bosonic systems

Consider a system consisting of a variable number of indistinguishable particles, and suppose these particles are bosons. Let $\mathscr{S}$ be the state space for one particle. Then the space of states in which two particles are present is the symmetric subspace of $\mathscr{S} \otimes \mathscr{S}$, which we will denote by $\mathscr{S} \vee \mathscr{S}$ or $\vee^2\mathscr{S}$; in general, the space of r-particle states is $\vee^r\mathscr{S}$, the subspace of $\otimes^r\mathscr{S}$ consisting of symmetric states. The full state space for our system of a variable number of particles is

$$\mathscr{U} = \mathscr{V} \oplus \mathscr{S} \oplus \vee^2\mathscr{S} \oplus \vee^3\mathscr{S} \oplus \cdots \tag{4.153}$$

Fig. 4.6.
(*a*) The hydrogen atom; (*b*) the harmonic oscillator.

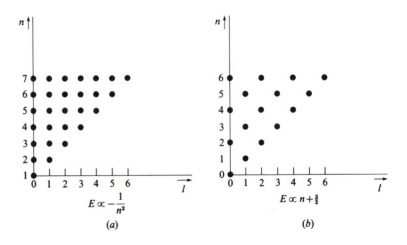

where $\mathscr{V}$ is a one-dimensional subspace containing a single state $|0\rangle$ in which no particles are present (the **vacuum state**). Note that $|0\rangle$ is a non-zero state vector and should not be confused with the zero vector 0.

Let $|\psi\rangle \in \mathscr{S}$ be any one-particle state, and let $\mathscr{U}_\psi$ be the subspace of $\mathscr{U}$ consisting of many-particle states in which every particle is in the state $|\psi\rangle$. Then $\mathscr{U}_\psi$ has a basis $|0\rangle, |\psi\rangle, |2\psi\rangle, \ldots$ where

$$|n\psi\rangle = |\psi\rangle|\psi\rangle \cdots |\psi\rangle \in \vee^n \mathscr{S}$$

is the state containing n particles. Thus $\mathscr{U}_\psi$ has the same structure as the state space of the harmonic oscillator, and we can introduce operators $a_\psi, a_\psi^\dagger$ to correspond to the raising and lowering operators of the harmonic oscillator:

$$a_\psi |n\psi\rangle = \sqrt{n}|(n-1)\psi\rangle, \quad a_\psi^\dagger |n\psi\rangle = \sqrt{(n+1)}|(n+1)\psi\rangle. \tag{4.154}$$

Then, as for the harmonic oscillator,

$$[a_\psi, a_\psi^\dagger] = 1. \tag{4.155}$$

a_ψ is called an **annihilation operator**; $a_\psi^\dagger$ is called a **creation operator**.

These operators can be extended so as to act on the whole of $\mathscr{U}$ as follows. There is a complete set of states like $|n\psi, \psi_1', \ldots, \psi_m'\rangle$, which contains n particles in the state $|\psi\rangle$ and m particles in states $|\psi_i'\rangle$ orthogonal to $|\psi\rangle$. The annihilation and creation operators a_ψ and $a_\psi^\dagger$ act on such a state by ignoring the particles in states other than $|\psi\rangle$:

$$a_\psi |n\psi, \psi_1', \ldots, \psi_m'\rangle = \sqrt{n}|(n-1)\psi, \psi_1', \ldots, \psi_m'\rangle;$$
$$a_\psi^\dagger |n\psi, \psi_1', \ldots, \psi_m'\rangle = \sqrt{(n+1)}|(n+1)\psi, \psi_1', \ldots, \psi_m'\rangle. \tag{4.156}$$

There is a more general definition of the annihilation operator a_ψ as an operator on the product space $\otimes^n \mathscr{S}$:

$$a_\psi |\phi_1\rangle \cdots |\phi_n\rangle = \frac{1}{\sqrt{n}} \{\langle\psi|\phi_1\rangle|\phi_2\rangle \cdots |\phi_n\rangle + \langle\psi|\phi_2\rangle|\phi_1\rangle|\phi_3\rangle\cdots|\phi_n\rangle$$
$$+ \cdots + \langle\psi|\phi_n\rangle|\phi_1\rangle \cdots |\phi_{n-1}\rangle\}. \tag{4.157}$$

The operator a_ψ of (4.156) is obtained by restricting this to the symmetric subspace of $\otimes^n \mathscr{S}$ (for a proof see problem 4.28).

Now let $|\psi_1\rangle, |\psi_2\rangle, \ldots$ be a complete set of states for $\mathscr{S}$. For each $|\psi_i\rangle$ there is a harmonic-oscillator space $\mathscr{U}_i$ and annihilation and creation operators $a_i, a_i^\dagger$. Let $|n_1 n_2 \cdots\rangle$ be the state containing n_i particles in state $|\psi_i\rangle$; these form a complete set of states for $\mathscr{U}$, and we have

$$\left.\begin{array}{l} a_i |\cdots n_i \cdots\rangle = \sqrt{n_i}|\cdots n_i - 1 \cdots\rangle \\ a_i^\dagger |\cdots n_i \cdots\rangle = \sqrt{(n_i+1)}|\cdots n_i + 1 \cdots\rangle \end{array}\right\}. \tag{4.158}$$

It follows from this that

$$\left.\begin{array}{l} [a_i, a_j^\dagger] = \delta_{ij} \\ [a_i, a_j] = 0 = [a_i^\dagger, a_j^\dagger] \end{array}\right\}. \tag{4.159}$$

By identifying the state $|n_1 n_2 \cdots\rangle$ with the product state $|n_1\rangle|n_2\rangle \cdots$, where

$|n_i\rangle \in \mathscr{U}_i$, we have an isomorphism

$$\mathscr{U} \cong \mathscr{U}_1 \otimes \mathscr{U}_2 \otimes \cdots. \tag{4.160}$$

In other words

● **4.12**   The state space of a variable number of bosons of a given type can be identified with the state space of a fixed number of distinguishable harmonic oscillators, one for each independent state of the bosons. ∎

If the particles do not interact with each other, the energy of a state of n particles is the sum of the energies of the individual particles. Thus if $|\psi\rangle$ is an eigenstate of the single-particle Hamiltonian, with energy E, the state $|n\psi\rangle$ is an eigenstate of the total Hamiltonian with eigenvalue nE. Comparing this with the harmonic oscillator, which has eigenvalues $(n+\frac{1}{2})\hbar\omega$ and a Hamiltonian $(a^\dagger a + \frac{1}{2})\hbar\omega$, we see that the Hamiltonian in the subspace $\mathscr{U}_\psi$ can be written

$$H = E a_\psi^\dagger a_\psi. \tag{4.161}$$

If the one-particle states $|\psi_i\rangle$ are all eigenstates of the one-particle Hamiltonian, with energies E_i, then the many-particle state $|n_1 n_2 \cdots\rangle$ has energy $n_1 E_1 + n_2 E_2 + \cdots$; it follows that the full Hamiltonian is

$$H = \sum_i E_i a_i^\dagger a_i. \tag{4.162}$$

Fermionic systems   If the particles are fermions, the space of r-particle states is the antisymmetric subspace of $\otimes^r \mathscr{S}$, denoted by $\wedge^r \mathscr{S}$, and the full state space is

$$\mathscr{U} = \mathscr{V}_0 \oplus \mathscr{S} \oplus \wedge^2 \mathscr{S} \oplus \cdots. \tag{4.163}$$

In this case the subspace $\mathscr{U}_\psi$, consisting of states in which all the particles are in the state $|\psi\rangle$, is only two-dimensional: since at most one particle can be in the state $|\psi\rangle$, the only possible states are the vacuum $|0\rangle$ and the one-particle state $|\psi\rangle$. We can still define annihilation and creation operators a_ψ and $a_\psi{}^\dagger$, avoiding the danger of creating two particles in the same state by making $a_\psi{}^\dagger$ annihilate $|\psi\rangle$. Thus for the general state, if $|\psi_1'\rangle, |\psi_2'\rangle, \ldots$ are all orthogonal to $|\psi\rangle$,

$$a_\psi|\psi\,\psi_1'\psi_2'\cdots\rangle = |\psi_1'\psi_2'\cdots\rangle, \quad a_\psi|\psi_1'\psi_2'\cdots\rangle = 0,$$
$$a_\psi{}^\dagger|\psi\,\psi_1'\psi_2'\cdots\rangle = 0, \quad a_\psi{}^\dagger|\psi_1'\psi_2'\cdots\rangle = |\psi\,\psi_1'\psi_2'\cdots\rangle. \tag{4.164}$$

These annihilation and creation operators do not satisfy the harmonic-oscillator commutation relation (4.127), but they do satisfy a similar *anti*commutation relation

$$\{a_\psi, a_\psi{}^\dagger\} = 1. \tag{4.165}$$

They also satisfy

$$a_\psi{}^2 = (a_\psi{}^\dagger)^2 = 0. \tag{4.166}$$

The fermion annihilation operator a_ψ can be defined as an operator on the product space $\otimes^n \mathscr{S}$ by a formula similar to (4.157):

$$a_\psi |\phi_1\rangle |\phi_2\rangle \cdots |\phi_n\rangle = \langle\psi|\phi_1\rangle|\phi_2\rangle\cdots|\phi_n\rangle - \langle\psi|\phi_2\rangle|\phi_1\rangle|\phi_3\rangle\cdots|\phi_n\rangle$$
$$+ \cdots + (-1)^{n+1}\langle\psi|\phi_n\rangle|\phi_2\rangle\cdots|\phi_{n-1}\rangle. \qquad (4.167)$$

Now let $|\psi_1\rangle$, $|\psi_2\rangle$, ... be an orthonormal complete set of one-particle states. There are corresponding pairs of annihilation and creation operators a_i, $a_i{}^\dagger$, each of which satisfies (4.165)–(4,166). We will now show that the a_i all anticommute with each other. Let

$$|\phi\rangle = \frac{1}{\sqrt{2}}(|\psi_i\rangle + |\psi_j\rangle), \quad |\theta\rangle = \frac{1}{\sqrt{2}}(|\psi_i\rangle + i|\psi_j\rangle),$$

so that

$$a_\phi = \frac{1}{\sqrt{2}}(a_i + a_j), \quad a_\theta = \frac{1}{\sqrt{2}}(a_i - ia_j)$$

(the minus sign occurs in a_θ because $a_\theta{}^\dagger$ involves $a_i{}^\dagger + ia_j{}^\dagger$). Then a_ϕ and a_θ also satisfy (4.165)–(4.166). These four sets of equations (for a_i, a_j, a_ϕ and a_θ) yield

$$\{a_i{}^\dagger, a_j{}^\dagger\} = \delta_{ij}, \qquad (4.168)$$

$$\{a_i, a_j\} = \{a_i{}^\dagger, a_j{}^\dagger\} = 0. \qquad (4.169)$$

The fact that $a_i{}^\dagger$ and $a_j{}^\dagger$ anticommute is to be expected for fermions: $a_i{}^\dagger a_j{}^\dagger|\Psi\rangle$ and $a_j{}^\dagger a_i{}^\dagger|\Psi\rangle$ should differ by the interchange of two particles, and therefore by a factor of -1.

The formula (4.162) for the Hamiltonian of a system of non-interacting particles holds for fermions as well as for bosons. The isomorphism (4.160) also holds, except that each space $\mathscr{U}_i$ is not a harmonic-oscillator space but a two-dimensional space spanned by $|0\rangle$ and $|\psi_i\rangle$.

The anticommutation relations (4.168)–(4.169) mean that there is a complete symmetry between the creation operators and the annihilation operators for a fermion (at least if the state space is finite-dimensional), which is not true of the boson operators. For a boson the annihilation operators have a simultaneous null vector, namely the vacuum state; the creation operators do not. For a fermion, on the other hand, the creation operators also have a simultaneous null vector $|\Omega\rangle = a_1{}^\dagger \cdots a_N{}^\dagger|0\rangle$, where N is the dimension of the one-particle state space $\mathscr{S}$; $|\Omega\rangle$ is the unique state in which all single-particle states are occupied. This means that any state can be obtained by applying annihilation operators to the full state $|\Omega\rangle$ just as well as by applying creation operators to the empty state $|0\rangle$. The annihilation operators can be regarded as creating holes in $|\Omega\rangle$, and in some circumstances (e.g. in semiconductors) these holes behave like particles. If the original particles were electrically charged (as in the case of electrons in a semiconductor), the holes behave like particles with the opposite charge.

Several types of particle   States containing several different types of particle can be described by putting together many-particle states of the sort we have just been considering. If the particles are A, B, ..., with state spaces $\mathscr{S}_A$, $\mathscr{S}_B$, ..., this gives a total state space

$$\mathscr{U} = \mathscr{U}_A \otimes \mathscr{U}_B \otimes \cdots \tag{4.170}$$

where

$$\mathscr{U}_X = \sum_{r=0}^{\infty} \vee^r \mathscr{S}_X \quad \text{if X is a boson,} \tag{4.171}$$

$$\mathscr{U}_X = \sum_{r=0}^{\infty} \wedge^r \mathscr{S}_X \quad \text{if X is a fermion} \tag{4.172}$$

(the summation sign denotes a direct sum of vector spaces).

For each particle X and each state $|\psi\rangle \in \mathscr{S}_X$ there are annihilation and creation operators $a_{X\psi}$, $a_{X\psi}^\dagger$ which act on $\mathscr{U}$ by acting on $\mathscr{U}_X$. Then $\mathscr{U}$ has a complete set of states of the form

$$a_{A1}^\dagger a_{A2}^\dagger \cdots a_{B1}^\dagger a_{B2}^\dagger \cdots |0\rangle \tag{4.173}$$

where $a_{X1}^\dagger, a_{X2}^\dagger, \ldots$ create different states of particle X and the vacuum state $|0\rangle$ in $\mathscr{U}$ is the product of the individual states in $\mathscr{U}_A$, $\mathscr{U}_B$, ... In this construction creation operators referring to different particles commute with each other.

There is an alternative description which can be used when the particles are all fermions or all bosons: the space $\mathscr{U}$ of (4.170) can be replaced by

$$\mathscr{U}' = \sum_{r=0}^{\infty} \vee^r (\mathscr{S}_A \oplus \mathscr{S}_B \oplus \cdots) \quad \text{for bosons} \tag{4.174}$$

(replace $\vee$ by $\wedge$ if the particles are fermions). In other words, the one-particle state spaces are put together to form a big one-particle space, so that the different particles A, B, ... appear as different states of a single underlying particle, which is then treated as a boson (or a fermion) in forming many-particle states. There is a natural correspondence between states in $\mathscr{U}$ and states in $\mathscr{U}'$: the basic state (4.173) in $\mathscr{U}$, which can be written as

$$S(|A1\rangle|A2\rangle \cdots) S(|B1\rangle|B2\rangle \cdots) \tag{4.175}$$

where S is the symmetrisation operator (2.142), corresponds to the state

$$S(|A1\rangle|A2\rangle \cdots |B1\rangle|B2\rangle \cdots) \tag{4.176}$$

in $\mathscr{U}'$ (replace S by the antisymmetrisation operator (2.143) for fermions). Thus $\mathscr{U}$ and $\mathscr{U}'$ are equivalent descriptions of the same physical system. However, in the case of fermions the two descriptions lead to different creation operators, for $\mathscr{U}'$ has a structure like (4.172) which suggests that all creation operators should *anti*commute (even when they refer to different particles). The choice of description ($\mathscr{U}$ or $\mathscr{U}'$) will therefore affect the physical content of a statement involving annihilation and creation operators.

The $\mathscr{U}'$ description (leading to anticommuting creation operators) is always used for the states of a fermion and its antiparticle. Thus if the fermion A has state space $\mathscr{S}_A$ and its antiparticle $\bar{A}$ has state space $\mathscr{S}_A$, the state space of variable numbers of A and $\bar{A}$ is

$$\mathscr{V}_A = \sum_{r=0}^{\infty} \wedge^r (\mathscr{S}_A \oplus \mathscr{P}_A). \tag{4.177}$$

The two types of description can be mixed: for example, in describing several particle–antiparticle pairs the creation operators for different pairs can commute, the total state space being $\mathscr{V}_A \otimes \mathscr{V}_B \otimes \cdots$ with each $\mathscr{V}_X$ given by (4.177).

Charge conjugation   Since a particle and its antiparticle have the same mass and spin, their state spaces are isomorphic: if $|X\psi\rangle$ is a state of the particle X, there is a corresponding state $|\bar{X}\psi\rangle$ of its antiparticle $\bar{X}$. **Charge conjugation** C is an operator which interchanges these two states, i.e. it changes every particle into its antiparticle without changing its spin state or wave function:

$$C|X\psi\rangle = |\bar{X}\psi\rangle, \quad C|\bar{X}\psi\rangle = |X\psi\rangle. \tag{4.178}$$

This defines C as an operator on $\mathscr{S}_A \oplus \bar{\mathscr{S}}_A$. The definition can be extended to the many-particle state space $\mathscr{V}_A$ in the obvious way, by making C act simultaneously on all the one-particle states in a product state, so that it changes all particles in a state into their antiparticles.

Like the parity operator, the charge conjugation operator satisfies $C^2 = 1$ and is both unitary and hermitian. For a particle X which is distinct from its antiparticle, (4.178) is the most general form of a charge conjugation operator with these properties, for the phase of the $\bar{X}$ state can be adjusted so as to make (4.178) true. If X is its own antiparticle, however, this no longer applies: the effect of charge conjugation on a state $|X\psi\rangle$ of X must be to give the same physical state, but this only means that the state vector $C|X\psi\rangle$ must be a multiple of $|X\psi\rangle$: thus $|X\psi\rangle$ is an eigenvector of C. Since $C^2 = 1$, the possible eigenvalues are ± 1. This eigenvalue is a property of X like its intrinsic parity; it is called the **charge conjugation parity** and denoted by η_C. It is only defined if X is **totally neutral**, i.e. if it has the value 0 for all additive quantum numbers like electric charge and baryon number, since otherwise it would have a distinct antiparticle. The values of η_C for some totally neutral particles are given in Table 4.2.

If there is symmetry between matter and antimatter (as we might expect from classical electrodynamics, in which positive and negative charge are both on the same footing), then charge conjugation will be an invariance: the operator C will commute with the Hamiltonian and η_C will be a conserved

Table 4.2. *Charge conjugation parity*

Particle	Photon	Z^0	π^0	η
η_C	-1	-1	$+1$	$+1$

quantity. This is true of strong, electromagnetic and gravitational interactions; it leads to restrictions on possible processes like those which can be deduced from parity conservation.

Example: positronium annihilation

Positronium is a bound state of an electron and a positron – a hydrogen atom with the proton replaced by a positron. Unlike the hydrogen atom, though, it has no stable ground state, as the electron and the positron can annihilate each other to form photons. This is an electromagnetic process, and is invariant under charge conjugation; hence η_C is conserved.

The state space of positronium is part of the two-particle space built from the one-particle space $\mathscr{S} \oplus \mathscr{\bar{S}}$ where $\mathscr{S}$ is the state space of an electron and $\mathscr{\bar{S}}$ that of a positron; this contains states $|e^-\psi\rangle$ and $|e^+\psi\rangle$. Since $\mathscr{S}$ and $\mathscr{\bar{S}}$ are isomorphic, $\mathscr{S} \oplus \mathscr{\bar{S}}$ can be identified with $\mathscr{C} \otimes \mathscr{S}$ where $\mathscr{C}$ is a two-dimensional space (the 'charge space') with basic states $|e^-\rangle$ and $|e^+\rangle$. Then charge conjugation acts on $\mathscr{C}$ by interchanging $|e^-\rangle$ and $|e^+\rangle$. Thus two-particle states which are symmetric or antisymmetric in charge space are eigenstates of C with eigenvalue $\eta_C = +1$ (for symmetric states) or -1 (for antisymmetric ones). The two-particle state must be antisymmetric overall; hence if the spin/orbital state is symmetric (l even and $s = 1$, or l odd and $s = 0$, where l is the relative orbital angular momentum and s is the total spin) then η_C must be -1, and if the spin/orbital state is antisymmetric (l odd and $s = 1$, or l even and $s = 0$) η_C must be $+1$. If the positronium is in its ground state we have $l = 0$ from the theory of the hydrogen atom; hence $\eta_C = -1$ if the spins are parallel ($s = 1$) and $\eta_C = +1$ if they are antiparallel ($s = 0$).

If the electron and positron annihilate each other to form n photons, the charge conjugation parity afterwards will be $(-1)^n$, since the photon has $\eta_C = -1$. Energy–momentum conservation makes $n = 1$ impossible (in the rest frame of the positronium the photon would have zero momentum), so the two smallest possibilities are $n = 2$ or 3. Charge conjugation invariance then shows that the positronium will decay into two photons when the spins are antiparallel but into three photons when the spins are parallel. Since the phase space factor is greater for the two-photon decay, this decay is faster and so the ground state of positronium with parallel spins is longer-lived than the state with antiparallel spins.

Parity is also conserved in positronium annihilation; this can be applied to the two-photon process to obtain information about the polarisation of the photons. The parity of the ground state ($l = 0$, $s = 0$) of positronium is $-(-1)^l = -1$, since the electron and positron, being fermionic antiparticles, have opposite intrinsic parities. Thus the two-photon state must have parity -1 and angular momentum 0. Suppose one of the photons is in an eigenstate of momentum with eigenvalue $\mathbf{k}$; then, taking the positronium to be at rest, the other photon must have momentum $-\mathbf{k}$. The two photons must have opposite values for the component of angular momentum in the direction of $\mathbf{k}$, since the total angular momentum is 0; this means that their helicities must be equal. Hence, labelling the photon states as $|\mathbf{p}\pm\rangle$ where $\mathbf{p}$ is the momentum and $\pm$ is the helicity of the photon, and taking into account the fact that photons are bosons, we have a two-photon state of the form

$$\alpha(|\mathbf{k}+\rangle|-\mathbf{k}+\rangle+|-\mathbf{k}+\rangle|\mathbf{k}+\rangle)+\beta(|\mathbf{k}-\rangle|-\mathbf{k}-\rangle+|-\mathbf{k}-\rangle|\mathbf{k}-\rangle) \tag{4.179}$$

The effect of the parity operator on a one-photon state is to reverse its momentum and to leave the components of angular momentum unchanged (since P commutes with $\mathbf{J}$), which is to reverse the helicity:

$$P|\mathbf{p}\pm\rangle=-|-\mathbf{p}\mp\rangle \tag{4.180}$$

(the minus sign is the intrinsic parity of the photon). Hence if the two-photon state (4.179) is to have negative parity we must have $\alpha=-\beta$.

The helicity of a photon is related to its polarisation. Eq. (3.101) defines the operator J_z for a photon moving in the z-direction, i.e. its helicity, in terms of polarisation states $|\phi_x\rangle$ and $|\phi_y\rangle$. From this the eigenstates of helicity can be calculated as

$$|\mathbf{k}\pm\rangle=\frac{1}{\sqrt{2}}(|\mathbf{k}\,x\rangle\pm i|\mathbf{k}\,y\rangle),\quad |-\mathbf{k}\pm\rangle=\frac{1}{\sqrt{2}}(|-\mathbf{k}x\rangle\mp i|-\mathbf{k}y\rangle).$$

In terms of these the two-photon state (4.179), with $\alpha=-\beta$, is

$$|\mathbf{k}x\rangle|-\mathbf{k}y\rangle-|\mathbf{k}y\rangle|-\mathbf{k}x\rangle+|-\mathbf{k}y\rangle|\mathbf{k}x\rangle-|-\mathbf{k}x\rangle|\mathbf{k}y\rangle. \tag{4.181}$$

Thus the two photons are polarised in perpendicular directions.

Particle-changing interactions

Creation and annihilation operators are used to construct Hamiltonians which change the number of particles present. To illustrate this, we will suppose that each particle has a one-dimensional state space (i.e. we ignore the spin state and wave function of each particle. We will refer to these suppressed degrees of freedom as **kinematical**). Then for particles A, B, ... there are creation operators $a_A^\dagger$, $a_B^\dagger$, ..., and there is a complete set of states like

$$|m\mathrm{A}, n\mathrm{B}, \ldots\rangle=\frac{1}{\sqrt{(m!\,n!\ldots)}}(a_A^\dagger)^m(a_B^\dagger)^n\cdots|0\rangle. \tag{4.182}$$

Now suppose that the single state of particle X has energy E_X. Then if the particles do not interact, the Hamiltonian is

$$H_0=E_A a_A^\dagger a_A+E_B a_B^\dagger a_B+\cdots \tag{4.183}$$

(cf. (4.162)). We will describe interactions between the particles by adding an extra term εV to this Hamiltonian and using time-dependent perturbation theory.

Suppose there are five particles A, B, C, D, α, and let

$$V=a_A a_C^\dagger a_\alpha^\dagger+a_B a_D^\dagger a_\alpha+a_A^\dagger a_C a_\alpha+a_B^\dagger a_D a_\alpha^\dagger. \tag{4.184}$$

(this is dimensionless, so the expansion parameter ε, which in this context is called a **coupling constant**, has the dimensions of energy). Then V is hermitian, since it is of the form $W + W^\dagger$. Suppose the initial state is $|\psi_0\rangle = |A\rangle$; then from (3.182) and (3.176) the first-order term in the perturbation expansion is

$$|\tilde{\psi}(t)\rangle = \frac{1}{i\hbar} \int_0^t dt_1 e^{iH_0 t_1/\hbar} V e^{-iH_0 t_1/\hbar} |\psi_0\rangle$$

$$= \frac{1}{i\hbar} \int_0^t dt_1 e^{i(E_C + E_\alpha - E_A)t_1/\hbar} |C\alpha\rangle. \tag{4.185}$$

Thus the effect of the interaction (4.184) is to induce the decay

$$A \rightarrow C + \alpha. \tag{4.186}$$

It is clear that the part of (4.184) responsible for this is the first term, destroying A and creating C and α. From the theory of §3.5 we know that if $E_A \neq E_C + E_\alpha$ the integral in (4.185) is very small and so the decay is highly unlikely to occur. Suppose the particle A is at rest, so that $E_A = m_A c^2$; then, since the energy of a moving particle is at least mc^2, the condition for this decay to be possible is

$$m_A \geqslant m_C + m_\alpha. \tag{4.187}$$

If this condition is not satisfied, the term in V describing the process (4.186) is still significant because of its contribution to higher-order processes. Suppose the initial state is $|AB\rangle$ and consider the second-order term in the perturbation expansion, namely

$$|\tilde{\psi}_2(t)\rangle = \frac{1}{(i\hbar)^2} \int_0^t dt_1 \int_0^{t_1} dt_2 e^{iH_0 t_1} V e^{-iH_0(t_1 - t_2)} V e^{-iH_0 t_2} |AB\rangle \tag{4.188}$$

(see (3.183)). From (4.184) we have

$$V|AB\rangle = |BC\alpha\rangle \tag{4.189}$$

and

$$V|BC\alpha\rangle = |AB\rangle + |CD\rangle; \tag{4.190}$$

thus the second-order term (4.188) contains a multiple of $|CD\rangle$, and so the interaction induces the process

$$A + B \rightarrow C + D. \tag{4.191}$$

These calculations can be represented by diagrams as follows. The terms in V can be drawn as in Fig. 4.7, which shows the annihilations and creations performed by the operators in each term (the diagrams are to be read from left to right). These provide four types of vertex which can be put together to form composite diagrams like Fig. 4.8, which represents the process (4.191). The two vertices of this diagram show that the process it represents takes place in second order of perturbation theory; the two halves of the diagram, on either side of the dotted line, illustrate the two equations (4.189)–(4.190) which contribute to the perturbation theory calculation. The dotted line itself intersects the virtual intermediate state $|BC\alpha\rangle$ which occurs in (4.189)–(4.190).

This procedure can be continued to draw diagrams which illustrate processes occurring in any order of perturbation theory. These are the Feynman diagrams introduced in Chapter 1.

Calculating the integral (4.188), we find that the amplitude for the process $A + B \rightarrow C + D$, to second order in ε, is

$$\langle CD | e^{-iH_0 t} | \tilde{\psi}_2(t) \rangle$$

$$= \varepsilon^2 \left[\frac{e^{-i(E_A + E_B)t} - e^{-i(E_C + E_D)t}}{(E_A - E_C - E_\alpha)(E_A + E_B - E_C - E_D)} - \frac{e^{-i(E_B + E_\alpha + E_C)t} - e^{-i(E_C + E_D)t}}{(E_A - E_C - E_\alpha)(E_B - E_D + E_\alpha)} \right].$$

(4.192)

If E_α is large compared with both ε and the energies of A, B, C and D, so that E_A/E_α is a small quantity of the order of ε/E_A, then the second term in (4.192) is of third order and the first term is approximately

$$-\frac{\varepsilon^2}{E_\alpha} \left[\frac{e^{-i(E_A + E_B)t} - e^{-i(E_C + E_D)t}}{E_A + E_B - E_C - E_D} \right].$$

(4.193)

This is the same (apart from sign) as the *first*-order amplitude that would be obtained from the Hamiltonian $H = H_0 + \varepsilon' V'$, where

$$\varepsilon' = \frac{\varepsilon^2}{E_\alpha}, \quad V' = a_A a_B a_C^\dagger a_D^\dagger + a_A^\dagger a_B^\dagger a_C a_D.$$

(4.194)

V' is called an **effective** Hamiltonian, and ε' is called an effective coupling

Fig. 4.7.
The terms of V in (4.184).

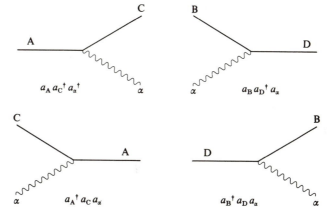

Fig. 4.8.
$A + B \rightarrow C + D$.

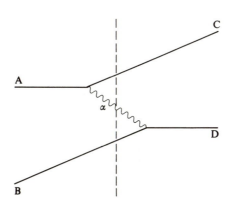

constant. The terms in V' are depicted by the vertices shown in Fig. 4.9. They appear to describe processes occurring by direct contact between the four particles.

The equivalence between the Hamiltonian (4.184) and the effective Hamiltonian (4.194) when the energies of the particles are small compared with E_α is an aspect of the inverse relation between the range of a force and the mass of its associated particle α, which was mentioned in Chapter 1. The energy E_α is of the order of $m_\alpha c^2$. An energy E which is much less than this is associated with momentum of the order of E/c, and therefore with lengths of the order of $\hbar c/E$, which are much greater than the length $h/m_\alpha c$ which was given in (1.29) as the range of the force. On this length scale, therefore, the force appears to operate only when the particles are in contact, which is the picture given by Fig. 4.9.

When the kinematical states of the particles are taken into account, it becomes of interest to consider Hamiltonians like (4.184) in which A = C and B = D; for now we have an annihilation operator a_{Ai} for each state $|\psi_i\rangle$ of particle A, and V can include terms

$$a_{Ai}a_{Aj}^\dagger a_\alpha^\dagger \tag{4.195}$$

describing the emission of a field quantum α by particle A, which is left in a different state. The diagram corresponding to Fig. 4.8 then depicts a process in which A and B both change state, i.e. the operation of a *force* between them. If A and B are both electrons and α is a photon, the force is the electromagnetic force between the electrons.

This method of describing interactions between particles constitutes **quantum field theory**, which will be introduced a little more fully in Chapter 7, though a proper development is beyond the scope of this book. This theory requires that annihilation and creation operators must always occur in the form of a **quantum field**, which is a combination

$$\phi_{Xi} = a_{Xi} + a_{\bar{X}i}^\dagger \tag{4.196}$$

referring to a state $|\psi_i\rangle$ of particle X and a state $|\psi_i'\rangle$ of its antiparticle $\bar{X}$. The states $|\psi_i\rangle$ and $|\psi_i'\rangle$ are the same for bosons, but for fermions they differ in their spin parts; $|\psi_i'\rangle$ is obtained from $|\psi_i\rangle$ by replacing each eigenstate of helicity by the eigenstate with opposite eigenvalue.

Fig. 4.9.
The terms of V' in (4.194).

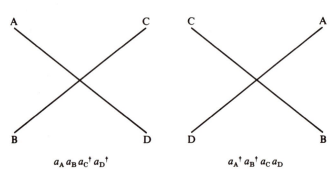

$$a_A a_B a_C^\dagger a_D^\dagger \qquad\qquad a_A^\dagger a_B^\dagger a_C a_D$$

In Chapter 6 we will give a description of the fundamental interactions between particles ignoring kinematical states, for which the appropriate model will be that introduced on p. 169, each particle having just one state. The quantum field for particle X will then be

$$\phi_X = a_X + a_X^\dagger,$$ (4.197)

which we will call a **reduced** quantum field. All interaction Hamiltonians are to be constructed from such operators.† The effect of this is that if an interaction admits a Feynman-diagram vertex involving the emission of a certain particle, it also admits a vertex in which this is replaced by absorption of the antiparticle, and vice versa.

The interaction Hamiltonian (4.184), which we have used for illustrative purposes, does not fulfil the requirement that it should be constructed out of quantum fields. To be physically acceptable, it would have to be replaced by

$$V = J\phi_\alpha + J^\dagger \phi_\alpha^\dagger$$ (4.198)

where

$$J = \phi_A^\dagger \phi_C + \phi_B \phi_D^\dagger.$$ (4.199)

The operator J is called the **current** associated with the force carried by the particle α. The low-energy effective Hamiltonian derived from (4.198) (corresponding to (4.194)) can be expressed entirely in terms of J:

$$V' = JJ^\dagger.$$ (4.200)

The last three chapters of this book are largely independent of each other. The reader who is irritated by the loose ends in the last section (§4.6), and is interested in the further formal development of quantum mechanics, can proceed straight to the account of quantum field theory in Chapter 7. A reader who wants to know about elementary particles should turn to Chapter 6, which contains applications of the theory of the last three chapters. However, this theory has a number of puzzling features, and so Chapter 5 is devoted to a deeper look at the concepts of quantum mechanics.

† The free-particle Hamiltonian (4.183) cannot be constructed in this way, but this difference between H_0 and V is an artificial feature of our simplified formalism; it can be removed if the particles are allowed more than one state. See problem 4.31.

Further reading   The topics in this chapter are covered in most textbooks on quantum mechanics. See the 'Further reading' for Chapters 2 and 3.

Problems on Chapter 4

1. For a simple particle moving in space, show that the wave function $\psi_{lm}(\mathbf{r}) = x^2 + y^2 - 2z^2$ represents a simultaneous eigenstate of $\mathbf{J}^2$ and J_z, with eigenvalues $l(l+1)\hbar^2$ and $m\hbar$ where l and m are to be determined. Find a function with the same eigenvalue for $\mathbf{J}^2$ and the maximum possible eigenvalue for J_z.

2. Show that there are $2l+1$ independent totally symmetric tensors $t_{i_1 \cdots i_l}$ ($i_r = 1, 2, 3$) satisfying $\sum_j t_{jji_3 \cdots i_l} = 0$, and that the functions $f(\mathbf{r}) = t_{i_1 \cdots i_l} x_{i_1} \cdots x_{i_l}$ are eigenfunctions of $\mathbf{J}^2$ with eigenvalue $l(l+1)\hbar^2$. [Consider $\nabla^2 f$.]

3. Show that for a system with angular momentum j the eigenstates of $\mathbf{n} \cdot \mathbf{J}$ are $U(R)|j\,m\rangle$ where R is a rotation which takes the z-axis to the axis $\mathbf{n}$. Deduce that $|d^j_{mn}(R)|^2$ is the probability that a measurement of $\mathbf{n} \cdot \mathbf{J}$ will give the value $n\hbar$ after a measurement of J_z has given the value $m\hbar$.

   How are these results compatible with the fact that R is not uniquely defined?

4. A beam of electrons in an eigenstate of J_z with eigenvalue $\frac{1}{2}\hbar$ is fed into a Stern–Gerlach apparatus, which measures the component of spin along an axis at an angle θ to the z-axis and separates the particles into distinct beams according to the value of this component. Find the ratio of the intensities of the emerging beams.

5. Let U and A be the 2×2 matrices $U = \cos \frac{1}{2}\theta + i\mathbf{n} \cdot \boldsymbol{\sigma} \sin \frac{1}{2}\theta$, $A = \mathbf{a} \cdot \boldsymbol{\sigma}$ where $\mathbf{n}$ is a unit vector. Show that $UAU^\dagger = \mathbf{a}' \cdot \boldsymbol{\sigma}$ where $\mathbf{a}'$ is obtained by rotating $\mathbf{a}$ about $\mathbf{n}$ through the angle θ. Deduce that the homomorphism ϕ of (4.61) maps SU(2) onto a group isomorphic to SO(3). Show that ϕ has kernel $\{\pm 1\}$.

6. In the spin state space $\mathscr{D}_{j_1} \otimes \mathscr{D}_{j_2}$ of two particles with spins j_1 and j_2, show that $|j_1 j_1\rangle |j_2 j_2\rangle$ is an eigenstate $|J\,M\rangle$ of the total angular momentum operators $\mathbf{J}^2$ and J_z. By applying the lowering operator $J_- = J_{1-} + J_{2-}$, find the Clebsch–Gordan coefficients $\langle J\,J-1|j_1 m_1, j_2 m_2\rangle$ for all relevant values of m_1 and m_2. Find the state $|J-1\,J-1\rangle$ (which must be orthogonal to $|J\,J-1\rangle$), and hence find the Clebsch–Gordan coefficients $\langle J-1\,J-1|j_1 m_1, j_2 m_2\rangle$. Consider how this process can be continued, and show that

   $$r(J, M)\langle J\,M-1|j_1\,m_1, j_2 m_2\rangle$$
   $$= r(j_1, m_1+1)\langle JM|j_1\,m_1+1, j_2 m_2\rangle + r(j_2, m_2+1)\langle JM|j_1\,m_1, j_2 m_2+1\rangle$$

   where $r(j, m) = \sqrt{[j(j+1) - m(m-1)]}$.

   Calculate the Clebsch–Gordan coefficients $\langle \frac{1}{2} m_1 | 1\,m_2, \frac{3}{2} m_3\rangle$ for all relevant values of m_1, m_2 and m_3.

7. In a system of three particles, each with angular momentum j and no other properties, how many independent states are there with total angular momentum l, where $0 \le l \le 3j$, and a given value of J_z?

8. Show that the five operators

$$T_0 = 2\frac{\partial^2}{\partial z^2} - \frac{\partial^2}{\partial x^2} - \frac{\partial^2}{\partial z^2}, \quad T_{\pm 1} = \mp\sqrt{6}\frac{\partial^2}{\partial x \, \partial z} - \sqrt{6i}\frac{\partial^2}{\partial y \, \partial z},$$

$$T_{\pm 2} = \frac{\sqrt{3}}{\sqrt{2}}\left(\frac{\partial}{\partial x} \pm 2i\frac{\partial^2}{\partial x \, \partial y} + \frac{\partial^2}{\partial y^2}\right)$$

form an irreducible set of operators of spin type 2. Find the ratio of the expectation values of T_0 in the five eigenstates of J_z of a particle with orbital angular momentum 2.

9. An electron in a state of a hydrogen atom with $n = 3$, $l = 2$ has total angular momentum $\frac{5}{2}$. Calculating any Clebsch–Gordan coefficients you need, find the ratio of the expectation values of z in the eigenstates of J_z with eigenvalues $\frac{5}{2}$ and $\frac{3}{2}$. What are the expectation values of x and y in these states?

10. Show that in a state with angular momentum 0, the expectation value of any vector operator is 0. Deduce that a spin-0 particle has no magnetic moment or electric dipole moment.

11. Show that in any irreducible representation of the rotation group, every vector operator is a multiple of $\mathbf{J}$.

12. Let $|\pm 1\rangle, |0\rangle$ be the three eigenstates of J_z of a particle with orbital angular momentum 1. Show that the expectation values of z^2 and r^2 are related by

$$\langle -1|r^2|-1\rangle = \langle 0|r^2|0\rangle = \langle 1|r^2|1\rangle = \langle 0|z^2|0\rangle + 2\langle -1|z^2|-1\rangle.$$

13. A hydrogen atom is placed in a uniform electric field with magnitude E in the z-direction. Labelling the states by the usual quantum numbers n, l, m, find the ratios of the first-order transition probabilities P_n from $|n \ l = 2, m\rangle$ to $|n, l = 1, m\rangle$ for $m = 0, \pm 1$.

14. If $S_{m_1}(-j_1 \le m_1 \le j_1)$ and $T_{m_2}(-j_2 \le m_2 \le j_2)$ are irreducible sets of operators of spin types j_1 and j_2, show that

$$U^j{}_m = \sum_{m_1, m_2} \langle j \, m|j_1 m_1, j_2 m_2\rangle S_{m_1} T_{m_2}$$

is an irreducible set of operators of spin type j.

15. A particle A with spin $\frac{1}{2}$ decays into two particles B and C, where B has spin $\frac{1}{2}$ and C has spin 1. What are the possible values of the relative orbital angular momentum of B and C? If this relative orbital angular momentum is 0, and if A is in an eigenstate of the z-component of spin with eigenvalue $+\frac{1}{2}$, find the probability that the z-component of the spin of B will also have the value $+\frac{1}{2}$.

16. The η-meson is a neutral particle with negative parity. Show that it cannot decay into two π^0 mesons if parity is conserved.

17. If two neutrons could form a bound state with total angular momentum 1, show that it would have negative parity.

18. Show that a particle which decays into two π^0 mesons has even integer spin.

19. The decay $\pi^0 \to 2\gamma$ takes place by the (parity-conserving) electromagnetic

interaction. Treating the photons as ordinary (massive) spin-1 particles, show that their relative orbital angular momentum is 1.

20. Show that in the ρ-meson ($J^P = 1^-$) the quark and antiquark have parallel spins.

21. Prove (4.107), (4.108) and (4.104).

22. **The Zeeman effect**. The Hamiltonian for an atom in a magnetic field of strength B along the z-axis is $H_0 + \mu B L_z$ where H_0 is the Hamiltonian of §4.4, **L** is the relative orbital angular momentum and μ is a constant. What are the stationary states and energy levels of this Hamiltonian?

23. Show that when a hydrogen atom decays from an excited state by emitting a single photon, the value of l changes by ± 1.

24. An experimenter has carefully prepared a particle of mass m in the first excited state (energy $\frac{3}{2}\hbar\omega$) of a one-dimensional harmonic potential $V(x) = \frac{1}{2}m\omega^2 x^2$, when he sneezes and knocks the centre of the potential a small distance a to one side. It takes him a time τ to blow his nose, and when he has done so he immediately puts the centre back where it was. Find, to lowest order in a, the probabilities P_0 and P_2 that the oscillator will now be in its ground state and its second excited state. Show that the probability that it will be in its nth excited state is of order $a^{2(n-1)}$.

25. For any complex number z, a state $|z\rangle$ of a harmonic oscillator is defined by $|z\rangle = \exp(za^\dagger)|0\rangle$. Show that $\langle z_1|z_2\rangle = \exp(\bar{z}_1 z_2)$ and that the expectation value of the energy in the state $|z\rangle$ is $(|z|^2 + \frac{1}{2})\hbar\omega$ where ω is the frequency of the oscillator. If the oscillator is in the state $|z_0\rangle$ at time $t = 0$, find the probability that it will again be found in the state $|z_0\rangle$ at time t.

26. Show that the nine operators $a_i^\dagger a_j + a_j^\dagger a_i$, $i(a_i^\dagger a_j - a_j^\dagger a_i)$ represent conserved observables of the three-dimensional harmonic oscillator, and that when multiplied by i they form a Lie algebra isomorphic to that of U(3).

27. Let $\mathcal{S}$ be a single-particle state space, with complete set of states $|\psi_i\rangle$, and let a_i be the corresponding annihilation operators on the space $\mathcal{U}$ of an indefinite number of particles. If H is the single-particle Hamiltonian, show that the Hamiltonian on $\mathcal{U}$ for a system of non-interacting particles is $\sum_{ij} \langle \psi_i | H | \psi_j \rangle a_i^\dagger a_j$.

28. Let $|\Psi\rangle$ be an n-particle state $|\psi_{i_1}\rangle \cdots |\psi_{i_n}\rangle$ where the $|\psi_i\rangle$ make up an orthonormal set and $\{i_1, \ldots, i_n\}$ contains r different indices in groups of $k_1, \ldots, k_r$. Let $S = \sum X_\rho$ be the symmetrisation operator. Show that $\langle \Psi | S^\dagger S | \Psi \rangle = n! \, k_1! \ldots k_r!$. Hence show that the annihilation operator a_ψ of (4.156) is the restriction of (4.157) to the symmetric subspace of $\otimes^n \mathcal{S}$.

29. Prove (4.192).

30. If the Hamiltonian is $H_0 + \varepsilon V$ where H_0 is given by (4.183) and $V = W^\dagger + W$ where $W = a_A a_C^\dagger a_\alpha^\dagger + a_B a_D^\dagger a_\alpha + a_D a_E^\dagger a_F^\dagger$, show that the probability of the process $A + B \to C + E + F$ is of order ε^6, and draw the corresponding Feynman diagram.

31. Let a_{Xi} be the annihilation operator for particle X in state i, and suppose the definition of quantum fields is extended to include operators like $\phi_{Xi} =$

$\sum_j (\alpha_{ij}a_{Xj} + \beta_{ij}a_{\bar{X}j}^\dagger)$. Show that it is possible to find matrices α_{ij} and β_{ij} so that the free-particle Hamiltonian H_0 of (4.183) can be written as $\sum_i (\phi_{Xi}^\dagger \phi_{Xi})$.

32. Consider a model of a charged particle X, with antiparticle $\bar{X}$, interacting with a photon γ, in which each particle is assumed to have just one state and the Hamiltonian is $H_0 + \varepsilon V$ with

$$H_0 = E_X(a_X^\dagger a_X + a_{\bar{X}}^\dagger a_{\bar{X}}) + E_\gamma a_\gamma^\dagger a_\gamma, \quad V = \phi_X^\dagger \phi_X \phi_\gamma$$

where a and $a^\dagger$ denote annihilation and creation operators and ϕ_X and ϕ_γ are the field operators $\phi_X = a_X + a_{\bar{X}}^\dagger$, $\phi_\gamma = a_\gamma + a_\gamma^\dagger$. Draw the lowest-order Feynman diagram for the process $X + \bar{X} \to 2\gamma$, and write down the amplitude for this process to occur in time t.

5

Quantum metaphysics

The conceptual structure of quantum mechanics today is as unhealthy as the conceptual structure of the calculus was at the time Berkeley's famous criticism was issued. Hilary Putnam (1965)

The fact that an adequate philosophical presentation [of quantum mechanics] has been so long delayed is no doubt caused by the fact that Niels Bohr brainwashed a whole generation of theorists into thinking that the job was done fifty years ago. Murray Gell-Mann (1979)

As these quotations show, the reader who finds quantum mechanics hard to understand is in good company. In this chapter we will examine some of the main difficulties and describe some proposals for how they should be understood. The chapter also contains some alternative formulations of quantum mechanics which are equivalent (at least mathematically) to the formulation of Chapters 2 and 3.

5.1. Statistical formulations of classical and quantum mechanics
(a) Classical mechanics

In classical mechanics the state of a system is completely specified by the values of the coordinates and momenta $(q_1, \ldots, q_n, p_1, \ldots, p_n)$, which we will denote collectively by (q, p). The set of all possible values of these is called the **phase space** of the system; this is the classical counterpart of the state space in quantum mechanics. We will denote it by $\mathscr{P}$.

In order to compare the statements of classical mechanics with the essentially probabilistic statements of quantum mechanics we will consider a more general specification of the system in which the information one has is not so precise as to determine the state with certainty, but only gives a probability distribution of the values of (q, p). This is a more realistic description of the knowledge available in practice about the state of the system, for any actual method of preparing the system will not lead to infinitely accurate values of q and p. Thus we define an **experimental status** (or simply

status)† to be a probability measure on phase space; by definition this is a non-negative real-valued function defined on appropriate subsets of $\mathscr{P}$, but it will be precise enough to think of it as a non-negative real-valued function ρ on $\mathscr{P}$ satisfying

$$\int_{\mathscr{P}} \rho(q, p) \, dq \, dp = 1 \tag{5.1}$$

if we allow ρ to be concentrated on a single point, i.e. to be a δ-function. Then the measure of a subset $E \subseteq \mathscr{P}$ is

$$\rho(E) = \int_{E} \rho \, dq \, dp. \tag{5.2}$$

The observables in classical mechanics are real-valued functions of q and p. The status ρ determines the probabilities of the values of an observable $A(q, p)$: the probability that the value of A lies in an interval Δ of the real line is

$$P(A \in \Delta) = \int_{A^{-1}(\Delta)} \rho(q, p) \, dq \, dp. \tag{5.3}$$

It follows that the expectation value of A is

$$\langle A \rangle_{\rho} = \int_{\mathscr{P}} A(q, p) \rho(q, p) \, dq \, dp. \tag{5.4}$$

This formula for the expectation value of any observable can be taken as the fundamental statement, for the probability statement (5.3) follows from it by replacing A by $\chi_{\Delta}(A)$, where χ_{Δ} is the characteristic function of Δ:

$$\chi_{\Delta}(A) = \begin{cases} 1 & \text{if } A \in \Delta \\ 0 & \text{if } A \notin \Delta. \end{cases} \tag{5.5}$$

For future reference, we note that the probability distribution ρ is uniquely determined by the expectation values $\langle A \rangle_{\rho}$ for all observables, for by taking A to be the characteristic function χ_{E} of a subset $E \subseteq \mathscr{P}$ one obtains from (5.4) the probability that $(q, p) \in E$. Thus an experimental status could be defined as a function which assigns to each observable A its expectation value $\langle A \rangle_{\rho}$.

If ρ_1 and ρ_2 are two statuses and w_1 and w_2 are two real numbers satisfying

$$0 \leqslant w_1, w_2 \leqslant 1, \quad w_1 + w_2 = 1, \tag{5.6}$$

then

$$\rho = w_1 \rho_1 + w_2 \rho_2 \tag{5.7}$$

is another status. If E_1 and E_2 are the experimental procedures which produce the statuses ρ_1 and ρ_2, the status $w_1 \rho_1 + w_1 \rho_2$ can be produced by tossing a biased coin which has probability w_1 of coming down heads, and following the procedure E_1 if the coin shows heads, E_2 if it shows tails. Such a status is called a **mixed state**. A **pure state** is one which cannot be written in the form (5.7). It is not hard to see that a pure state must be concentrated on a single point of

---

† In mathematical discussions of the foundations of mechanics this is often called a 'state', but this assumes particular answers to some questions which we want to leave open.

phase space, i.e. it must be a δ-function; thus a pure state, corresponding to a point of phase space, is what we previously called simply a 'state'.

A set of elements of a vector space which contains all linear combinations of its members like (5.7), with coefficients satisfying (5.6), is called a **convex set**. An element which cannot be written as such a linear combination is called an **extreme point** of the set. Thus pure states are extreme points of the convex set of experimental statuses.

The time development of a classical system is governed by the motion of the point (q, p) in phase space according to Hamilton's equations (3.5). This causes the probability distribution ρ to change according to the equation

$$\frac{\partial \rho}{\partial t} = \{H, \rho\} \tag{5.8}$$

where $H(q, p)$ is the Hamiltonian of the system and the curly brackets denote the Poisson bracket (3.171).

(b) Quantum mechanics: the statistical operator

The maximum available information about a system, according to quantum mechanics, is contained in the state vector $|\psi\rangle$. To parallel the statistical discussion of classical mechanics, we now consider how to describe less than maximum information in quantum mechanics. Suppose the normalised state vector of a system is one of a number of states $|\psi_1\rangle, \ldots, |\psi_n\rangle$, the probability that it is $|\psi_i\rangle$ being p_i. The **statistical operator** of the system (also known as its **density matrix**) is

$$\rho = \sum_{i=1}^{n} p_i |\psi_i\rangle\langle\psi_i|. \tag{5.9}$$

Since the p_i are real, ρ is a hermitian operator. It yields the probabilities of the values of observables as follows:

●**5.1**   If the statistical operator of a system is ρ, the probability that a measurement of an observable A gives the value α is

$$p_A(\alpha | \rho) = \mathrm{tr}\,(\rho P_\alpha) \tag{5.10}$$

where P_α is the projection onto the eigenspace of A with eigenvalue α. The expectation value of A is

$$\langle A \rangle_\rho = \mathrm{tr}\,(A\rho). \tag{5.11}$$

Proof. Let $|\psi\rangle$ be any normalised state vector, let $X = |\psi\rangle\langle\psi|$, and let Y be any other operator. Then if $\{|\phi_i\rangle\}$ is a complete orthonormal set of states we have

$$\mathrm{tr}\,(XY) = \sum_i \langle\phi_i|XY|\phi_i\rangle = \sum_i \langle\psi|Y|\phi_i\rangle\langle\phi_i|\psi\rangle = \langle\psi|Y|\psi\rangle, \tag{5.12}$$

using (2.68). Now the probability that a measurement of A will give the result α when the statistical operator is ρ is given in terms of the probabilities when the state vector is $|\psi_i\rangle$ by

$$p_A(\alpha \mid \rho) = \sum_i \text{(probability that the state is } |\psi_i\rangle) p_A(\alpha \mid \psi_i)$$

$$= \sum_i p_i \langle \psi_i | P_\alpha | \psi_i \rangle$$

$$= \sum_i p_i \, \text{tr} \, (|\psi_i\rangle\langle\psi_i| P_\alpha) \quad \text{by (5.12)}$$

$$= \text{tr} \, (\rho P_\alpha).$$

Similarly, the expectation value of A is

$$\langle A \rangle_\rho = \sum_i p_i \langle \psi_i | A | \psi_i \rangle = \text{tr} \, (\rho A). \quad \blacksquare$$

By putting $A = 1$ in (5.11) and $P_\alpha = |\psi\rangle\langle\psi|$ in (5.10) we obtain

$$\text{tr} \, \rho = 1 \tag{5.13}$$

and

$$\langle \psi | \rho | \psi \rangle \geqslant 0 \quad \text{for all } |\psi\rangle. \tag{5.14}$$

An operator ρ satisfying (5.14) is said to be **positive**. The set of all positive hermitian operators with trace 1 forms a convex set. A system is said to be in a **pure state** if its statistical operator is an extreme point of this convex set; otherwise it is in a **mixed state**. The meaning of convex linear combinations of statistical operators, and the significance of pure states, is the same as in classical mechanics:

●**5.2**   (i) A situation in which there is probability w_1 that the statistical operator is ρ_1, and probability w_2 that it is ρ_2, is described by the statistical operator $w_1\rho_1 + w_2\rho_2$.

(ii) A system is in a pure state if and only if its statistical operator is of the form $|\psi\rangle\langle\psi|$.

Proof. (i) Suppose

$$\rho_1 = \sum_i p_i |\psi_i\rangle\langle\psi_i|, \quad \rho_2 = \sum_j q_j |\phi_j\rangle\langle\phi_j|.$$

In the situation considered, the probability of the state of the system being $|\psi_i\rangle$ is $w_1 p_i$ and the probability of it being $|\phi_j\rangle$ is $w_2 q_j$. Hence the statistical operator is

$$\rho = \sum_i w_1 p_i |\psi_i\rangle\langle\psi_i| + \sum_j w_2 q_j |\phi_j\rangle\langle\phi_j|$$

$$= w_1\rho_1 + w_2\rho_2. \quad \blacksquare$$

(ii) Since ρ is hermitian, it has a complete set of eigenvectors $|\phi_i\rangle$ and so can be written as

$$\rho = \sum_i p_i |\phi_i\rangle\langle\phi_i| \tag{5.15}$$

with $\sum p_i = 1$ since tr $\rho = 1$. If the state is pure, only one of the p_i can be non-zero, so ρ is of the form $|\psi\rangle\langle\psi|$. Conversely, suppose

$$\rho = |\psi\rangle\langle\psi| = w_1\rho_1 + w_2\rho_2. \tag{5.16}$$

Writing ρ_1 as in (5.15), we have

$$\langle\psi|\rho_1|\psi\rangle = \sum_i p_i |\langle\phi_i|\psi\rangle|^2 \leqslant \left[\sum_i p_i\right]\left[\sum_j |\langle\phi_j|\psi\rangle|^2\right] = 1 \tag{5.17}$$

with equality at the second step only if

$$p_i = |\langle\phi_i|\psi\rangle|^2 = 1 \quad \text{for some } i \tag{5.18}$$

since otherwise there will be positive cross-terms in the product. If (5.18) is true, $|\phi_i\rangle = |\psi\rangle$ and so $\rho_1 = \rho$. Thus

$$\text{Either} \quad \langle\psi|\rho_1|\psi\rangle < 1 \quad \text{or} \quad \rho_1 = \rho. \tag{5.19}$$

Similarly, either $\langle\psi|\rho_2|\psi\rangle < 1$ or $\rho_2 = \rho$. But

$$1 = \langle\psi|\rho|\psi\rangle = w_1\langle\psi|\rho_1|\psi\rangle + w_2\langle\psi|\rho_2|\psi\rangle. \tag{5.20}$$

Since $w_1 + w_2 = 1$, we must have $\langle\psi|\rho_1|\psi\rangle = \langle\psi|\rho_2|\psi\rangle = 1$ and therefore $\rho_1 = \rho_2 = \rho$. Thus $\rho = |\psi\rangle\langle\psi|$ describes a pure state. ∎

From the Schrödinger equation and its hermitian conjugate, we obtain the equation of motion of the statistical operator as

$$i\hbar\frac{\partial\rho}{\partial t} = \sum_i p_i\left\{\left(i\hbar\frac{d}{dt}|\psi_i\rangle\right)\langle\psi_i| + |\psi_i\rangle\left(i\hbar\frac{d}{dt}\langle\psi_i|\right)\right\}$$

$$= \sum_i p_i\{H|\psi_i\rangle\langle\psi_i| + |\psi_i\rangle(-\langle\psi_i|H)\}$$

$$= [H, \rho]. \tag{5.21}$$

The solution (3.2)–(3.3) of the Schrödinger equation gives

$$\rho(t) = e^{-iHt/\hbar}\rho(0)e^{iHt/\hbar}. \tag{5.22}$$

The comparison between classical and quantum mechanics, in this statistical form, is summarised in Table 5.1.

One feature of quantum mechanics which has no counterpart in classical mechanics is the effect of a measurement on the state of the system. In both theories it is permissible to take account of the result of a measurement by changing the status ρ, for the probabilities contained in ρ reflect inadequate knowledge of the system, and by its nature a measurement improves this knowledge. Thus in classical mechanics an exact measurement of q and p increases one's knowledge of the system to the point where it can be described by a δ-function rather than the broader density ρ which was appropriate before the measurement. We suppose, however, that it is not the state of the system that has changed, but only the state of our knowledge: the experimental *status* of the system has changed because the result of the measurement is an extra element in the experimental preparation of the system. It is possible to perform the measurement but not take its result into account: this does not affect any probabilities and the status ρ is unaffected by the measurement.

In quantum mechanics, on the other hand, the measurement has a physical effect on the state of the system, and the status will change even if the result of the measurement is not taken into account. The change in the statistical operator is given by

●**5.3** If the statistical operator of a system is ρ just before a measurement of an observable A, then immediately after the measurement its statistical operator is

$$\rho' = \sum_\alpha P_\alpha \rho P_\alpha \tag{5.23}$$

where P_α is the projection operator onto the eigenspace of A with eigenvalue α, and the sum is over all eigenvalues of A.

Proof. Suppose first of all that the system is in a pure state $|\psi\rangle$ before the measurement, so that $\rho = |\psi\rangle\langle\psi|$. After the measurement, according to Postulates II and III (p. 50–51), it will be in the eigenstate $P_\alpha|\psi\rangle$ with probability $\langle\psi|P_\alpha|\psi\rangle$. The state vector obtained by normalising $P_\alpha|\psi\rangle$ is

$$|\psi_\alpha\rangle = \frac{P_\alpha|\psi\rangle}{\sqrt{\langle\psi|P_\alpha|\psi\rangle}}; \tag{5.24}$$

hence the statistical operator after the measurement is

$$\rho' = \sum_\alpha \langle\psi|P_\alpha|\psi\rangle |\psi_\alpha\rangle\langle\psi_\alpha|$$
$$= \sum_\alpha P_\alpha|\psi\rangle\langle\psi|P_\alpha \tag{5.25}$$

since P_α is hermitian.

Now consider the general case $\rho = \sum p_i|\psi_i\rangle\langle\psi_i|$. With probability p_i the

Table 5.1. *Classical and quantum mechanics*

	Classical	Quantum
Pure state	Phase space point (q, p)	State vector $|\psi\rangle$
General status	Probability density $\rho(q, p)$	Positive hermitian operator ρ
	$\int \rho \, dq \, dp = 1$	$\mathrm{tr}\,\rho = 1$
Condition for pure state	$\rho = \delta$-function	$\rho = |\psi\rangle\langle\psi|$ (ρ has rank 1)
Equation of motion	$\dfrac{\partial\rho}{\partial t} = \{H, p\}$	$ih\dfrac{\partial\rho}{\partial t} = [H, p]$
Observable	Function $A(q, p)$	Hermitian operator A
Expectation value	$\int A\rho \, dq \, dp$	$\mathrm{tr}\,(A\rho)$

statistical operator after the measurement will be

$$\rho_i' = \sum_\alpha P_\alpha |\psi_i\rangle\langle\psi_i| P_\alpha, \tag{5.26}$$

and so the full statistical operator is

$$\rho' = \sum_i p_i \rho_i' = \sum_\alpha P_\alpha \rho P_\alpha. \quad \blacksquare$$

●5.1 and ●5.3 show that all experimental probabilities are determined by ρ. The probability that the system is in a given state, however, is not determined by ρ, for different probability distributions over states may give the same statistical operator. For example, suppose the state space is n-dimensional, and let $|\psi_1\rangle, \ldots, |\psi_n\rangle$ be a complete orthonormal set of states. Suppose the system is known to be in one of these states, but they are all equally probable; then the statistical operator is

$$\rho = \frac{1}{n} \sum |\psi_i\rangle\langle\psi_i| = \frac{1}{n}\mathbf{1}, \tag{5.27}$$

by (2.68). Thus this same statistical operator could be obtained from any orthonormal complete set. It can also be taken to describe complete ignorance of what state the system is in.

The statistical operator can be used to prove

●5.4   Let A and B be two observables such that A commutes with $e^{iHt/\hbar}Be^{-iHt/\hbar}$. Suppose A is measured at time $t=0$ and B is measured at time t. Then the probabilities of the various results of measurement of B are the same as if the measurement of A did not occur.

Proof. Let ρ be the initial statistical operator. If there is no measurement of A the statistical operator at time t is $e^{-iHt/\hbar}\rho e^{iHt/\hbar}$ and the probability that the measurement of B gives the result β is

$$p_1(\beta, t) = \text{tr}\left[e^{-iHt/\hbar}\rho e^{iHt/\hbar}P_\beta\right]$$

where P_β is the projection operator associated with β. On the other hand, if there is a measurement of A at $t=0$ the statistical operator becomes $\sum P_\alpha \rho P_\alpha$ immediately after the measurement (where the sum is over the eigenvalues α of A), and this evolves to $e^{-iHt/\hbar}\sum P_\alpha \rho P_\alpha e^{iHt/\hbar}$ so that the probability of the result β at time t is

$$p_2(\beta, t) = \text{tr}\left[\sum_\alpha P_\beta e^{-iHt/\hbar}P_\alpha \rho P_\alpha e^{iHt/\hbar}\right] = \text{tr}\left[\sum_\alpha P_\alpha e^{iHt/\hbar}P_\beta e^{-iHt/\hbar}P_\alpha \rho\right].$$

Now $e^{iHt/\hbar}P_\beta e^{-iHt/\hbar}$ is a projection operator onto an eigenspace of $e^{iHt/\hbar}Be^{-iHt/\hbar}$ and therefore commutes with A, since the eigenspaces of $e^{iHt/\hbar}Be^{-iHt/\hbar}$ are invariant under the commuting operator A. Hence

$$p_2(\beta, t) = \text{tr}\left[\sum_\alpha P_\alpha{}^2 e^{iHt/\hbar}P_\beta e^{-iHt/\hbar}\rho\right]$$
$$= \text{tr}\left[P_\beta e^{-iHt/\hbar}\rho e^{iHt/\hbar}\right]$$

(since $\sum P_\alpha^2 = \sum P_\alpha = 1$), which is the same as if there had been no measurement of A at $t = 0$. ∎

Combined systems Consider a system composed of two subsystems S and T, with state space $\mathscr{S} \otimes \mathscr{T}$. The statistical operator ρ for the combined system will then be a positive hermitian operator on $\mathscr{S} \otimes \mathscr{T}$. If $\{|\phi_i\rangle\}$ is a complete orthonormal set of states for $\mathscr{S}$ and $\{|\psi_i\rangle\}$ is a similar set for $\mathscr{T}$, the trace of any operator Ω on $\mathscr{S} \otimes \mathscr{T}$ is

$$\mathrm{tr}\,\Omega = \sum_{i,j} \langle\psi_j|\langle\phi_i|\Omega|\phi_i\rangle|\psi_j\rangle. \tag{5.28}$$

Let A be an observable of system T. Regarded as an observable of the combined system, it is represented by an operator $1 \otimes A$ on $\mathscr{S} \otimes \mathscr{T}$; hence its expectation value is

$$\langle A\rangle = \mathrm{tr}\,[\rho(1 \otimes A)] = \sum_{i,j}\langle\psi_j|\langle\phi_i|\rho(1 \otimes A)|\phi_i\rangle|\psi_j\rangle$$

$$= \sum_j \langle\psi_j|\langle\phi_i|\rho|\phi_i\rangle A|\psi_j\rangle = \sum_j \langle\psi_j|\,\mathrm{tr}_S\,(\rho)\cdot A|\psi_j\rangle$$

$$= \mathrm{tr}_T\,[\mathrm{tr}_S\,(\rho)\cdot A] \tag{5.29}$$

where $\mathrm{tr}_S\,(\rho)$ is the operator on $\mathscr{T}$ defined by

$$\mathrm{tr}_S\,(\rho)|\psi\rangle = \sum_i \langle\phi_i|\rho(|\phi_i\rangle|\psi\rangle), \tag{5.30}$$

a bra vector of $\mathscr{S}$ being regarded as a map $\langle\phi|: \mathscr{S} \otimes \mathscr{T} \to \mathscr{T}$ in the obvious way (i.e. taking $|\chi\rangle|\psi\rangle$ to $\langle\phi|\chi\rangle|\psi\rangle$). This can be written as

$$\mathrm{tr}_S\,(\rho) = \sum_i \langle\phi_i|\rho|\phi_i\rangle. \tag{5.31}$$

The operator $\mathrm{tr}_S\,(\rho)$ is called the **partial trace** of ρ. Eq. (5.29) shows that it acts as a statistical operator for system T considered on its own.

5.2. Quantum theory of measurement

The most radical difference between quantum and classical mechanics is the special role played by the process of measurement in quantum mechanics. It is not just that the process of observing a physical system unavoidably changes the properties of the system – that is true in classical physics also, since any observation entails an interaction between the observing apparatus and the observed system – but that the change in question is specially described in the fundamental postulates of the theory. The result is that there are two distinct laws governing the change of state of a system. The first is the Schrödinger equation (Postulate VI), whose writ runs as long as the system is not disturbed by an experiment; this is completely deterministic, giving a unique prediction for the future state of the system if its present state is known. The second is the projection postulate, which operates whenever the system is subjected to an experiment; this is a probabilistic statement, and describes an unpredictable change in the system brought about by the experiment.

This seems very unsatisfactory for a fundamental physical theory: it is at best ill-defined, at worst inconsistent. An apparatus is a physical system, and an experiment is a physical process; it ought therefore to be subject to the Schrödinger equation like any other physical process. How is this to be reconciled with the projection postulate? Alternatively, what defines a physical process as an experiment which obeys the projection postulate rather than the Schrödinger equation?

Bohr's answer to this difficulty was that quantum mechanics only applies to microscopic systems, on a much smaller scale than the apparatus which is used to perform experiments upon them. The apparatus must be macroscopic, large enough for its properties to be directly apprehended by human observers, in the way that is assumed by classical physics. Thus the apparatus is always described by classical concepts. Both the Schrödinger equation and the projection postulate apply only to quantum systems, and the latter comes into play when the quantum system interacts with a classical apparatus.

This view divides the physical world into two sorts of object, quantum and classical, each obeying its own characteristic laws. It is still open to objections like those raised above. How are we to decide whether a particular object is of the classical type or the quantum type? A classical apparatus can be described as a collection of quantum objects; why should quantum mechanics not apply to it?

Let us pursue the idea that as a fundamental theory quantum mechanics should apply to all physical systems, and investigate the quantum mechanics of the measurement process so as to clarify the relation between the Schrödinger equation and the projection postulate. Let $\mathscr{S}$ be the state space of the quantum object on which an experiment is being performed, and let $\mathscr{A}$ be the state space of the experimental apparatus (also regarded as a quantum system). We will consider the development of the combined state of the object and apparatus in the state space $\mathscr{S} \otimes \mathscr{A}$. Let $|\psi_1\rangle, |\psi_2\rangle \in \mathscr{S}$ be two eigenstates of the object corresponding to two different results of the experiment; these results must leave the apparatus in different states $|\alpha_1\rangle$ and $|\alpha_2\rangle$ (describing, say, different positions of a pointer). Suppose the apparatus is initially in a state $|\alpha_0\rangle$. The experiment consists of allowing the object and the apparatus to interact in such a way that if the object state is $|\psi_1\rangle$, then after the experiment the object state will still be $|\psi_1\rangle$ and the apparatus will record the appropriate result, i.e. will be in the state $|\alpha_1\rangle$; and similarly if the state is initially $|\psi_2\rangle$ then the apparatus state changes to $|\alpha_2\rangle$. Thus during the experiment the Hamiltonian must be such that

$$\left. \begin{aligned} e^{-iH\tau/h}(|\psi_1\rangle|\alpha_0\rangle) &= e^{i\theta_1}|\psi_1\rangle|\alpha_1\rangle \\ e^{-iH\tau/h}(|\psi_2\rangle|\alpha_0\rangle) &= e^{i\theta_2}|\psi_2\rangle|\alpha_2\rangle \end{aligned} \right\}, \tag{5.32}$$

where τ is the time taken by the experiment, and θ_1 and θ_2 are phases that may be introduced by the experiment. Now suppose that before the experiment the object is in a state

$$|\psi_0\rangle = c_1|\psi_1\rangle + c_2|\psi_2\rangle; \tag{5.33}$$

then after it the object and apparatus together will be in the state

$$e^{-iH\tau/\hbar}(|\psi_0\rangle|\alpha_0\rangle) = e^{-iH\tau/\hbar}(c_1|\psi_1\rangle|\alpha_0\rangle + c_2|\psi_2\rangle|\alpha_0\rangle)$$
$$= e^{i\theta_1}c_1|\psi_1\rangle|\alpha_1\rangle + e^{i\theta_2}c_2|\psi_2\rangle|\alpha_2\rangle. \tag{5.34}$$

Postulate II applied to this state yields the statement that if an experiment is done on the apparatus to determine the position of the pointer – for example, by photographing it – the result will be α_1 with probability $|c_1|^2$ and α_2 with probability $|c_2|^2$. If the result is α_1 then, according to the projection postulate, the state of the object and the apparatus after the apparatus has been examined is $|\psi_1\rangle|\alpha_1\rangle$ and so the state of the object is $|\psi_1\rangle$.

Thus applying the Schrödinger equation to the combined system of the object and apparatus is equivalent to applying the projection postulate to the object alone. This does not, however, resolve the difficulty attached to the projection postulate, because in order to interpret the result of applying the Schrödinger equation, namely the state (5.34), we had to apply the projection postulate at the level of the apparatus. This could in turn be replaced by a Schrödinger evolution applied to whatever observes the apparatus – the camera, if it is photographed – but again the projection postulate would have to be applied to the state of the camera, the apparatus and the object. This can be continued indefinitely – we can include the developing equipment, and then the eye and then the brain of the experimenter who looks at the film – but clearly we will never get away from the necessity of invoking the projection postulate at some stage, the Schrödinger equation having been applied at all previous stages. We still have to divide the world into a quantum realm and a classical realm, with different laws applying in the two realms. Part of the objection to this division has been removed by this analysis, for it shows that there is no need to give a precise definition of the boundary between quantum and classical – wherever one places the boundary, the results will be the same.

Schrödinger's cat paradox One proposal is that the boundary between the two realms should be placed at the boundary of human consciousness, so that the division between quantum and classical is identified with the division between body and mind. On this view, the full quantum-mechanical state vector should include the states of the quantum object, the experimental apparatus and the brain of the experimenter. The projection postulate is applied when the experimenter becomes aware of a particular brain state. This view assigns a special status to human brains, and is attuned to the philosophical opinion called **Cartesian dualism**, according to which mind and matter are two separate substances, mind having a particular relationship to human brains.

Schrödinger's cat paradox is designed to illustrate the strangeness of this view. Suppose a cat is shut up in a box containing the following 'diabolical device': a single atom of a radioactive substance, with a half-life of one hour, is placed next to a Geiger counter which is wired up so that if it discharges a sealed glass tube is smashed, releasing poison gas which kills the cat. Let $|\psi\rangle$ be the state of the atom before it decays and $|\psi'\rangle$ its state after decay, so that in isolation the state $|\psi\rangle$ would develop in time t to

$$e^{-\gamma t}|\psi\rangle + \sqrt{(1 - e^{-2\gamma t})}|\psi'\rangle \quad (\gamma = \tfrac{1}{2}\log 2 \text{ hr}^{-1}). \tag{5.35}$$

The diabolical Hamiltonian is such that in time t the state of the whole system develops from $|\psi\rangle|\text{cat alive}\rangle$ to

$$e^{-\gamma t}|\psi\rangle|\text{cat alive}\rangle + \sqrt{(1 - e^{-2\gamma t})}|\psi'\rangle|\text{cat dead}\rangle \tag{5.36}$$

(the state $|\text{cat dead}\rangle$ includes a discharged Geiger counter, some broken glass and a smell of burnt almonds). Unless we apply the projection postulate, we cannot say that the cat is either alive or dead at any particular time. According to the view described in the previous paragraph, the projection postulate is not to be applied until a human observer interacts with the system; thus it is only when a would-be rescuer opens the box that the cat has a definite state. Even if *rigor mortis* has set in and the cat-lover deduces that the atom decayed at the beginning of the hour, this version of quantum mechanics insists that the cat only entered the state of death when the box was opened.

The paradox is compounded in an elaboration due to Wigner. Suppose that when another hour has passed after the box was opened, you go into the room to find out the fate of the cat. If you find out by looking for yourself, then you would seem to be in the same position as the other investigator: you will regard the cat as being in a superposition state like (5.36), which is only projected to a state of life or death when the information reaches your consciousness. However, if you find out by asking the other person what they found when they opened the box, you may regard the projection as having occurred at that earlier time: the cat is already either alive or dead, and you are simply finding out which. But you may alternatively regard the other person as part of the physical universe, to be described by a state vector like anything else; in that case they, and the cat, and the atom are all in a superposition state

$$\frac{1}{\sqrt{2}}|\psi\rangle|\text{cat alive}\rangle|\text{observer happy}\rangle$$

$$+ \frac{1}{\sqrt{2}}|\psi'\rangle|\text{cat dead}\rangle|\text{observer sad}\rangle. \tag{5.37}$$

The actual state of the cat then depends on whether the projection postulate should be applied to states of your consciousness or to states of any human consciousness. But then why not feline consciousness?

Clearly this discussion is quite unreal. It is tempting to conclude that a superposition state $|\Phi\rangle + |\Psi\rangle$ means simply that the state is either $|\Phi\rangle$ or $|\Psi\rangle$, and that an experiment just consists of finding out which of these is true. However, this would not allow the interference effects which are typical of quantum mechanics; in the two-slit experiment, for example, where the states

$|\Phi\rangle$ and $|\Psi\rangle$ could be the wave functions describing the passage of the electron through the two slits, the assumption that the state is either $|\Phi\rangle$ or $|\Psi\rangle$ ('the electron went through either hole A or hole B') would not give the interference pattern which results from the superposition $|\Phi\rangle + |\Psi\rangle$. In general, interpreting the state $|\Phi\rangle + |\Psi\rangle$ as 'either $|\Phi\rangle$ or $|\Psi\rangle$' would lose the distinction between $|\Phi\rangle + e^{i\theta}|\Psi\rangle$ for different phases θ. The difference shows clearly in the statistical operators for the two situations: if the state is $2^{-\frac{1}{2}}(|\Phi\rangle + e^{i\theta}|\Psi\rangle)$ the statistical operator is

$$\rho = \tfrac{1}{2}(|\Phi\rangle + e^{i\theta}|\Psi\rangle)(\langle\Phi| + e^{-i\theta}\langle\Psi|)$$
$$= \tfrac{1}{2}|\Phi\rangle\langle\Phi| + \tfrac{1}{2}|\Psi\rangle\langle\Psi| + \tfrac{1}{2}(e^{i\theta}|\Psi\rangle\langle\Phi| + e^{-i\theta}|\Phi\rangle\langle\Psi|) \qquad (5.38)$$

whereas if the state is either $|\Phi\rangle$ or $|\Psi\rangle$ (with equal probabilities) the statistical operator is

$$\rho' = \tfrac{1}{2}|\Phi\rangle\langle\Phi| + \tfrac{1}{2}|\Psi\rangle\langle\Psi|. \qquad (5.39)$$

The interference effects lie in the extra terms in (5.38). This represents a **coherent** superposition of the two states, (5.39) an **incoherent** one, in the same sense as on p. 81.

Nevertheless, it is clearly true that the difference between $|\Phi\rangle + |\Psi\rangle$ and '$|\Phi\rangle$ or $|\Psi\rangle$' is not manifested by macroscopic systems of the kind we have just been considering; interference phenomena are not observed in cats, and if the state of the world is $|\psi\rangle|\text{cat alive}\rangle + |\psi'\rangle|\text{cat dead}\rangle$ then, at least for all practical purposes, we can take this as meaning that either the cat is alive or the cat is dead. We will now see how this can be justified by arguments within quantum mechanics.

Properties of macroscopic apparatus Consider the experiment described on p. 186, in which the apparatus is initially in the state $|\alpha_0\rangle$ and the object is in a superposition state $|\psi\rangle = \sum c_i|\psi_i\rangle$ where $|\psi_i\rangle$ are eigenstates of the experiment. The statistical operator is initially

$$\rho_0 = |\psi\rangle|\alpha_0\rangle\langle\alpha_0|\langle\psi| = |\psi\rangle\langle\psi| \otimes |\alpha_0\rangle\langle\alpha_0|. \qquad (5.40)$$

The partial trace of this, representing the statistical operator of the object alone, is

$$\text{tr}_A(\rho_0) = \sum_n \langle\phi_n|\rho_0|\phi_n\rangle$$

(where $|\phi_n\rangle$ is a complete set of apparatus states)

$$= \sum |\psi\rangle\langle\phi_n|\alpha_0\rangle\langle\alpha_0|\phi_n\rangle\langle\psi|$$
$$= |\psi\rangle\langle\psi|, \qquad (5.41)$$

using $\sum|\phi_n\rangle\langle\phi_n| = 1$ and the fact that $|\alpha_0\rangle$ is normalised. This is just the statistical operator we expect for the object when it is in the state $|\psi\rangle$. After the experiment, when the apparatus state has become correlated to the object state so that the combined state is

$$|\Psi\rangle = \sum c_i e^{i\theta_i}|\psi_i\rangle|\alpha_i\rangle, \qquad (5.42)$$

the statistical operator is

$$\rho = |\Psi\rangle\langle\Psi| = \sum c_i \bar{c}_j e^{i(\theta_i - \theta_j)} |\psi_i\rangle |\alpha_i\rangle\langle\alpha_j|\langle\psi_j| \tag{5.43}$$

whose partial trace is

$$\text{tr}_A(\rho) = \sum_n \langle\phi_n|\rho|\phi_n\rangle = \sum_{ij} c_i \bar{c}_j e^{i(\theta_i - \theta_j)} |\psi_i\rangle\langle\phi_n|\alpha_i\rangle\langle\alpha_j|\phi_n\rangle\langle\psi_j|$$

$$\tag{5.44}$$

$$= \sum c_i \bar{c}_j \delta_{ij} |\psi_i\rangle\langle\psi_j|$$

(since the different apparatus states $|\alpha_i\rangle$ must be orthogonal)

$$= \sum |c_i|^2 |\psi_i\rangle\langle\psi_i|. \tag{5.45}$$

This is the appropriate statistical operator for the object when it has probability $|c_i|^2$ of being in the state $|\psi_i\rangle$. Thus we have

> ●**5.5**   If two systems S and A interact so that states $|\psi_i\rangle$ of S become associated with states $|\alpha_i\rangle$ of A, the statistical operator $\text{tr}_A(\rho)$ of S reproduces the effect of the projection postulate after an experiment on S with eigenstates $|\psi_i\rangle$. ∎

Note that the statistical operator (5.43) does not give licence to conclude that the state of S is one of the $|\psi_i\rangle$, the probabilities being $|c_i|^2$. This can only be done if the projection postulate is applied to the state of the apparatus. (In this connection, it is salutary to recall the comments about the ambiguity of the statistical operator on p. 184.) This status of S, in which it does not have a definite state but is part of a larger system which is in a pure state, is called an **improper mixture**.

Now let us take account of the fact that the apparatus is a macroscopic system. This means that each distinguishable configuration of the apparatus (for example, each position of the pointer) is not a single quantum state but corresponds to a vast number of different quantum states (to say that the pointer is against a certain mark on the scale does not by any means determine the state of motion of each molecule in the pointer). Thus in the above analysis the single apparatus state $|\alpha_0\rangle$ should be replaced by a statistical distribution over microscopic quantum states $|\alpha_{0,s}\rangle$; the initial statistical operator should be not (5.40) but

$$\rho_0 = \sum_s p_s |\psi\rangle|\alpha_{0,s}\rangle\langle\alpha_{0,s}|\langle\psi|. \tag{5.46}$$

Each apparatus state $|\alpha_{0,s}\rangle$ will respond to an eigenstate $|\psi_i\rangle$ of the object by changing to a state $|\alpha_{i,s}\rangle$ which is one of the quantum states whose macroscopic description is that the pointer is in position i; to be precise,

$$e^{iH\tau/\hbar}(|\psi_i\rangle|\alpha_{0,s}\rangle) = e^{i\theta_{i,s}}|\psi_i\rangle|\alpha_{i,s}\rangle. \tag{5.47}$$

The important feature here is the phase factor, which depends on the index s. The differences between the energies of the quantum states $|\alpha_{0,s}\rangle$, in relation to the time τ, are likely to be such that the phases $\theta_{i,s}$ (mod 2π) are randomly distributed between 0 and 2π.

It follows from (5.46) and (5.47), with $|\psi\rangle = \sum c_i |\psi_i\rangle$, that the statistical operator after the experiment is

$$\rho = \sum_{s,i,j} p_s c_i \bar{c}_j e^{i(\theta_{i,s} - \theta_{j,s})} |\psi_i\rangle |\alpha_{i,s}\rangle \langle\alpha_{j,s}| \langle\psi_j|. \tag{5.48}$$

As well as satisfying (5.45), so that it reproduces the projection postulate for the object, this also shows that the projection postulate is effectively satisfied for the apparatus. 'Effectively' here means 'in its consequences for macroscopic observables'. Such an observable will be unable to distinguish between different quantum states with the same macroscopic description, so its matrix elements between $|\psi_i\rangle |\alpha_{i,s}\rangle$ and $|\psi_j\rangle |\alpha_{j,r}\rangle$ will be independent of r and s. If this is true of an observable A, its expectation value will be

$$\mathrm{tr}(\rho A) = \sum_{s,i,j} p_s c_i \bar{c}_j e^{i(\theta_{i,s} - \theta_{j,s})} \langle\alpha_{j,s}| \langle\psi_j| A |\psi_i\rangle |\alpha_{i,s}\rangle$$

$$= \sum_{i,j} c_i \bar{c}_j a_{ij} \sum_s p_s e^{i(\theta_{i,s} - \theta_{j,s})}. \tag{5.49}$$

Because of the random distribution of the phases $\theta_{i,s}$, the sum over s will vanish if $i \neq j$; hence

$$\mathrm{tr}(\rho A) = \sum_i |c_i|^2 a_{ii} = \mathrm{tr}(\rho' A) \tag{5.50}$$

where

$$\rho' = \sum_{i,s} |c_i|^2 p_s |\psi_i\rangle |\alpha_{i,s}\rangle \langle\alpha_{i,s}| \langle\psi_i|. \tag{5.51}$$

This is the statistical operator which is appropriate if the projection postulate is applied to the apparatus in the sense that if the pointer is observed to be in position i, its state will be $|\alpha_{i,s}\rangle$ for some s, and the probability that it is $|\alpha_{i,s}\rangle$ is the same as the probability that its original state was $|\alpha_{0,s}\rangle$. Thus we have

> ●5.6 Suppose a quantum system interacts with a macroscopic apparatus so as to introduce random phases in the apparatus states. Let ρ be the statistical operator of the apparatus after the experiment, calculated according to the Schrödinger equation, and let ρ' be the result of applying the projection postulate to ρ. Then it is not possible to perform an experiment with macroscopic apparatus which will detect the difference between ρ and ρ'. ■

For a wide class of possible apparatus, the randomness of the phases required in ●5.6 has been shown to apply if the apparatus evolves irreversibly so as to form a permanent record of the experiment. This is the **Daneri–Loinger–Prosperi theorem** (Daneri *et al.* 1962).

Continuous observation So far in this section we have concentrated on a single experiment or measurement, which takes place in a short time compared with the natural time development of the quantum system. If this time development is studied by means of an experiment which covers a period during which the state of the system changes appreciably, as in determining the rate of decay of an unstable particle or system, the last part of Postulate III (p. 114) applies. This raises new questions concerning compatibility with the Schrödinger equation.

A period of continuous observation cannot be reduced to a succession of short measurements, as is shown by the following:

> ●**5.7 A watched pot never boils.** Let A be an observable of a quantum system with eigenvalues 0 and 1. Suppose that measurements of A are made at times $0 = t_0, t_1, \ldots, t_N = T$ in the interval $[0, T]$, and that the projection postulate is applied after each measurement. Let p_n be the probability that the measurement at time t_n gives the result 0. Then if $N \to \infty$ in such a way that $\max(p_{n+1} - p_n) \to 0$,
>
> $$p_N - p_0 \to 0 \tag{5.52}$$
>
> (so that if the system is in an eigenstate of A at $t = 0$, it will still have the same value of A at time T).

Proof. Let P_0 be the projection operator onto the eigenspace of A with eigenvalue 0, and let $P_1 = 1 - P_0$ be the projection onto the subspace with eigenvalue 1. Let ρ_n be the statistical operator just before the measurement at time t_n; then by ●5.3 the statistical operator after this measurement is

$$\rho_n' = P_0 \rho_n P_0 + P_1 \rho_n P_1, \tag{5.53}$$

so the statistical operator just before the measurement at time t_{n+1} is

$$\rho_{n+1} = e^{-iH\tau_n} \rho_n' e^{iH\tau_n} \tag{5.54}$$

where H is the Hamiltonian of the system and $\tau_n = t_{n+1} - t_n$. Note that if ρ_n is a sum of k terms of the form $|\psi\rangle\langle\psi|$, ρ_{n+1} is a sum of at most $2k$ such terms; since $\rho_0 = |\psi_0\rangle\langle\psi_0|$, it follows that ρ_n contains a finite number of such terms. From (5.54) we have

$$\rho_{n+1} = \rho_n' - i\tau_n[H, \rho_n'] + O(\tau_n^2). \tag{5.55}$$

Since $P_0^2 = P_0$ and $P_0 P_1 = 0$, this gives

$$P_0 \rho_{n+1} P_0 = P_0 \rho_n P_0 - i\tau_n[P_0 H P_0, P_0 \rho_n P_0] + O(\tau_n^2). \tag{5.56}$$

Hence the probability that the measurement at time t_n gives the result 0 is

$$p_{n+1} = \operatorname{tr}(\rho_{n+1} P_0) = \operatorname{tr}(P_0 \rho_{n+1} P_0) \quad \text{(because } P_0^2 = P_0\text{)}$$
$$= \operatorname{tr}(P_0 \rho_n P_0) - i\tau_n \operatorname{tr}[P_0 H P_0, P_0 \rho_n P_0] + O(\tau_n^2). \tag{5.57}$$

But $P_0 \rho_n P_0$ is the sum of a finite number of terms $|\psi\rangle\langle\psi|$, and for any operator X

$$\operatorname{tr}(X|\psi\rangle\langle\psi|) = \langle\psi|X|\psi\rangle = \operatorname{tr}(|\psi\rangle\langle\psi|X). \tag{5.58}$$

Hence the trace of the commutator in (5.57) vanishes, and

$$p_{n+1} = p_n + O(\tau_n^2). \tag{5.59}$$

Let $\tau = \max(\tau_n)$; then there is a constant k such that

$$p_{n+1} - p_n \leqslant k\tau_n^2 \leqslant k\tau\tau_n \tag{5.60}$$

and therefore

$$p_N - p_0 = \sum_{n=0}^{N-1}(p_{n+1} - p_n) \leqslant k\tau \sum_{n=0}^{N-1} \tau_n = k\tau T \to 0 \quad \text{as } \tau \to 0. \quad \blacksquare$$

In some situations a system is subjected to genuine repeated interactions which determine whether it has changed or not; for example, an unstable particle in a bubble chamber undergoes such an interaction on each encounter with a molecule of the liquid in the bubble chamber. In these circumstances ●5.6 indicates that the decay of the unstable particle will be slowed down by the continual projection of its state vector. The effect is too small to be experimentally detectable.

Aharonov & Vardi (1980) have shown that given any sequence of states there is a continuous sequence of measurements which will force the system to follow the given sequence of states. An interesting feature of this process is that if the sequence of states is a sequence of eigenstates of position, so that it corresponds to a trajectory in space, then the final state acquires the phase e^{iS} of Feynman's postulate, where S is the action of the trajectory.

In situations where the system is continuously coupled to an apparatus which will respond to its decay the second form of Postulate III (p. 114) is used. In terms of the statistical operator, this can be expressed by saying that if P_i are the projection operators onto the eigenspaces of the experiment, the statistical operator at time t is

$$\rho(t) = \sum_i P_i e^{-iHt/\hbar} \rho(0) e^{iHt/\hbar} P_i. \tag{5.61}$$

If this is applied to the apparatus and object together, rather than to the object alone, the effect is approximately the same (i.e. an approximate version of the measurement theorem ●5.5 can be proved; see Sudbery 1984).

The force of ●5.7 is to show that the continuous part of Postulate III (namely (5.61)) cannot be deduced from the first part (the projection postulate, p. 51). It can be argued that ●5.5 shows that (5.61) should be taken as fundamental and the projection postulate deduced from it, since an apparatus can be regarded as continuously observed. However, (5.61) has some unsatisfactory features: it cannot be expressed as a differential equation and (as a result of this) if $\rho(s)$ is substituted for $\rho(0)$ in (5.61), the result is not the same as $\rho(s+t)$. Thus the development of the statistical operator for a time s followed by development for a time t is not the same as development for a time $s+t$. Another aspect of this is that the development of the system from time t is not determined purely by the state at time t, but is also affected by what the state was before time t.

Clearly Postulate III is in a mess. In §5.5 we will discuss proposals that it should be eliminated entirely.

The perplexities discussed in the previous section could be taken as an indication that the quantum-mechanical state vector does not tell us everything there is to know about the state of a system: that there are some further variables, at present hidden from us, whose values will completely specify the state of the system and determine its future behaviour more definitely than quantum mechanics allows. Another argument that such further variables must exist was put forward by Einstein, Podolsky and Rosen in 1935.

The Einstein–Podolsky–Rosen paradox

Consider an electron and a positron which are created together (as in Fig. 1.3) in a state with total spin 0. According to ●4.6 this is the antisymmetric combination of two spin-$\frac{1}{2}$ particles, so the spin state is

$$|\Psi\rangle = \frac{1}{\sqrt{2}}(|\uparrow\rangle|\downarrow\rangle - |\downarrow\rangle|\uparrow\rangle) \tag{5.62}$$

where $|\uparrow\rangle$ and $|\downarrow\rangle$ denote the one-particle eigenstates of the spin component s_z with eigenvalues $+\frac{1}{2}$ and $-\frac{1}{2}$ respectively, and in the two-particle spin state the state of the electron is written as the first factor.

Since a state with zero angular momentum is invariant under rotations, it must retain the form (5.62) whatever axis is used to define the basis of one-particle spin states. Thus we can also write

$$|\Psi\rangle = \frac{1}{\sqrt{2}}(|\rightarrow\rangle|\leftarrow\rangle - |\leftarrow\rangle|\rightarrow\rangle) \tag{5.63}$$

where $|\rightarrow\rangle$ and $|\leftarrow\rangle$ are one-particle eigenstates of s_x.

Now suppose the electron and the positron move in opposite directions until they are separated from each other by a large distance, and then the z-component of the spin of the electron is measured. This is an observable $s_z(e^-)$ of the whole system, and after the measurement the state of the system will be projected onto an eigenstate of this observable: if the measurement gave the value $+\frac{1}{2}$, then the result of the projection will be that the system is in the state $|\uparrow\rangle|\downarrow\rangle$. This means that the positron is in the state $|\downarrow\rangle$, and a measurement of the z-component of its spin, $s_z(e^+)$, will certainly give the value $-\frac{1}{2}$. Now this information about the positron has been obtained by means of an experiment conducted a long way away from the positron, without any possibility of affecting it. Einstein, Podolsky and Rosen argued that this implied that the fact about the positron discovered by the experiment, namely $s_z(e^+) = -\frac{1}{2}$, must have been a real objective fact which was already true before the experiment on the electron.

But now suppose the experiment on the electron measures not the z-component but the x-component of its spin. Then from (5.63) it follows that the state of the system is projected to either $|\rightarrow\rangle|\leftarrow\rangle$ or $|\leftarrow\rangle|\rightarrow\rangle$, so that the positron now has a definite value for the x-component $s_x(e^+)$. Again, this must have been true before the experiment. Hence before the experiment the positron had definite values for both $s_z(e^+)$ and $s_x(e^+)$. But these are incompatible observables, and have no simultaneous eigenstates; so no quantum-mechanical state can assign definite values to both of them. The conclusion drawn by Einstein, Podolsky and Rosen was that the quantum-mechanical description is incomplete, and that there are 'elements of reality' which it does not include.

Before going on to consider the possibility of making quantum mechanics into a more complete theory as this argument requires, let us look more closely at the orthodox quantum-mechanical description of the EPR (Einstein–Podolsky–Rosen) situation. The experiment on the electron leaves the whole system in an eigenstate: $|\uparrow\rangle|\downarrow\rangle$ if $s_z(e^-)$ is measured and the answer is $+\frac{1}{2}$, $|\rightarrow\rangle|\leftarrow\rangle$ if $s_x(e^-)$ is measured and the value is $+\frac{1}{2}$. This means that the positron is now in a definite state, $|\downarrow\rangle$ or $|\leftarrow\rangle$ in these cases, which it was not in before the experiment. It does not mean, however, that the state of the positron has been changed by the experiment on the electron, because the positron did not have a definite state before the experiment. If we insist on describing the positron separately, we can only do so by means of its statistical operator: from (5.62) or (5.63) this (before the experiment) is

$$\rho_{e^+} = \text{tr}_{e^-} |\Psi\rangle\langle\Psi| = \tfrac{1}{2}(|\uparrow\rangle\langle\uparrow| + |\downarrow\rangle\langle\downarrow|) \tag{5.64}$$

$$= \tfrac{1}{2}(|\rightarrow\rangle\langle\rightarrow| + |\leftarrow\rangle\langle\leftarrow|), \tag{5.65}$$

which is $\frac{1}{2}\times$ the identity operator on the two-dimensional spin space. Now consider the statistical operator of the positron immediately after the experiment, before information about its result has had time to reach the vicinity of the positron. If the experiment measured s_z, the state of the positron is either $|\uparrow\rangle$ or $|\downarrow\rangle$, with equal probability, and the statistical operator is (5.64); if the experiment measured s_x, the state of the positron is either $|\rightarrow\rangle$ or $|\leftarrow\rangle$, with equal probability, and the statistical operator is (5.65) – which is the same as in the other case, and the same as before the experiment. Although the three situations – before the experiment, after the s_z experiment, and after the s_x experiment – have different descriptions in terms of states of the positron, they all have the same statistical operator, and there is no observable difference between them. Thus there is no observable action at a distance between the experiment on the electron and the distant positron; in particular, it is not possible to use the EPR experiment to send information faster than light.

The EPR paradox was originally formulated not in terms of spin states but in terms of wave functions. Suppose two simple particles have a wave function

$$\psi(\mathbf{r}_1, \mathbf{r}_2) = \delta(x_1 - x_2)\phi_1(y_1, z_1)\phi_2(y_2, z_2) \tag{5.66}$$

where ϕ_1 and ϕ_2 are well separated wave packets. The x_1-dependence can be written as a superposition of the eigenfunctions $\delta(x_1 - a)$ of $\hat{x}_1$:

$$\delta(x_1 - x_2) = \int_{-\infty}^{\infty} \delta(x_1 - a)\delta(x_2 - a)\, da. \tag{5.67}$$

Then if x_1 is measured and found to have the value a, the projection postulate implies that the wave function of the two-particle system becomes $\delta(x_1-a)\delta(x_2-a)\phi_1\phi_2$, so that x_2 certainly has the value a. On the other hand, we can also write $\delta(x_1-x_2)$ as a superposition of the eigenfunctions $e^{ipx_1/\hbar}$ of $\hat{p}_1$:

$$\delta(x_1-x_2)=\frac{1}{2\pi}\int_{-\infty}^{\infty}e^{ik(x_1-x_2)}\,dk. \tag{5.68}$$

This implies that if p_1 is measured and found to have the value $\hbar k$, the wave function ψ becomes $e^{ikx_1}e^{-ikx_2}\phi_1\phi_2$, and so p_2 certainly has the value $-\hbar k$. The same argument as before then leads to the conclusion that particle 2 has definite position and momentum, in contradiction to the uncertainty principle.

Now let us consider in what ways quantum mechanics could be supplemented so as to make it a more complete theory in the sense of Einstein, Podolsky and Rosen: what hidden variables could there be? The most obvious form for such a complete theory, perhaps, would be to describe quantum particles as made up of smaller constituents, whose positions and velocities might constitute the hidden variables. The behaviour of the quantum particle might be rigidly determined by the precise configuration of these constituents; the reason for the probabilistic nature of our present laws would then be our lack of knowledge of this configuration. However, although constituents of quantum particles have indeed been discovered (namely quarks), there is no experimental indication that they behave in any way differently from quantum particles themselves.

The characteristic features of quantum mechanics (particularly the interference effects between probabilities) make it difficult to construct hidden-variable theories along the lines just suggested. Indeed, for some time it was widely thought to have been proved (by von Neumann) that no such theory could reproduce all the consequences of quantum mechanics. This proof, however, was mistaken, as the following counter-example shows.

The de Broglie/Bohm pilot wave theory   Consider a single simple particle moving in space in a potential $V(\mathbf{r})$. Suppose that the particle is described at time t not only by a wave function $\psi(\mathbf{r},t)$ but also by a vector $\mathbf{q}(t)$, and that the wave function satisfies the usual Schrödinger equation

$$i\hbar\frac{\partial\psi}{\partial t}=-\frac{\hbar^2}{2m}\nabla^2\psi+V\psi, \tag{5.69}$$

while the vector $\mathbf{q}$ satisfies

$$\frac{d\mathbf{q}}{dt}=\frac{\mathbf{j}(\mathbf{q},t)}{\rho(\mathbf{q},t)} \tag{5.70}$$

where $\mathbf{j}$ and ρ are the probability current and density introduced in §3.1:

$$\mathbf{j}=\frac{\hbar}{m}\,\mathrm{Im}\,[\bar{\psi}\,\nabla\psi],\quad \rho=|\psi|^2. \tag{5.71}$$

Now suppose that at time $t=0$ we have a large number of such particles all with the same wave function $\psi(\mathbf{r},0)$ but with varying values of $\mathbf{q}$, the proportion of the particles for which this value lies in a region dV containing $\mathbf{q}$ being $\rho(\mathbf{q}, 0)\, dV$. Let this proportion at time t be $\sigma(\mathbf{q}, t)$; then if we think of $\mathbf{q}$ as the position of a particle we can regard the whole collection of particles as a fluid with density σ and velocity distribution $\mathbf{u} = \mathbf{j}/\rho$, according to (5.70). These must satisfy the equation of continuity

$$\frac{\partial \sigma}{\partial t} + \mathbf{V} \cdot (\sigma \mathbf{u}) = 0, \tag{5.72}$$

i.e.

$$\frac{\partial \sigma}{\partial t} = -\mathbf{V} \cdot \left(\frac{\sigma \mathbf{j}}{\rho} \right). \tag{5.73}$$

This has a unique solution $\sigma(\mathbf{r}, t)$ if $\sigma(\mathbf{r}, 0)$, $\mathbf{j}(\mathbf{r}, t)$ and $\rho(\mathbf{r}, t)$ are given. But $\sigma = \rho$ satisfies the equation, since then it becomes the continuity equation (3.42) which was shown to be a consequence of the Schrödinger equation (5.69). Hence if the distribution of the values of $\mathbf{q}$ among the particles is given by ρ at time $t = 0$, it is given by ρ at all later times.

Thus we can suppose that every particle whose wave function satisfies the Schrödinger equation (5.69) has a definite position vector $\mathbf{q}$, and that all our experimental arrangements happen to produce particles with a particular distribution of positions: of those particles whose wave function is ψ, a proportion $|\psi(\mathbf{q})|^2\, dV$ lie in the volume dV at the point $\mathbf{q}$. This will be true if the experimental arrangement that produces particles with wave function ψ has probability[†] $|\psi|^2\, dV$ of producing a particle in the volume dV. If ψ and $\mathbf{q}$ develop according to the deterministic equations (5.69)–(5.70), this statement of distribution will remain true at all times if it is true at any one time.

If this is to be taken seriously as a theory of quantum particles, it must show how to realise the possibility that it admits of having a collection of particles with the same wave function ψ but with a different distribution in space from the usual $|\psi|^2$. There are no experimental indications whatsoever that this can be done. However, the significance of this idea is not so much that it is a serious deterministic theory as that it shows that such a theory can be compatible with quantum mechanics.

The theory can be extended to discrete quantum numbers like spin, essentially by linking these to position variables in the apparatus by which they are measured (Bell 1982). It can also be extended to deal with several particles. When this is done a strange feature becomes apparent. Consider the case of two particles: the variables are $\mathbf{q}_1$, $\mathbf{q}_2$ and a two-particle wave function $\psi(\mathbf{r}_1, \mathbf{r}_2)$, and the equations of motion are the Schrödinger equation together with

$$\frac{d\mathbf{q}_1}{dt} = \frac{\mathbf{j}_1}{\rho}, \quad \frac{d\mathbf{q}_2}{dt} = \frac{\mathbf{j}_2}{\rho} \tag{5.74}$$

† The argument can be framed entirely in terms of probability rather than proportions of a large number of particles, but the latter idea is very useful as an aid to thought in dealing with probability.

where
$$\mathbf{j}_1 = (\hbar/m) \operatorname{Im} [\bar{\psi} \, \mathbf{V}_1 \psi], \quad \mathbf{j}_2 = (\hbar/m) \operatorname{Im} [\bar{\psi} \, \mathbf{V}_2 \psi], \quad \rho = |\psi|^2.$$

Here $\mathbf{j}_1$, and therefore $d\mathbf{q}_1/dt$, can be a function of $\mathbf{q}_2$: the motion of the first particle depends on the position of the second particle. Thus there is an action at a distance between the two particles, and this is true even if the potential $V(\mathbf{r}_1, \mathbf{r}_2)$ contains no forces between the particles. It is a reflection of a correlation between the particles produced purely by the formalism of quantum mechanics, if the wave function is suitably chosen. In particular, the EPR wave function (5.66) will produce such a correlation between separated particles.

We will now see that this action at a distance is an inescapable feature of hidden-variable theories which reproduce the predictions of quantum mechanics.

Bell's inequalities   We will consider a situation in which experiments are performed on two separated particles, and draw consequences from the assumption that the result of an experiment on one of the particles is determined by the nature of that experiment alone, and is not affected by any experiment that may be performed on the other particle. This is the assumption of **locality**. We will find that it leads to restrictions on the possible correlations between experiments on the two particles which are not satisfied by some of the predictions of quantum mechanics.

In principle there is no connection between locality and determinism. Locality could be a property of a theory which only gave probabilities for the results of experiments, in the following way. Suppose that all probabilities are determined by a number of variables which we will denote collectively by λ (in the case of two separated particles, these could include variables describing the particles individually and also variables describing general conditions affecting them both). Then for any experiment E there will be a probability $p_E(\alpha \mid \lambda)$ of getting the result α when the variables have the values λ. The theory is **local** if experiments E and F which are separated in space are probabilistically independent; according to **P3** (p. 42), this implies that

$$p_{E\,\&\,F}(\alpha \, \& \, \beta \mid \lambda) = p_E(\alpha \mid \lambda) p_F(\beta \mid \lambda). \tag{5.75}$$

However, any local theory which reproduces the predictions of quantum mechanics for the separated spin-$\frac{1}{2}$ particles in the EPR experiment is equivalent to a deterministic theory for this situation. Let E be the measurement of the spin component of the electron in a certain direction, and let F be the measurement of the spin component of the distant positron in the same direction. Let $\uparrow$ and $\downarrow$ denote the two possible results. Then, since the total spin is zero, we know that E and F always have different results; in probabilistic terms,

$$p_{E\,\&\,F}(\uparrow \, \& \, \uparrow) = p_{E\,\&\,F}(\downarrow \, \& \, \downarrow) = 0. \tag{5.76}$$

Let $\rho(\lambda)$ be a probability density giving the probability that the variables have the value λ; then the total probability of (5.76) is given by

$$p_{E\&F}(\uparrow \& \uparrow) = \int p_{E\&F}(\uparrow \& \uparrow|\lambda)\rho(\lambda)\,d\lambda$$

$$= \int p_E(\uparrow|\lambda)p_F(\uparrow|\lambda)\rho(\lambda)\,d\lambda. \tag{5.77}$$

If this is 0, the integrand, being positive, must vanish everywhere:

$$\text{either} \quad \rho(\lambda)=0 \quad \text{or} \quad p_E(\uparrow|\lambda)=0 \quad \text{or} \quad p_F(\uparrow|\lambda)=0. \tag{5.78}$$

Similarly,

$$\text{either} \quad \rho(\lambda)=0 \quad \text{or} \quad p_E(\downarrow|\lambda)=0 \quad \text{or} \quad p_F(\downarrow|\lambda)=0. \tag{5.79}$$

Since E has only the two possible results $\uparrow$ and $\downarrow$,

$$p_E(\uparrow|\lambda)=0 \;\Leftrightarrow\; p_E(\downarrow|\lambda)=1. \tag{5.80}$$

It follows from (5.78)–(5.80) that if $\rho(\lambda)\neq 0$, all four probabilities must be 0 or 1. So for all values of λ which are actually possible, the results of the experiments are completely determined by λ.

Thus, if we assume that the probability distribution of the hidden variables is not affected by what experiments are to be performed on the particles, we need only consider deterministic theories. Suppose that the two separated particles can each be subjected to one of three experiments A, B, C, each of which has two possible results (say, 'positive' and 'negative'). Then in a deterministic local theory the outcome of experiment A on particle 1 is determined by a property of the system which we will denote by a_1; this is a variable which can take values $+$ and $-$. Similarly we have variables b_1, c_1, a_2, b_2, c_2. Suppose now that experiment A always gives opposite values for the two particles; then $a_1 = -a_2$. Similarly, if B and C always give opposite results for the two particles we have $b_1 = -b_2$ and $c_1 = -c_2$.

Now consider particles which are produced with a fixed probability that they will have a particular set of values of a, b and c. Let $P(a=1, b=1)$ denote the probability that a particle has the specified values of a and b. Then

$$P(b=1, c=-1) = P(a=1, b=1, c=-1) + P(a=-1, b=1, c=-1)$$

$$\leqslant P(a=1, b=1) + P(a=-1, c=-1). \tag{5.81}$$

Hence when pairs of particles are produced with opposite values of a, b and c,

$$P(b_1=1, c_2=1) \leqslant P(a_1=1, b_2=-1) + P(a_1=-1, c_2=1). \tag{5.82}$$

Each of the terms in this inequality is a probability for the outcomes of experiments on different particles, and so the inequality can be tested even if the experiments A, B, C cannot be performed simultaneously on a single particle.

The inequality (5.82) is violated by the probabilities calculated from quantum mechanics in the following case. Suppose the two particles are spin-$\frac{1}{2}$ particles produced in a state of total spin 0, like the electron and positron considered at the beginning of this section; then we know that a measurement of the component of spin in any given direction will give opposite results for the two particles. Let A, B, C be measurements of the components of spin along three coplanar axes, with an angle θ between the axes of A and B and an angle ϕ between those of B and C. Let us calculate the term $P(b_1 = 1, c_2 = 1)$ on the left-hand side of (5.82): this is to be interpreted as the probability that measurements of spin components of particles 1 and 2, along axes at an angle ϕ to each other, both give the result $+\frac{1}{2}$. Take the axis for particle 1 to be the z-axis; then if the measurement of its spin component gives the result $+\frac{1}{2}$, particle 1 is left in the eigenstate $|\uparrow\rangle$ and particle 2 in the eigenstate $|\downarrow\rangle$. The eigenstates of the measurement on particle 2 are obtained by rotating $|\uparrow\rangle$ and $|\downarrow\rangle$ through the angle ϕ (about the x-axis, say); thus the eigenstate with eigenvalue $+\frac{1}{2}$ is

$$|+(\phi)\rangle = e^{-i\phi J_x}|\uparrow\rangle = (\cos \tfrac{1}{2}\phi + 2iJ_x \sin \tfrac{1}{2}\phi)|\uparrow\rangle$$
$$= \cos \tfrac{1}{2}\phi|\uparrow\rangle + i \sin \tfrac{1}{2}\phi|\downarrow\rangle \qquad (5.83)$$

using (4.46) and (4.38). Hence the probability we are looking for is

$$P(b_1 = 1, c_2 = 1) = \tfrac{1}{2}|\langle +(\phi)|\downarrow\rangle|^2 = \tfrac{1}{2}\sin^2 \tfrac{1}{2}\phi \qquad (5.84)$$

(since the probability of the result $+\frac{1}{2}$ for particle 1 is $\frac{1}{2}$). Similarly,

$$P(a_1 = 1, b_2 = -1) = \tfrac{1}{2}\cos^2 \tfrac{1}{2}\theta$$

and

$$P(a_1 = 1, c_2 = 1) = \tfrac{1}{2}\cos^2 \tfrac{1}{2}(\theta + \phi).$$

Thus (5.82) becomes

$$\sin^2 \tfrac{1}{2}\phi \leqslant \cos^2 \tfrac{1}{2}\theta + \cos^2 \tfrac{1}{2}(\theta + \phi)$$

or

$$\cos \theta + \cos \phi + \cos (\theta + \phi) \geqslant -1. \qquad (5.85)$$

This is violated if $\theta = \phi = 3\pi/4$.

To summarise,

> ●**5.8   Bell's theorem.** Suppose two separated particles can each be subjected to one of three two-valued experiments, and that when the same experiment is performed on both particles it always gives opposite results. If the particles are described by a local theory, and if the probabilities of their properties are not affected by what experiments are going to be performed on them, the probabilities of the results of the experiment satisfy the inequality (5.82).
>
> This inequality is violated in quantum mechanics by the system of two spin-$\frac{1}{2}$ particles having total spin 0. ■

Bell's inequality (5.82) has been tested by a number of experiments, all but one of which have shown (with one mild extra assumption) that it is violated (see Clauser and Shimony 1978). Note that quantum mechanics predicts greater correlation between the particles than local theories; the effect of any experimental inefficiency would be to destroy the correlation, so the observation of a violation of (5.82) is more significant than a failure to observe it.

The experiments have been performed with electron–positron pairs (as described in this section), with protons, and with polarised photons (see problem 5.9). The condition that the state of the particles should not be affected by the measurements to be made on them has been guaranteed by an experiment on pairs of photons by Aspect, Dalibard & Roger (1982) in which one photon may encounter either apparatus A or apparatus B; the change from one to the other occurs while the photons are in flight. This is shown in Fig. 5.1.

5.4. Alternative formulations of quantum mechanics

The mathematical apparatus of quantum mechanics contains an assortment of objects – state vectors, inner products, hermitian operators, unitary operators, groups of invariances, ... All this ironmongery is connected together in various ways. From the usual point of view, which is the one adopted in this book, these connections are seen as definitions of all the other objects in terms of state vectors, which are regarded as basic. It is possible, however, to move around the apparatus and look at it from a different direction; then one of the other objects might appear as basic. Any of the objects in the above list can in fact be taken as fundamental and the others defined in terms of it. The problems of interpretation posed by quantum mechanics can take on a slightly different complexion if the mathematical formalism is swung round, so we will briefly describe some of these other formulations before discussing possible interpretations.

Algebraic formulations

Instead of starting with state vectors as the basic concept, it is possible to start with operators. There are already elements of this in the approach to particular quantum systems in this book. In Postulate IV, for example, describing the system of a single simple particle, we did not describe the state space of the system and specify how the operators $\hat{x}_i$ and $\hat{p}_i$ act on this space; instead we postulated some equations satisfied by the operators (namely the

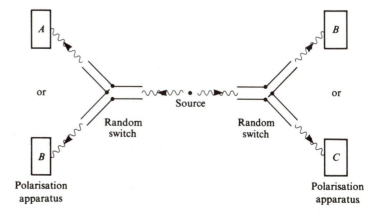

Fig. 5.1.
The Aspect experiment.

canonical commutation relations) and deduced the properties of the state vectors from these. In this case, because of the Stone/von Neumann theorem, the state space is essentially uniquely determined by the commutation relations of the operators; in other cases (such as angular momentum operators) this is not so, but it is still true that the fundamental statements take the form of relations between operators which do not refer to the space they act on. They may say that the operators form a certain group like the rotation group, or that they form a certain Lie algebra like the canonical commutation relations or the angular momentum commutation relations (as we have seen, these are closely related). In either case the algebraic relations between the operators pose a problem which is solved by finding a linear representation of the algebraic system.

An approach due to Jordan takes the basic objects to be hermitian operators in their capacity of observables. The basic algebraic operations between them are modelled on the operations of addition and multiplication which are defined also for classical observables; this makes it possible to formulate a common theory of mechanics which includes both classical and quantum mechanics, the difference between them appearing as that between different structures in the same algebraic theory. In terms of orthodox quantum mechanics the product of two observables A and B cannot be described by the operator product $\hat{A}\hat{B}$, since this is not hermitian if A and B do not commute. Because of this Jordan took the product of two observables A and B to be the observable whose corresponding operator is

$$\widehat{A * B} = \tfrac{1}{2}(\hat{A}\hat{B} + \hat{B}\hat{A}). \tag{5.86}$$

This product is commutative and satisfies

$$A * (A^2 * B) = A^2 * (A * B) \tag{5.87}$$

where $A^2 = A * A$. A **Jordan algebra** is a vector space on which is defined a commutative (but not necessarily associative) bilinear multiplication $A * B$ satisfying (5.87); thus it is an algebra whose product is modelled on the anticommutator of operators in the same way as the product in a Lie algebra is modelled on the commutator†.

In a Jordan algebra of hermitian operators it is also true that

$$A^2 + B^2 = 0 \;\Rightarrow\; A = B = 0. \tag{5.88}$$

A Jordan algebra satisfying this condition is called **formally real**. In a formally real Jordan algebra the powers of a single element obey the associative law; this means that it is possible to define functions of an observable to correspond to functions of a real variable.

A theory of mechanics can now be characterised in terms of a set of observables Ω which form a Jordan algebra with an identity element 1. The

---

† Not all Jordan algebras can be represented by linear operators with the Jordan product being given by the anticommutator. But those that cannot are exceptional: there is only one such algebra among the finite-dimensional formally real Jordan algebras which are algebraically simple (have no non-trivial ideals).

values of the observables are introduced through the notion of a status (usually called a 'state', but this can be confusing). A status is defined to be linear map σ: $\Omega \to \mathbb{R}$ satisfying

$$\sigma(A^2) \geqslant 0, \quad \sigma(1) = 1; \tag{5.89}$$

$\sigma(A)$ is to be interpreted as the expectation value of A in the status σ. Then the statuses form a convex set. A pure state is defined to be an extreme point of this set, as in §5.1.

Suppose for the moment that the Jordan algebra Ω is finite-dimensional. Then if it is isomorphic to an algebra of hermitian operators on some vector space, a status σ is necessarily of the form

$$\sigma(A) = \operatorname{tr}(A\rho) \tag{5.90}$$

where ρ is a hermitian operator satisfying $\operatorname{tr}\rho = 1$, and σ is a pure state if and only if $\rho = |\psi\rangle\langle\psi|$ for some vector $|\psi\rangle$. Thus the state vectors can be recovered from the structure of the algebra of observables. On the other hand, if the Jordan algebra Ω is associative it must be isomorphic to an algebra of functions on some finite set X, and a status must be of the form

$$\sigma(A) = \sum_{x \in X} A(x)\rho(x) \tag{5.91}$$

for some function ρ on X; σ is a pure state if and only if $\rho(x)$ is non-zero for just one element $x \in X$.

In the infinite-dimensional case topological conditions must be imposed on the Jordan algebra. Then (5.91) can be replaced by

$$\sigma(A) = \int_X A(x)\rho(x)\, dx \tag{5.92}$$

where $\rho(x)\, dx$ is a probability measure on the set X; thus the phase space of classical mechanics can be recovered from an associative algebra of observables. If the Jordan algebra Ω is isomorphic to an algebra of operators on a state space, the pure states on Ω are not in one-to-one correspondence with state vectors but can include extra elements corresponding to.eigenbras of an operator with a continuous spectrum.

Modern work along these lines tends to concentrate on C^*-algebras, in which the product is modelled on the simple operator product. These are particularly useful in dealing with systems with infinitely many degrees of freedom, i.e. field theories.

Quantum logic An awkward feature of the use of state vectors to describe physical states is that the correspondence between them is not one-to-one. If we want a single mathematical object to describe a physical state we must declare all multiples of a particular state vector to be equivalent to each other, and work with the equivalence classes so formed. These are the one-dimensional subspaces, or **rays**, of state space, and they form the objects of a well-established mathematical theory, namely **projective geometry**.

A **projective space** can be defined to be the set of one-dimensional subspaces of a vector space, which in this context are called **points**. A **line** in the projective space is the set of rays contained in a two-dimensional subspace of the vector space, a **plane** is the set of rays contained in a three-dimensional vector subspace, and so on. This can be visualised by taking the vector space to be $\mathbb{R}^3$ and identifying each ray with the point in which it intersects a fixed plane in $\mathbb{R}^3$ which does not contain the origin (this omits some rays, namely those which lie in the two-dimensional vector subspace parallel to the fixed plane; these make up the 'line at infinity').

Alternatively, a projective space can be characterised intrinsically, without starting from a vector space, by means of the axioms of projective geometry. For a projective plane, these are:

Π1   Any two points lie on a unique line.
Π2   Any two lines meet in a unique point.
Π3   There is a line and a point not on it.
Π4   Every line contains at least three points.

These axioms contain only the undefined terms 'line', 'point' and 'lie on' ('meet' and 'contains' are to be regarded as defined in terms of 'lie on'). Thus the essential structure of a projective plane can be given by listing all the points and all the lines and saying which points lie on which lines. Let us write $P < l$ if P is a point and l is a line, and P lies on l; then a projective plane consists of a set $\mathscr{L}'$ (points and lines) together with a relation $<$ which holds between some pairs of elements of the set. We can use $\leqslant$ to mean '$<$ or $=$', as usual. Let us add two more elements 0 and Π to $\mathscr{L}'$ to get a set $\mathscr{L}$, and specify $0 \leqslant x$ and $x \leqslant \Pi$ for every element x of $\mathscr{L}$; then the relation $\leqslant$ satisfies

L1 $x \leqslant x$ for all $x \in \mathscr{L}$;

L2 $x \leqslant y \ \& \ y \leqslant z \ \Rightarrow \ x \leqslant z$;

L3 $x \leqslant y \ \& \ y \leqslant x \ \Rightarrow \ x = y$;

L4   Given $x, y \in \mathscr{L}$ there is an element $x \vee y$ such that

$$x \leqslant x \vee y, \quad y \leqslant x \vee y$$

and

$$x \leqslant z \ \& \ y \leqslant z \ \Rightarrow \ x \vee y \leqslant z;$$

L5   Given $x, y \in \mathscr{L}$ there is an element $x \wedge y$ such that

$$x \wedge y \leqslant x, \quad x \wedge y \leqslant y$$

and

$$z \leqslant x \ \& \ z \leqslant y \ \Rightarrow \ z \leqslant x \wedge y.$$

A set $\mathscr{L}$ with a relation $\leqslant$ satisfying **L1–L5** is called a **lattice**. The elements $x \vee y$ and $x \wedge y$ are called respectively the **join** (or **least upper bound** or **l.u.b.**) and the **meet** (or **greatest lower bound** or **g.l.b.**) of x and y. In a projective plane $x \wedge y$ is the line containing x and y if they are distinct points; if x is a point and y is a line, $x \vee y$ is y if x lies on y, otherwise it is Π; and if x and y are distinct lines, $x \vee y$ is Π. Similarly, $x \wedge y$ is the point of intersection of x and y if they are distinct lines; if x is a point and y is a line, $x \wedge y$ is x if x lies on y, otherwise it is 0; and if x and y are distinct points, $x \wedge y$ is 0.

A possible set of axioms for a projective space (of any dimension) can be obtained by replacing **Π2** by

Π2′ A line and a point not on it lie in a unique plane.

If the projective space S is isomorphic to the space of rays in a vector space over a field (or other algebraic structure) F, then both F and the dimension of the vector space can be recovered from the geometric structure of S. (The algebraic nature of F is determined by the geometric properties of S: for example, if Desargues's theorem is true in S then F is associative, and if Pappus's theorem is true in S then F is a field.)

The lattice description of a projective plane can be extended to a projective space S of any dimension by adding the planes, 3-spaces, . . . of S as elements of the lattice. If S is the set of rays in a vector space V, the lattice can then be described as the lattice of subspaces of V, ordered by inclusion, the meet of two subspaces being their intersection and the join being their linear sum:

$$M \leqslant N \Leftrightarrow M \subseteq N, \quad M \wedge N = M \cap N, \quad M \vee N = M + N \tag{5.93}$$

where $M + N = \{u + v : u \in M, v \in N\}$, and M and N are subspaces of V.

Lattices also occur in the contexts of set theory and logic, which are closely related to each other. The set of all subsets of a set forms a lattice if it is ordered by inclusion, the meet and join being the intersection and union of subsets:

$$S \leqslant T \Leftrightarrow S \subseteq T, \quad S \wedge T = S \cap T, \quad S \vee T = S \cup T. \tag{5.94}$$

A set of propositions, ordered by implication, forms a lattice if it is closed under conjunction ('and') and disjunction ('or'), which give the meet and join:

$$P \leqslant Q \Leftrightarrow P \text{ implies } Q, \quad P \wedge Q = P \,\&\, Q, \quad P \vee Q = P \text{ or } Q. \tag{5.95}$$

(In order to satisfy **L3**, the elements of the lattice must be taken to be equivalence classes of propositions, with P equivalent to Q if $P \Leftrightarrow Q$.)

The lattices (5.94) and (5.95) differ from the subspace lattice (5.93) in that they satisfy the distributive law

$$x \wedge (y \vee z) = (x \wedge y) \vee (x \wedge z). \tag{5.96}$$

This fails for subspaces of a vector space, as can be seen by taking M, N, P to be three one-dimensional subspaces of a two-dimensional space, as in Fig. 5.2. In this situation we have $M \wedge N = M \wedge P = 0$, so the right-hand side of (5.96) is 0; but $N + P$ is the whole space, so the left-hand side is M.

The lattices of subsets and propositions both have elements 0 and 1 satisfying $0 \leqslant x$ and $x \leqslant 1$ for all x; for subsets of X the empty set is 0 and the whole set X is 1, while for propositions the class of contradictions is 0 and the class of tautologies is 1. It is also true in these lattices that every element x is associated with an element x' such that

OL1 $(x')' = x$;
OL2 $x \vee x' = 1$;
OL3 $x \leqslant y \Leftrightarrow y' \leqslant x'$

(for subsets of X, S' is the complement $X - S$; for propositions P' is the negation 'not P'). Such a lattice is said to be **orthocomplemented**. The lattice of subspaces of a vector space V is orthocomplemented if V has an inner product: M' is the **orthogonal complement** $M^{\perp} = \{u : \langle u, v \rangle = 0, \forall v \in M\}$.

Now propositions can be associated with subspaces of the state space of a quantum system by considering, for any subspace M, the orthogonal projection onto M. This is a hermitian operator P_M with eigenvalues 0 and 1, and therefore represents an observable which can take these values; we associate M with the proposition 'P_M takes the value 1' ('$P_M = 1$' for short). This proposition is true when the system is in a state belonging to the subspace M. If N is another subspace, the intersection $M \cap N$ contains simultaneous eigenstates of P_M and P_N, for which the propositions '$P_M = 1$' and '$P_N = 1$' are both true. These similarities between the lattice of subspaces and a lattice of propositions led Birkhoff and von Neumann to suggest that the two lattices should have the same interpretation in all respects, so that the join of two subspaces should correspond to the disjunction of the corresponding propositions: the subspace $M + N$ corresponds to the proposition '$P_M = 1$ or $P_N = 1$'. The failure of the distributive law then shows that propositions in quantum mechanics do not obey classical logic, and this is held to account for our difficulties in understanding quantum mechanics.

To see the relevance of this approach to some of the problems we have been discussing in this chapter, consider two state vectors $|\phi\rangle$ and $|\psi\rangle$. The one-dimensional subspaces P and Q containing these are identified with propositions which we can state as 'The system is in the state $|\phi\rangle$' or '...$|\psi\rangle$'. The two-dimensional subspace N spanned by $|\phi\rangle$ and $|\psi\rangle$ is the lattice join $P \vee Q$, which is to be identified with the disjunction of the corresponding propositions: 'The system is in the state $|\phi\rangle$ or it is in the state $|\psi\rangle$'. This is to be regarded as true whenever the state of the system belongs to N, i.e. whenever it

Fig. 5.2.
Failure of the distributive law.

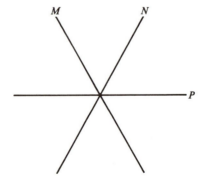

is a superposition of $|\phi\rangle$ and $|\psi\rangle$. In §5.1 we rejected this interpretation of superposition on the grounds that it would lead to the disappearance of the interference effects of quantum mechanics, but this argument fails if the distributive law is not used. Let us examine the argument concerning the two-slit experiment. Here $|\phi\rangle$ and $|\psi\rangle$ are the wave functions $\phi(\mathbf{r})$ and $\psi(\mathbf{r})$ describing waves radiating from the slits A and B respectively. When both slits are open, and the electron is intercepted on a fluorescent screen, an interference pattern is observed on the screen. In this situation the wave function is $\phi + \psi$; the interpretation being questioned is that the electron is either in the state $|\phi\rangle$ or in the state $|\psi\rangle$, i.e. it passes through either slit A or slit B. The ensuing argument can be phrased as follows:

$$\text{The wave function is } \phi + \psi \text{ and there is interference.} \tag{1}$$

∴ By assumption, the electron passes through slit A or slit B and there is interference; \hfill (2)

∴ Either the electron passes through slit A and there is interference

or it passes through slit B and there is interference. \hfill (3)

But both alternatives of (3) are false;

∴ The assumption in (2) must be false.

The step from (2) to (3) clearly uses the distributive law

$$(P \text{ or } Q) \text{ and } R \quad \equiv \quad (P \text{ and } R) \text{ or } (Q \text{ and } R). \tag{5.97}$$

The quantum-mechanical lattice of subspaces of state space has, instead of the distributive law, the weaker law

$$x \leqslant y \;\Rightarrow\; (x \vee y') \wedge y = x. \tag{5.98}$$

A lattice in which this holds is called **orthomodular**. Two elements of an orthomodular lattice are **compatible** if

$$(x \wedge y) \vee (x \wedge y') = x. \tag{5.99}$$

If this holds the sublattice generated by x and y is distributive. A similar result is true for the sublattice generated by any number of compatible elements. The significance of this in quantum mechanics is that if a number of propositions can have their truth or falsity decided by the same experimental arrangement, then they obey classical logic. The **centre** of a lattice is the set of elements which are compatible with all elements of the lattice. A lattice is **irreducible** if its centre is $\{0, 1\}$.

An **atom** in a lattice is an element a such that

$$x \leqslant a \;\Rightarrow\; x = 0 \text{ or } x = a. \tag{5.100}$$

A lattice is **atomic** if for every element x there is an atom a with $a \leqslant x$. Thus the lattice of a projective geometry is atomic, the atoms being points; so is the lattice of subspaces of a vector space; so is the lattice of subsets of a set. If the elements of a lattice are regarded as propositions, the atoms are the mutually exclusive propositions which constitute complete statements about the system.

The theory of orthomodular orthocomplemented lattices provides a general framework for the discussion of physical theories, including both classical and quantum mechanics. A **status** can be defined as a function σ from the lattice to the unit interval $[0, 1]$ satisfying

(i) $\sigma(0) = 0, \quad \sigma(1) = 1;$ (5.101)

(ii) $x \leqslant y' \;\Rightarrow\; \sigma(x \lor y) = \sigma(x) + \sigma(y);$ (5.102)

(iii) $\sigma(x) = \sigma(y) = 1 \;\Rightarrow\; \sigma(x \land y) = 1.$ (5.103)

These requirements (and their extensions to countable infinite subsets) imply that σ is a status of the kind we have already met. If the lattice is distributive, with the completeness property that every countable subset has a least upper bound, then it is a lattice of subsets of some space and σ is a probability measure on that space. Thus a distributive lattice corresponds to classical mechanics. There is no simple logical (i.e. lattice–theoretic) characterisation of the lattice of subspaces of state space, but **Gleason's theorem** states that a status σ on such a lattice† must be of the form

$$\sigma(M) = \mathrm{tr}\,(P_M \rho)$$ (5.104)

where ρ is a hermitian operator satisfying $\mathrm{tr}\,\rho = 1$.

Superselection rules   The lattice–theoretic approach leads to a generalisation of quantum mechanics which is physically significant. The lattice of subspaces of state space is irreducible; considered as a logic, it is as non-classical as possible – every proposition is incompatible with some other proposition. It is a simple matter to construct a reducible lattice, which contains propositions obeying classical logic as well as a non-classical part. Given two lattices $\mathscr{L}_1$ and $\mathscr{L}_2$, we define their **direct sum** to be the Cartesian produce $\mathscr{L}_1 \times \mathscr{L}_2$ with the ordering

$$(x_1, x_2) \leqslant (y_1, y_2) \;\Leftrightarrow\; x_1 \leqslant y_1 \text{ and } x_2 \leqslant y_2.$$ (5.105)

This direct sum has a non-trivial centre consisting of the four elements $(0, 0)$, $(0, 1), (1, 0)$ and $(1, 1)$. The two non-trivial propositions $(0, 1)$ and $(1, 0)$ obey a classical logic in relation to all other propositions.

If $\mathscr{L}_1$ and $\mathscr{L}_2$ are the lattices of subspaces of two vector spaces $\mathscr{S}_1$ and $\mathscr{S}_2$, so that their atoms are all the one-dimensional subspaces of $\mathscr{S}_1$ and $\mathscr{S}_2$, then the direct sum has these atoms but no others: in particular, it does not have elements corresponding to superpositions of a vector in $\mathscr{S}_1$ and one in $\mathscr{S}_2$. Thus the direct sum describes a system in which the superposition principle does not hold: the state vectors of the system belong to the space $\mathscr{S}_1 \oplus \mathscr{S}_2$, but they are restricted to the subspaces $\mathscr{S}_1$ and $\mathscr{S}_2$. Superpositions of vectors from different subspaces do not describe possible physical states.

A rule restricting possible states in this way is called a **superselection rule**.

---

† More precisely, a lattice of closed subspaces of a Hilbert space.

There is a superselection rule for each of the absolutely conserved discrete quantum numbers, namely electric charge, baryon number and the three kinds of lepton number: for example, no superposition of states with different electric charges has ever been observed.

Three-valued logic A lattice whose elements are regarded as propositions is sometimes called a logic; an orthomodular orthocomplemented lattice is a **quantum logic**, a distributive orthocomplemented lattice is a **classical logic**. This should not be confused with other systems of logic, in particular the propositional calculus. In a lattice of propositions, if x and y are propositions then $x \vee y$ and $x \wedge y$ are also propositions ('x and y' and 'x or y') but $x \leqslant y$ ('x implies y') is a statement about propositions. In the propositional calculus, if x and y are propositions then so are $x \vee y$, $x \wedge y$ and $x \supset y$ ('x materially implies y'). In this system the notion of **truth value** plays a central role; this is an assignment to every proposition of a value 'true' or 'false', with rules giving the truth values of $x \vee y$, $x \wedge y$, and $x \supset y$ in terms of those of x and y. The relationship between the two systems can be seen by writing 1 for 'true' and 0 for 'false'; then an assignment of truth values to a set of propositions can be regarded as a status which only takes the values 0 and 1. It can be shown (problem 5.13) that (5.101)–(5.103) then imply the usual rules for the truth values of $x \vee y$ and $x \wedge y$.

However, it is easy to see that it is impossible to assign truth values to the lattice of subspaces of a state space. (*Proof*: Let x be a one-dimensional subspace with truth value $\sigma(x) = 1$, X a two-dimensional subspace containing x. Then $\sigma(X) = 1$. Any other one-dimensional subspace $y \leqslant X$ must have $\sigma(y) = 0$, for $\sigma(y) = 1$ would imply $\sigma(0) = \sigma(x \wedge y) = 1$. But we can find orthogonal y and z, different from x, such that $y \vee z = X$; hence $\sigma(X) = \sigma(y \vee z) = 0$, which is a contradiction.) In other words, if a system is in a state which is not an eigenstate of an observable A, a statement 'A has the value α' is neither true nor false. To cater for this it has been suggested that the classical propositional calculus should be extended to admit a third truth value 'undecided'. The rules giving the truth values of $x \vee y$ and $x \wedge y$ are extended as follows:

Table 5.2. *Truth table for three-valued logic*

x	t	t	t	f	f	f	u	u	u
y	t	f	u	t	f	u	t	f	u
$x \wedge y$	t	f	u	f	f	f	u	f	u
$x \vee y$	t	t	t	t	f	u	t	u	u

For further details see Wallace Garden 1984 or Reichenbach 1944.

5.5. **Interpretations of quantum mechanics**

This section consists of a summary of the main answers that have been proposed to the questions of what quantum mechanics means and what it can tell us about the physical world. First we will look cursorily at some general views on what any scientific theory can mean; this is just a list of isms. Then, after reminding ourselves why quantum mechanics calls for interpretation so much more than classical mechanics, we will examine the various applications of these isms to quantum mechanics, paying particular attention to the question of whether they can be reconciled with each other.

Some meta-scientific vocabulary

(vocabulary for use in statements *about* scientific statements).

Empiricism is the doctrine that the justification for any belief can only, ultimately, come from sense experience. There is room for a considerable range of views about what beliefs are justified, and to what extent, by what experiences; an extreme position is **solipsism**, which is the belief that nothing exists outside oneself.

Positivism is the view that the meaning of any statement resides in the way it is to be verified (which must be by consulting sense experience – in the scientific context, this means doing an experiment). If a statement is not verifiable, it is meaningless. (This is the **verification principle**.) However, most scientific laws are not verifiable, since they are general statements and cover an infinite number of cases; according to the verification principle, this book is entirely meaningless. A more realistic account of the relation between meaningfulness and experiment is Popper's **falsification principle**, according to which a statement is meaningful if it can be *falsified* by experiment; that is to say, if the statement has logical consequences which can be found empirically to be false. (Popper actually proposed this not as a condition for meaningfulness, but as a criterion for a statement to be scientific.)

Operationalism is the view that individual terms in a scientific theory should be defined by reference to experimental procedures. The model theory for this view is the special theory of relativity, with its operational definitions of distance and time. A more permissive account is that a scientific theory is a **hypothetico–deductive system**, which may refer to unobservable quantities in its basic postulates, and need only give empirical meaning to quantities derived from the basic ones. Quantum mechanics provides a good example of this procedure: the state vector $|\psi\rangle$ is not defined in operational terms, and it is only the derived quantities $|\langle\phi|\psi\rangle|^2$ that relate to experiment.

Pragmatism is the general philosophical view that the meaning of a statement resides in the way in which it governs our actions; it is true if it is useful. **Instrumentalism** is the related view concerning scientific theories, that they are to be regarded as instruments for making predictions about the results of experiments.

The peculiarities of quantum mechanics

1. Indeterminism. Quantum mechanics differs radically from previous physical theories, not just because its assertions are probabilistic, but because of the fundamental status that is claimed for these assertions. Other uses of probability in physics arise in statements of partial knowledge about a situation; it is assumed that it would be possible to obtain further knowledge which would resolve a probability into certainties. If quantum mechanics is taken as a final theory, however, it must be accepted that there is no possibility of such further knowledge.

This feature of quantum mechanics is not hard to understand, but it is hard to accept. Until the advent of quantum mechanics it had been a basic assumption of physical science that every event had a cause; if the fundamental laws are probabilistic then some aspects of some events are uncaused. The feeling that to accept this is to do violence to the scientific spirit was expressed by Einstein in his famous saying 'I cannot believe that the good Lord plays dice'. To put this in a slightly different light, a general statement like 'every event has a cause' can be regarded not a statement of *fact* (it can never be falsified), but as a statement of *intention*: we are going to look for a cause for every event. Then quantum mechanics constitutes an admission of failure.

It is sometimes argued that the indeterminism of quantum mechanics is to be welcomed as allowing free will. Readers wishing to pursue this line of thought should be aware of the argument that 'free will' neither means nor implies the existence of uncaused events. Many thinkers believe that free will is compatible with (indeed, requires) determinism; see Berofsky 1966.

However palatable or unpalatable it may be, indeterminism raises no conceptual problems which are peculiar to quantum mechanics. In examining any proposed interpretation of quantum mechanics, it is important to consider to what extent it serves to elucidate probability statements in general and to what extent it specifically attends to quantum mechanics.

2. Indeterminacy. The way in which properties are ascribed to particles and systems in quantum mechanics is a more puzzling departure from the procedures of classical mechanics. This has two aspects. First, there is the denial of definite values for properties which every particle must have in classical mechanics – the fact that a particle need not have definite position or momentum (and cannot have both simultaneously). It is particularly puzzling that although a particle can have a definite position at one time, it cannot have definite positions at all times in an interval (i.e. a definite trajectory), since that would give it also a definite momentum. One might argue that there is nothing special about the properties of position and momentum, and we should not necessarily expect a particle to have them any more than we expect it to have a definite shape, smell or sense of humour, but there is a deep-seated feeling that position and momentum are 'essential' properties (as opposed to 'accidental' ones) and that a particle is inconceivable without them.

Secondly, and more seriously, there is the indeterminate status of a property of a system (i.e. an observable) when the system is not in an eigenstate of it. Does the system have this property or not? Since the observable can be measured and will be found to have a definite value, we cannot say that statements about it are meaningless; on the other hand, any statement that the observable has a particular value will be falsified by some experiment on the system when it is in this state.

Indeterminacy is related to indeterminism, since it is the fact that the results of measurements are not rigidly determined by the state vector which makes it impossible to ascribe definite values of an observable to a system; but they should be distinguished, since there are conceivable theories in which all observables have definite values but the development of the system is not uniquely determined by those values.

3. Inseparability. The peculiarities of state vectors of the form $|\phi_1\rangle|\psi_1\rangle + |\phi_2\rangle|\psi_2\rangle$, for a system composed of two subsystems, have been discussed in §5.2 in connection with the Einstein–Podolsky–Rosen paradox. When the whole system is in a state of this form, it is not possible to say that either subsystem is in a definite state, but it is possible to gain information about one subsystem from an experiment performed on the other.

In principle, therefore, quantum mechanics denies the possibility of describing the world by dividing it into small parts and completely describing each part – a procedure which is often regarded as essential for the progress of science. Because of this feature, quantum mechanics is sometimes called a **holistic** theory.

4. The projection postulate. The difficulties associated with the projection postulate were discussed in detail in §5.2. To summarise, they are:

1. The projection postulate is ill-defined: there is no precise definition of what constitutes a measurement, and no specification of the time at which projection is supposed to occur.

2. It is **dualistic**, requiring a division of the world into (microscopic) object and (macroscopic) apparatus. It also splits the law of time development into the deterministic Schrödinger equation and the probabilistic projection postulate.

3. It is **anticausal**: as Schrödinger's cat paradox illustrates, it makes physical events consequences of their observation, instead of saying that events are observed because they happen.

4. It gives no account of continuous observation.

In discussing the various interpretations of quantum mechanics we will pay particular attention to the way they explain the projection postulate.

Nine interpretations   An interpretation of quantum mechanics is essentially an answer to the question 'What is the state vector?' Different interpretations cannot be distinguished on scientific grounds – they do not have different experimental consequences; if they did they would constitute different *theories*. They can, however, be rationally compared with regard to their intelligibility and satisfactoriness as explanations. No one interpretation is generally accepted (although the phrase 'the Copenhagen interpretation' is often used as a synonym for 'orthodoxy', whatever the user thinks that is; it has been applied to at least four different interpretations). After each interpretation we will give the main objections to it. These are not meant to be definitive, or even necessarily meaningful; you may well decide that some of them are wrong or do not make sense. On the basis of these objections, however, it will be argued that many of the differences between interpretations are more apparent than real, and that on analysis a number of the interpretations turn out to be saying the same thing in different ways.

1. The minimal interpretation

On this view, which was strongly expressed by Bohr, one should not attempt to interpret the state vector in order to extract information about quantum objects; one should not even speak of quantum objects. The state vector is just a mathematical device used in calculating the results of experiments; to perform such calculations successfully is the sole purpose of any scientific theory. The experiments must (logically must) be described in the terms of classical physics, since the apparatus consists of macroscopic objects; we do not know how to describe our experiences with such objects to each other except in the terms of classical physics. If we mention microscopic objects it is only as a shorthand device to refer to some features of a calculation which relates different classically described states.

In following this interpretation it is important to distinguish between **preparation** and **measurement** of a system. A preparation occurs at the beginning of an experiment and is associated with a state vector $|\psi_0\rangle$; a measurement occurs at the end of an experiment and its various possible outcomes are associated with bra vectors $\langle\phi_1|$, $\langle\phi_2|$, ... The experimental arrangement is associated with a Hamiltonian H and a time t; then the probability that the measurement has the ith outcome is

$$p_i = |\langle\phi_i|e^{-iHt/\hbar}|\psi_0\rangle|^2. \tag{5.106}$$

Note that $|\psi_0\rangle$, H and $\langle\phi_i|$ are all defined in terms of macroscopic experimental arrangements.

This view dissolves all the puzzles concerning quantum objects, since there are no objects to be puzzled about. There is no projection postulate, for any calculation concerns only a preparation and a measurement, and cannot be extended to describe anything that happens after the measurement. If an experiment carries on after a measurement M which gave the result α, then one is embarking on a new experiment whose preparation procedure consists of performing the measurement M and selecting the cases in which the result was α. There is nothing surprising in the fact that this preparation has associated with it a state vector $|\phi_i\rangle$ which is different from $e^{-iHt/\hbar}|\psi_0\rangle$.

Some preparation procedures will be associated not with a state vector $|\psi_0\rangle$ but with a statistical operator ρ_0 (for example, one might prepare a beam of electrons without fixing their spins). In such a case the probability (5.106) must be replaced by

$$p_i = \text{tr} \left[P_i e^{-iHt/\hbar} \rho_0 e^{iHt/\hbar} \right] \tag{5.107}$$

where P_i is an appropriate projection operator, which reduces to (5.106) for a pure state $(\rho_0 = |\psi_0\rangle\langle\psi_0|)$. Thus in this interpretation there is no logical distinction between a pure state and a general statistical operator; both describe preparation procedures. Hence our term '(experimental) status'; hence also the usage of 'state' to refer to statistical operators.

Objections. This interpretation has been called 'extended solipsism'. A solipsist refuses to accept that the experience of seeing a tree is evidence that the tree exists; there are only sense experiences. Likewise, a follower of the minimal interpretation refuses to accept that the formation of a charged particle track in a bubble chamber is evidence for the existence of the charged particle; there are only macroscopic events. This is a solipsism on behalf of macroscopic apparatus towards the microscopic objects they perceive, and is as implausible as the human version of solipsism.

Moreover, it cannot be true that the sole purpose of a scientific theory is to predict the results of experiments. Why on earth would anyone want to predict the results of experiments? Most of them have no practical use; and, even if they had, practical usefulness has nothing to do with scientific inquiry. Predicting the results of experiments is not the *purpose* of a theory, it is a *test* to see if the theory is true. The purpose of a theory is to understand the physical world.

Although the instrumentalist philosophy which underlies the minimal interpretation is often expressed in the form given here, and is open to the above objection, Bohr's formulation was less crude: 'The task of science is both to extend the range of our experience and reduce it to order'. Heisenberg combined this with an operationalist view which derived from his own discovery of matrix mechanics, which he developed by considering matrices of frequencies of spectral lines. He taught that a theory should only contain experimentally observable quantities, and proposed that this principle should be applied to elementary particle physics by renouncing all mention of the time evolution of the state vector between preparation and measurement. This more radical form of quantum mechanics is called **S-matrix theory**, and stands in opposition to quantum field theory. As will be seen in Chapter 7, it has not been successful as a theory of elementary particles; in this area quantum field theory has been triumphant.

It is not necessary to be so austere as to renounce all belief in quantum objects to embrace the above solution of the problem of measurement, namely that being careful to distinguish between preparation and measurement makes the projection postulate unnecessary. In answer to the question 'What is the state of a system after a measurement?' it has been argued (particularly by Margenau) that true measurements on a quantum system always destroy the system. For example, to measure a component of polarisation of a photon one must not only pass the photon through a doubly refracting crystal (Fig. 2.2) but also detect it when it emerges from the crystal; and the act of detection (say, by making it impinge on a photographic plate) destroys the photon. This argument denies that there are any measurements of the first kind (p. 48). However, such measurements must exist for macroscopic apparatus, which must retain a permanent record of the results of experiment; thus this solution of the problem of measurement depends on refusing to extend quantum descriptions to macroscopic objects.

References: Bohr 1958, Stapp 1972, Peres 1984.

2. The literal interpretation

This is the interpretation which is implicit in most modern textbooks (including this one, so far). They speak as if the state vector (more precisely, the corresponding point of projective space) is an objective property of a system in the same sense as the values of coordinates and momentum are objective properties of a system in classical mechanics. The projection postulate is then a statement about an actual change in the state vector following a measurement.

In this interpretation indeterminism and indeterminacy are simply accepted as facts about the world. Inseparability means that it is not possible to apply this interpretation to subsystems; one cannot say that individual objects have state vectors, but is forced to consider the state vector of the universe.

Objections. The state vector cannot be an objective property of an individual system, for in general it is not possible to establish by experiment that the state vector is one vector rather than another. For example, if $|\psi\rangle$ is an eigenstate of an observable A with eigenvalue α, and if $|\phi\rangle$ is not orthogonal to $|\psi\rangle$, then a measurement of A which gives the value α does not prove that the state vector is $|\psi\rangle$ rather than $|\phi\rangle$, because the measurement might give this result when the state is $|\phi\rangle$. (This argument is valid if it is assumed that an objective statement must be capable of being proved true by experiment, but not if it is only assumed that there must be a possibility that it could be proved false. As Popper pointed out, the latter is the normal situation in science.)

All the unsatisfactory features of the projection postulate, as listed above, stand as objections to the literal interpretation of quantum mechanics.

3. The objective interpretation

The literal interpretation can be modified by supposing that the state vector is restricted to lie in certain subspaces of state space, and that it makes spontaneous and instantaneous transitions from one of these subspaces to another with probabilities determined by the solution of the Schrödinger equation. If the state space is

$$\mathscr{S} = \mathscr{S}_1 \oplus \mathscr{S}_2 \oplus \cdots \tag{5.108}$$

where $\mathscr{S}_i$ are the allowed subspaces, and if the solution of the Schrödinger equation is

$$|\psi(t)\rangle = |\psi_1(t)\rangle + |\psi_2(t)\rangle + \cdots \quad \text{with } |\psi_i(t)\rangle \in \mathscr{S}_i, \tag{5.109}$$

then the state of the system at time t is assumed to be one of these $|\psi_i(t)\rangle$, the probability that it is $|\psi_i(t)\rangle$ being

$$p_i(t) = \langle \psi_i(t) | \psi_i(t) \rangle. \tag{5.110}$$

This statement is the same as the continuous part of the projection postulate (p. 114; see also (5.61)). It can be derived from the following assignment of probabilities for the transition from one subspace to another:

> **Bell's postulate.** The probability that the state of the system is in the subspace $\mathscr{S}_i$ at time t and makes a transition to $\mathscr{S}_j$ between times t and $t + \delta t$ is $w_{ij}\, \delta t$, where
>
> $$w_{ij} = \begin{cases} 2 \, \text{Re} \, [(i\hbar)^{-1} \langle \psi_j(t)|H|\psi_i(t)\rangle] & \text{if this is} \geq 0, \\ 0 & \text{if it is negative.} \end{cases} \tag{5.111}$$

Then we can prove

⬤**5.9** The probabilities (5.110) follow from Bell's postulate.

Proof. First we cast the statement of probabilities (5.110) in the form of a differential equation. Let P_i be the projection operator onto $\mathscr{S}_i$, so that

$$|\psi_i(t)\rangle = P_i|\psi(t)\rangle \quad \text{and} \quad i\hbar \frac{d}{dt}|\psi_i(t)\rangle = P_i H|\psi(t)\rangle.$$

Then

$$i\hbar \frac{dp_i}{dt} = \langle \psi_i(t)|P_i H|\psi(t)\rangle - \langle \psi(t)|H P_i|\psi(t)\rangle$$

$$= \langle \psi_i(t)|H|\psi(t)\rangle - \langle \psi(t)|H|\psi_i(t)\rangle. \tag{5.112}$$

Now Bell's postulate gives the probability that the state of the system is in the subspace $\mathscr{S}_i$ at time $t + \delta t$ as the probability that it was in $\mathscr{S}_i$ at time t, minus the total probability that there was a transition out of $\mathscr{S}_i$ between t and $t + \delta t$, plus the probability that there was a transition into $\mathscr{S}_i$ in this time:

$$p_i(t + \delta t) = p_i(t) - \sum_{j \neq i} w_{ij}\, \delta t + \sum_{j \neq i} w_{ji}\, \delta t, \tag{5.113}$$

so that

$$\frac{dp_i}{dt} = \sum_j (w_{ji} - w_{ij}). \tag{5.114}$$

For a given i, let P be the set of j for which the first condition in (5.111) holds, so that the imaginary part of $\langle \psi_j(t)|H|\psi_i(t)\rangle$ is positive or zero, and let N be the set of j for which it is negative. Since H is hermitian, $\langle \psi_j|H|\psi_i\rangle$ and $\langle \psi_i|H|\psi_j\rangle$ are complex conjugates and so their imaginary parts have opposite signs. Hence

$$j \in P \;\Rightarrow\; w_{ij} \geq 0 \text{ and } w_{ji} = 0,$$
$$j \in N \;\Rightarrow\; w_{ij} = 0 \text{ and } w_{ji} \geq 0,$$

which gives

$$\frac{dp_i}{dt} = \sum_j (w_{ji} - w_{ij}) = -\sum_{j \in P} w_{ij} + \sum_{j \in N} w_{ji}$$

$$= \sum_{j \in P} 2 \operatorname{Im} \left[\hbar^{-1} \langle \psi_j | H | \psi_i \rangle \right] + \sum_{j \in N} 2 \operatorname{Im} \left[\hbar^{-1} \langle \psi_i | H | \psi_j \rangle \right]$$

$$= \sum_j 2 \operatorname{Im} \left[\hbar^{-1} \langle \psi_i | H | \psi_j \rangle \right]$$

$$= (i\hbar)^{-1} \sum_j \left(\langle \psi_i | H | \psi_j \rangle - \langle \psi_j | H | \psi_i \rangle \right) \tag{5.115}$$

which is the same as (5.112) since $|\psi(t)\rangle = \sum_j |\psi_j(t)\rangle$. For a given probability at $t = 0$, therefore, the solution $p_i(t)$ is the same as that of (5.112), namely $\langle \psi_i(t) | \psi_i(t) \rangle$. ∎

There are a number of possibilities for the subspaces $\mathscr{S}_i$. They can be taken as the eigenspaces of macroscopic observables, or they can be defined in microscopic terms, for example as the spaces with definite numbers of particles of some specified kind (e.g. photons, or fermions).

This interpretation eliminates all mention of measurement and projection, and thus avoids all the problems associated with the projection postulate.

Objections. Since the equation (5.110) for the probabilities involves all the states $|\psi_i(t)\rangle$, the development of the system is not solely determined by the state that it happens to be in. The whole state $|\psi(t)\rangle$ must be regarded as a property of the system as well as one of the $|\psi_i(t)\rangle$. Thus the interpretation involves a proliferation of properties of the system. Moreover, some of these properties cannot be determined by experiment. In an experiment, according to this interpretation, one determines the state $|\psi_i(t)\rangle$ in one of the subspaces $\mathscr{S}_i$; yet its future evolution is determined by the accompanying states $|\psi_j(t)\rangle$, which the experimenter cannot know about. (The proliferation of properties can be avoided by abandoning the use of a differential equation to describe the development of the system. In this case the experimenter's failure to determine future probabilities stems from their lack of knowledge of the past history of the system.)

It is not clear to what extent this interpretation is compatible with special relativity. Because of EPR effects, the state vector must be taken as describing the entire universe, and instantaneous transitions in this state vector seem to conflict with the fact that simultaneity is relative.

Finally, the freedom in the choice of the subspaces $\mathscr{S}_i$ casts some doubt on the objectivity claimed for the state vector which lies in one of these subspaces.

Reference: Bell 1984.

4. The epistemic ('subjective') interpretation

Instead of being taken as an intrinsic property of the system, the state vector can be regarded as a representation of the observer's knowledge of the system. Then indeterminacy in the values of observables becomes simply lack of knowledge of these values; and both inseparability and the projection postulate lose their mystery. There is nothing mysterious in the fact that the state vector of a system changes after a measurement if that just means that the observer's knowledge changes; the very purpose of a measurement is to increase one's knowledge. Similarly, in the EPR situation there is nothing mysterious in the fact of an experiment on one object changing the state (i.e. one's knowledge) of a distant object; I change my knowledge of distant objects every morning when I pick up the newspaper.

Objections. Because it refers to a particular observer, this interpretation is sometimes criticised for being subjective. However, the concept of knowledge contains both subjective and objective elements: a statement that a person N knows a proposition P is a statement both about the person (that they believe P) and about the proposition (that it is true). The subjective element can be removed from the epistemic interpretation by considering all possible observers and defining a unique state vector for the system as that which represents the maximum possible knowledge which any observer can have. This is then an intrinsic property of the system, so that we are back to the literal interpretation.

The attempt to explain away the projection of the state vector as simply an increase in knowledge is shown to be unsuccessful by considering the maximum obtainable knowledge. If this changes, it must be because of a change in the system itself; and the problems of when this happens, and why it should happen when it cannot be derived from the Schrödinger equation, remain unresolved. This can be clearly seen by considering the case of a decaying unstable particle; when the observer acquires knowledge that the particle has decayed, this is clearly because it *has* decayed. The idea is more plausible when applied to an instantaneous measurement, but it is hard to distinguish it from the classical idea that the function of a measurement is to find out something which is already true. This is to make the mistake of confusing a superposition $|a\rangle + |b\rangle$ with a mixture '$|a\rangle$ or $|b\rangle$'.

On the other hand, if one is prepared to accept the charge of subjectivity and insists that the state vector refers to the knowledge of a particular observer, then one faces the question 'What is it that that observer knows?' If it is something about the system, we are back to the literal or objective interpretation; if it is something about the results to be expected from future experiments, we are back to the minimal interpretation.

Reference: Heisenberg 1959.

5. The ensemble interpretation

Some authors (including Einstein) deny that the state vector describes the state of an individual system: it can be properly applied only to a large number of systems, all prepared in the same way. This collection of systems is called an **ensemble**. Then the probabilities $|\langle \psi | \psi_i \rangle|^2$ in Postulate II refer to the fraction of the ensemble in which an experiment has a particular result. Those systems for which a result α_i was obtained in the experiment constitute a subset of the original collection, and therefore form a different ensemble; naturally, this is described by a different state vector. Thus the process of projection is not an interruption to the Schrödinger evolution of the ensemble, but a shift of attention to a different ensemble.

If an ensemble E_1 containing N_1 systems is combined with a different ensemble E_2 containing N_2 systems, the resulting ensemble is called a mixture of E_1 and E_2. If E_1 and E_2 are described by state vectors $|\psi_1\rangle$ and $|\psi_2\rangle$, the mixture is described by the statistical operator

$$\rho = w_1 |\psi_1\rangle\langle\psi_1| + w_2 |\psi_2\rangle\langle\psi_2| \quad \text{where } w_i = \frac{N_i}{N_1 + N_2}. \tag{5.116}$$

Thus in this interpretation, as in the minimal interpretation (and also in the epistemic interpretation), there is no conceptual distinction between a status and a pure state (i.e. a description by a statistical operator and a description by a state vector).

Objections. This interpretation is not addressed to the specific problems of quantum mechanics, but is a way of understanding any probabilistic theory. As an account of probability its defects have been discussed in Chapter 2 (p. 43). The concept of an ensemble is vague, because it is not clear what is meant by 'a large number' of systems. If the statements about fractions of an ensemble are to be experimentally meaningful, the ensemble must consist of a finite number of systems. But then there is the possibility (remote but undeniable) that an experiment on the ensemble will yield results in proportions different from those given by the theory, and one cannot claim that these proportions are definite predictions of the theory.

On the other hand, if the ensemble is infinite and any finite collection of systems is just a sample from it, then the ensemble has no empirical reality: it is a theoretical entity associated with a particular system in exactly the same way as the state vector is associated with the system in the literal interpretation. The ensemble has an advantage over the state vector in that it exists (in this theoretical sense) when the state vector does not; for example, in an **EPR** experiment with two separated electrons in a state of total spin 0, neither electron has a definite state vector but each can be associated with an ensemble described by the statistical operator $\frac{1}{2}(|\uparrow\rangle|\downarrow\rangle + |\downarrow\rangle|\uparrow\rangle)$. The fullest possible description, however, must encompass the complete system of two electrons and assign it a definite state vector; this has the same conceptual status as an infinite ensemble of two-electron systems.

As with the epistemic interpretation, the account of the projection postulate offered by the ensemble interpretation does not resolve any problems. If an ensemble can be divided into experimentally distinguishable subensembles, it is a mixture; an ensemble described by a single state vector is homogeneous, in the sense that there is no detectable distinction between its members. In the course of an experiment an initially homogeneous ensemble becomes a mixture. This is an objective change in the ensemble which cannot be explained away in terms of a shift of attention on the part of the observer.

Reference: Ballentine 1970.

6. The relative-state and many-worlds interpretations

Everett's **relative-state** interpretation is a version of the literal interpretation which makes it possible to speak of the state of a subsystem. It insists, however, that the state of any system has no absolute meaning but is only defined relative to a given state of the rest of the universe. The only state which has an absolute meaning is that of the whole universe, including all observers and their consciousness. The idea can be demonstrated by considering a system S with two states $|\psi_1\rangle$ and $|\psi_2\rangle$; then the state of the universe can be written as

$$|\Psi\rangle = |\psi_1\rangle|\alpha_1\rangle + |\psi_2\rangle|\alpha_2\rangle \qquad (5.117)$$

where $|\alpha_1\rangle$ and $|\alpha_2\rangle$ are states of the rest of the universe which could include states of an apparatus showing different results of an experiment whose eigenstates are $|\psi_1\rangle$ and $|\psi_2\rangle$. Then the state of the system, *relative* to the state $|\alpha_1\rangle$ of the rest of the universe, is $|\psi_1\rangle$. This incorporates the projection postulate by emphasising that it is a conditional statement – if the result of the experiment was α_1, then the state of the system is $|\psi_1\rangle$ – and including this conditionality in the formalism. (The conventional formulation, as in Postulate III, has an antecedent which refers to experience and puts only the consequent in the formalism.) By developing von Neumann's theory of the measurement process (see §5.2), Everett (1957) proved the consistency of this procedure of retaining the full state vector (5.117) and showed that it could account for the agreement between different observers about what they thought had happened in a particular experiment (even though another part of the universal state vector described a different result). He also showed that Postulate II, giving the probabilities of the different results of an experiment, could be reduced to a natural probability distribution on state vectors. (This result is claimed to mean that 'the formalism yields its own interpretation'.)

The **many-worlds** interpretation is a picturesque account of the relative-state interpretation which describes the state of the universe given by (5.117) as a universe which has split into two branches, in one of which the state of the system is $|\psi_1\rangle$, the experiment has given the corresponding result α_1, and all observers are aware of that result; while in the other branch the course of events has gone according to the state $|\psi_2\rangle$. In general, wherever the conventional theory requires an application of the projection postulate the many-worlds interpretation says that the universe splits into parallel worlds of the kind familiar from science fiction stories such as Philip K. Dick's *The Man in the High Castle* (Dick 1962). It can accommodate possible interference between different terms in the universal state vector by assuming that the different strands of the universe can recombine.

Objections. The relative-state interpretation differs very little from the objective interpretation. By making the universal state vector develop purely according to the Schrödinger equation, it does not formally include the indeterminism which is nevertheless present in the reality experienced by any of the observers it describes. It might seem more honest to make the formalism describe that reality and no other, i.e. to drop the parts of the universal state vector which describe a situation which we actual observers know to be false. However, by not doing this the relative-state interpretation avoids the ambiguities of the projection postulate.

The many-worlds interpretation sells the pass. To say that an experiment had a result β in some parallel universe (when we observed it to have the result α) is surely just another form of words for saying that it might have had the result β, but didn't. We are perfectly entitled to define the 'real world' to be the one in which what we observed to happen did happen; then the splitting of the universe into several branches, only one of which is real, is exactly the same process as that described by the projection postulate, and is beset by exactly the same problems of defining when and under what circumstances it should happen.

In defence of the many-worlds interpretation, it can be claimed that it is justifiable to call an event 'real' if it can have an observable effect, and that this is true of the experimental results which we did not observe, because of possible interference between different parts of the universal state vector. These effects are present in the objective interpretation, in which the situation is described by saying that the future development of the system is affected by the unrealised possibilities for the results of past experiments. The difference between this and the statement that these possibilities have been working themselves out in an alternative universe is purely verbal.

Both the relative-state and the many-worlds interpretation are open to the objection that they do not make it possible to represent the knowledge obtained from experiment. If an experiment on a system has the result α_1, corresponding to the eigenstate $|\psi_1\rangle$, and if $|\alpha_1\rangle$ is the appropriate state of the apparatus and the experimenter and the rest of the universe, one cannot deduce that the state of the universe is $|\psi_1\rangle|\alpha_1\rangle$; it might be $|\psi_1\rangle|\alpha_1\rangle + |\psi_2\rangle|\alpha_2\rangle$ or $|\psi_1\rangle|\alpha_1\rangle + \frac{1}{2}|\psi_2\rangle|\alpha_2\rangle$. One can, however, deduce that the relative state of the system is $|\psi_1\rangle$; and it can be argued that there are logical reasons why one should not be able to represent oneself in a theory.

References: DeWitt & Graham 1973, Borges 1941.

7. The quantum-logical interpretation

Quantum logic (described in §5.4) was developed to support the contention that difficulties in the interpretation of quantum mechanics all stem from the use of classical logic in discussing physical systems. An analogy is drawn between logic and geometry in their relations to physics: just as Euclidean geometry is only one among several possible forms of geometry, and the question of which geometry applies to the physical world is to be decided empirically, so, it is argued, classical logic is only one among several possible forms of logic, and the laws of physics may show that it is not valid in the real world. This is what happens if we interpret subspaces of state space as propositions, with the lattice symbols $\leqslant$, $\vee$ and $\wedge$ interpreted as 'implies', 'and' and 'or' as described in §5.4.

This interpretation is directed mainly against the problem of indeterminacy. The idea is that it makes it possible to assert that a system has definite values for all observables, even though some of them are incompatible. For example, let X denote the position of a particle moving in one dimension, and let P denote its momentum. Let $p_1, p_2, \ldots$ denote the possible values of P (written as if they were countable for convenience). Then a proposition $X = x$ is incompatible with each of the propositions $P = p_1$. Nevertheless, $X = x$ is compatible with the proposition '$P = p_1$ or $P = p_2$ or $\ldots$' (i.e. 'P has *some* value') because, according to non-distributive quantum logic,

$$(X = x) \quad \text{and} \quad (P = p_1 \text{ or } P = p_2 \text{ or } \ldots)$$

is not equivalent to

$$(X = x \text{ and } P = p_1) \quad \text{or} \quad (X = x \text{ and } P = p_2) \quad \text{or} \ldots.$$

This solution of the problem of indeterminacy brings with it a solution of the problem of measurement. If every observable has a definite value, then the process of measurement simply reveals what that value is, and the projection of the state vector is a matter of refining the propositions that are true of the system (moving from '$P = p_1$ or $P = p_2$ or $\ldots$' to '$P = p_1$', say). The problems of inseparability can be resolved in a similar way; EPR correlations between two subsystems are explained as correlations between their separate properties dating from the time of their joint production. Bell's theorem, which normally shows that this explanation is incompatible with locality, can be discounted since it uses classical (distributive) logic.

Objections. This interpretation is based on nothing but a mathematical pun. To interpret $\vee$ (the linear sum of subspaces of state space) as the logical connective *or* is to change the meaning of 'or' too drastically to be acceptable:

among other things, in quantum logic the statement '$P \vee Q$ is true' does not imply 'either P is true or Q is true'. This means that it is only in a very weak sense that quantum logic makes it possible for every observable to have a value. We saw on p. 209 that it is not possible to suppose that every quantum proposition is either true or false.

The analogy between logic and geometry is superficial. It is possible to formulate non-Euclidean geometry without using or mentioning Euclidean geometry, but it is not possible to formulate quantum logic without using classical logic (in the meta-theory). Thus the solutions to the problems of measurement and inseparability are cheap; they depend on a selective ban on the distributive law. If quantum logic were consistently adopted as a logic in the true sense of the word (i.e. a method of reasoning), it would involve reconstructing the whole of mathematics – a herculean and probably impossible task.

If it is admitted that arguments about quantum propositions must be conducted according to classical logic – this is the normal situation in mathematical logic – then the apparatus of quantum logic becomes simply a reformulation of the mathematical machinery of quantum mechanics. It is then no longer an interpretation of the theory, but itself stands in need of an interpretation.

References: Birkhoff & von Neumann 1936, Putnam 1968.

8. Hidden-variable interpretations

The hypothesis that the behaviour of quantum systems is governed by hidden variables normally constitutes a new *theory*, whose purpose is to explain quantum mechanics and which can, in principle, be distinguished from it by differences in its empirical predictions. It becomes an *interpretation* if assumptions are added to make the differences unobservable even in principle. For example, the de Broglie/Bohm theory of §5.3 becomes such an interpretation if it is assumed that for any particle, whatever its source and past history, the probability of its position being $\mathbf{r}$ is $|\psi(\mathbf{r})|^2$. (Remember that both the position vector and the wave function are intrinsic properties of the particle in this theory.)

The de Broglie/Bohm interpretation is designed to remove the element of indeterminism from quantum mechanics. This objective is shared by many, but not all, hidden-variable interpretations; the hidden variables may be supposed to change unpredictably. The defining property of a hidden-variable interpretation is that all observables have precise values which are expressed in terms of the hidden variables; like the quantum-logical interpretation, their main target is indeterminacy.

Objections. If the hidden variables are so carefully hidden as to make them undetectable apart from the state vector, one has very little reason to believe in them. In the de Broglie/Bohm example, the assumption about the distribution of particle positions is highly implausible if these positions really do have a separate existence.

There is a continuum between hidden-variable interpretations and objective interpretations; if the special subspaces in the objective interpretation are taken to be eigenspaces of particle position, that interpretation becomes the same as a hidden-variable interpretation (see problem 5.14).

Bell's theorem shows that these interpretations, whether deterministic or not, must postulate instantaneous action at a distance and therefore conflict with special relativity.

Reference: Belinfante 1973.

9. The stochastic interpretation

There is a formal similarity between the Schrödinger equation and stochastic differential equations, which describe the unpredictable motion of a particle subject to random impulses, like a floating pollen particle undergoing Brownian motion. This leads to an interpretation of quantum mechanics in which a particle is supposed to have a definite position **r** at each time, and in each time interval δt there is a definite transition probability for this position to change by a given amount $\delta \mathbf{r}$.

Objections. Like the hidden-variables interpretations, this runs foul of Bell's theorem: if one thinks of the transition probabilities as being caused by impulses from a medium (as suggested by Brownian motion, or by a model of randomly fluctuating electromagnetic fields), then the properties of this medium depend on the instantaneous position of distant particles. The properties of the medium as it affects a particular particle also depend, at all times t, on the form of the wave function of that particle at $t = 0$. This is a strange and implausible feature of the interpretation.

Reference: Ghirardi, Omero, Rimini & Weber 1978, Nelson 1985.

Conclusion   In the absence of empirical indications, the interpretation of quantum mechanics is a matter of individual choice. The arguments sketched in this section suggest that the choice is between the tough-minded (but boring) minimal interpretation; the satisfying (but puzzling) objective interpretation; and the comprehensible (but implausible) hidden-variables interpretation.

Bones of Chapter 5   Bell's postulate 216

Further reading Thorough general discussions of conceptual problems in quantum mechanics can be found in Jammer 1974, d'Espagnat 1976 and Primas 1981. Wheeler & Zurek 1983 is an invaluable collection of reprints, which includes the original papers on many of the topics discussed in this chapter. For measurement theory see also Bohm 1951. On hidden variables and Bell's inequalities see d'Espagnat 1979 (for a non-technical account), Bell 1971, Clauser & Shimony 1978 (for details of experiments) and Belinfante 1973 (for a thorough survey). For quantum logic see Hughes 1981 (a non-technical account), Piron 1976 and Jauch 1968.

Problems on Chapter 5

1. Consider a system with a state space of finite dimension n. The statement that all states of the system are equally likely could be taken to mean that its statistical operator is $\rho = \int |\psi\rangle\langle\psi| \, d\psi$ where the integral is taken over the set of unit vectors (which form a sphere S^{2n-1}) and $d\psi$ is the usual measure on S^{2n-1}, which is invariant under unitary transformations of $|\psi\rangle$. Show that $\rho = n^{-1}1$.

2. Prove that for a system with a finite-dimensional state space, if all states are equally likely then the probability that an experiment will have a given result is proportional to the dimension of the eigenspace of that result.

3. A system is subjected to random repetitions of an experiment E, the probability that an experiment happens in a small time interval δt being $w \, \delta t$. If E has just two possible results, show that the statistical operator ρ satisfies

$$d\rho/dt = (i\hbar)^{-1}[H, \rho] + w(2\Pi\rho\Pi - \Pi\rho - \rho\Pi)$$

where Π is the projection operator onto one of the eigenspaces of E.

4. Let $\{|\psi_n\rangle\}$ be an orthonormal complete set of states for a system S, and let T be another system. Show that any state $|\psi\rangle$ of the combined system ST can be written as $|\Psi\rangle = \sum |\psi_n\rangle|\theta_n\rangle$ for some states $|\theta_n\rangle$ of T, and that when the combined system is in the state $|\Psi\rangle$ the statistical operator of S is

$$\text{tr}_S(|\Psi\rangle\langle\Psi|) = \sum \langle\theta_n|\theta_n\rangle|\psi_n\rangle\langle\psi_n|.$$

5. A system has two orthogonal states $|\Phi\rangle$ and $|\Psi\rangle$, and an operator A is defined by $A|\Phi\rangle = |\Psi\rangle$, $A|\Psi\rangle = |\Phi\rangle$. Calculate the expectation values of A in the situations described by the statistical operators ρ and ρ' of (5.38) and (5.39).

6. Let $\mathscr{P}$ be the phase space of a classical particle moving in one dimension, and let $\mathscr{W}$ be the space of wave functions for a quantum particle in one dimension.

For any function $g(x, p)$ on $\mathscr{P}$ an operator A_g on $\mathscr{W}$ can be defined by

$$(A_g \phi)(x) = (2\pi)^{-\frac{1}{2}} \int g(x + \tfrac{1}{2}ht, p) e^{-itp} \phi(x + ht) \, dt \, dp.$$

Given a wave function ψ, define a function g on $\mathscr{P}$ by

$$g(x, p) = \int \overline{\psi(x - \tfrac{1}{2}hs)} e^{-ips} \psi(x + \tfrac{1}{2}hs) \, ds.$$

Show that

(i) $\quad A_g = |\psi\rangle\langle\psi|$;

(ii) $\quad \int g(x, p) \, dp = 2\pi |\psi(x)|^2$;

(iii) $\quad \int g(x, p) \, dx = 2\pi |\tilde{\psi}(p)|^2$

where $\tilde{\psi}$ is the Fourier transform of ψ.

7. Let ST be a combined system with a Hamiltonian of the form $H_S \otimes 1 + 1 \otimes H_T$, where H_S and H_T are operators on the state spaces of S and T. Show that the result of an experiment on system S cannot be affected by performing an experiment on system T. Hence show that no EPR experiment can be used for faster-than-light signalling.

8. Describe the evolution of the full state vector, including the state of the apparatus, in the EPR experiment.

9. Let $|\phi_x\rangle$ and $|\phi_y\rangle$ be the polarisation states of a photon moving in the z-direction, as in Chapter 2. Show that the state of two such photons with total angular momentum 0 is $2^{-\frac{1}{2}}(|\phi_x\rangle|\phi_x\rangle + |\phi_y\rangle|\phi_y\rangle)$. If two photons are in this state, find the probability that one of them will pass through a polaroid at an angle θ to the x-axis if the other passes a polaroid aligned with the x-axis.

   Let A, B and C be the experiments of seeing whether a photon will pass through three polaroids, with an angle θ between the axes of A and B and an angle ϕ between those of B and C, and suppose these can be applied to either of two photons whose total angular momentum is 0. Find a version of Bell's inequality appropriate to this situation, and find values of θ and ϕ for which it is violated by the predictions of quantum mechanics.

10. Given a lattice of subsets of a set, construct a corresponding lattice of propositions.

11. Find the meet and join of two elements of the direct sum of two lattices.

12. Show that an orthomodular lattice is irreducible if and only if it is not a direct sum.

13. Show that if a lattice has a status σ taking only the values 0 and 1, it is distributive. Determine $\sigma(x \wedge y)$ and $\sigma(x \vee y)$ in terms of $\sigma(x)$ and $\sigma(y)$.

14. Consider a particle moving in one dimension in a potential $V(x)$, with wave function $\psi(x)$. Let x_n be a sequence of points on the line, labelled (in order) by an integer n, and let $\psi_n(x, t)$ be a wave function which coincides with $\psi(x, t)$ for $x_n \leqslant x \leqslant x_{n+1}$, and which vanishes for $x < x_n - \varepsilon$ and $x > x_{n+1} + \varepsilon$. Show that if Bell's postulate is used to find transition probabilities between the states $|\psi_n\rangle$, then in the limit as $\varepsilon \to 0$ the probability of transition from $|\psi_{n-1}\rangle$ to $|\psi_n\rangle$ in time δt is $j(x_n) \, \delta t$ if this is positive, where j is the probability current associated with ψ. Relate this to the de Broglie/Bohm pilot wave model.

6

Quantum numbers
THE PROPERTIES OF PARTICLES

In this chapter we return to the topic of elementary particles and apply the theory of Chapters 2, 3 and 4 to make the qualitative description of Chapter 1 into a quantitative one, at least as far as the intrinsic properties, or quantum numbers, are concerned. The description of the interactions between the particles given here is not fully quantitative, but is a simplified account of the type outlined in § 4.6.

Throughout this chapter we take $\hbar = 1$.

6.1. Isospin The proton and the neutron are distinguished by their electric charge, so that their responses to electromagnetic forces differ; but in their behaviour under strong forces they appear to be very similar, as is shown by the following pieces of evidence.

First, **mirror nuclei**. There are many pairs of nuclei in which the number of protons in one is equal to the number of neutrons in the other and vice versa, for example ^{3}H (tritium), which contains one proton and two neutrons, and the helium isotope ^{3}He, which contains two protons and one neutron. Other examples are ($^{19}_9$F$_{10}$, $^{19}_{10}$Ne$_9$) and ($^{14}_6$C$_8$, $^{14}_8$O$_6$). It is found that in such pairs the structures of the set of energy levels of the two nuclei are very similar, and the resemblance is enhanced if allowance is made for the extra electrostatic potential energy of the nucleus with more protons. If we assume that the Hamiltonian is the sum of several terms, one for each pair of particles in the nucleus, this suggests that the potential for two protons is the same as that for two neutrons. No conclusions can be drawn concerning the potential for a proton and a neutron, since the number of such pairs is the same in each of the two nuclei.

Scattering data, however, suggest that the n–p potential is the same as the n–n and p–p potentials. Scattering experiments determine the distribution of particles emerging from a collision, and are thus concerned with unbound states (whereas the states of a nucleus are bound states). It is found that the wave functions of these states for two protons, after allowing for the electrostatic repulsion, are similar to those for a neutron and a proton, provided one compares states with the same spin and orbital angular momentum. The point of this proviso is that, since protons are fermions, there are restrictions on the spin/orbital state of two protons that do not apply to a proton and a neutron: if the spin state is symmetric (total spin $s = 1$), the orbital state must be antisymmetric (relative orbital angular momentum l odd), and if the spin state is antisymmetric ($s = 0$), the orbital state must be symmetric (l even). In these states the n–p scattering is the same as the p–p scattering.

These facts can be accounted for by regarding the proton and neutron as two states of a single particle, the **nucleon**, and supposing that the Hamiltonian of the strong interaction does not discriminate between these states. Then the nucleon's state space is $(\mathscr{W} \otimes \mathscr{S}) \otimes \mathscr{I}$ where $\mathscr{W} \otimes \mathscr{S}$ is the familiar spin/orbital state space and $\mathscr{I}$ is a two-dimensional state space with a complete set of states $\{|p\rangle, |n\rangle\}$; the strong Hamiltonian is of the form $H_{st} = H' \otimes \mathbf{1}$ where H' acts on $\mathscr{W} \otimes \mathscr{S}$, so that H_{st} commutes with all operators on $\mathscr{I}$. In particular it commutes with the unitary operators U defined by†

$$\left.\begin{array}{l} U|p\rangle = \alpha|p\rangle + \beta|n\rangle \\ U|n\rangle = \gamma|p\rangle + \delta|n\rangle \end{array}\right\}, \tag{6.1}$$

where the matrix

$$\begin{pmatrix} \alpha & \beta \\ \gamma & \delta \end{pmatrix}$$

belongs to SU(2) (any unitary operator on $\mathscr{I}$ is a multiple of such a U). Thus the identical behaviour of the proton and the neutron under the strong force can be regarded as the result of invariance of the strong force under the group of operations represented by (6.1); this group is isomorphic to SU(2), and it acts on the state space $\mathscr{I}$ in the same way as rotation operators act on the spin space $\mathscr{S}$.

Since this is a continuous group of invariances, it has hermitian generators which, according to ●3.5, represent conserved observables. These operators act on the state space $\mathscr{I}$ in the same way as the generators of rotations – the angular momentum operators – act on the spin space $\mathscr{S}$. Thus $\mathscr{I}$ has three operators I_1, I_2, I_3 defined by the 2×2 matrices $\frac{1}{2}\sigma_1, \frac{1}{2}\sigma_2, \frac{1}{2}\sigma_3$ where σ_i are the three Pauli matrices (4.39) (but usually denoted by τ_i in this context). These represent conserved observables; they are called the components of **isospin**. The proton and neutron are in eigenstates of the third component of isospin

---

† This definition contains superpositions of states with different electric charge, whose existence is forbidden by a superselection rule. It might seem, therefore, that the operator U has no physical significance. However, we can consider a fictitious world in which the electromagnetic force does not operate; in such a world there would be no electric charge and therefore no superselection rule. Since the strong force is so much stronger than the electromagnetic force, it is reasonable to expect this fictitious world to be a good approximation to the real world. In particular, if we can use the transformations (6.1) to derive consequences which do not refer to the forbidden superpositions, we can expect these consequences to be (approximately) true in the real world.

with eigenvalues $+\frac{1}{2}$ and $-\frac{1}{2}$ respectively. The symmetry operations of (6.1) are called **isospin transformations**.

The components of isospin have the same commutation relations as the components of angular momentum. Hence the theory of §4.1, which was based solely on these commutation relations, can be applied to isospin. Thus the quantity $\mathbf{I}^2 = I_1{}^2 + I_2{}^2 + I_3{}^2$ has eigenvalues $I(I+1)$ where I is an integer or half-integer; the proton and the neutron, whose isospin was constructed on the model of a spin-$\frac{1}{2}$ particle, have $I = \frac{1}{2}$.

There are many consequences of isospin conservation in nuclear physics. The isospin of many-nucleon states is obtained by adding that of the individual nucleons, in the same way as angular momentum is added. The invariance of the Hamiltonian under isospin transformations then implies that the $2I + 1$ states with a given value of I should all have the same energy. Thus in the mirror nuclei (^{3}H, ^{3}He) each pair of corresponding energy levels constitutes a doublet with $I = \frac{1}{2}$, the ^{3}H state having $I_3 = -\frac{1}{2}$ and the ^{3}He state having $I_3 = \frac{1}{2}$. These states can be regarded as being formed by adding one nucleon (with $I = \frac{1}{2}$) to a two-nucleon state which has $I = 0$ (i.e. a state containing a proton and a neutron). Similarly, the pair (^{13}C, ^{13}N) have $I = \frac{1}{2}$ and are formed by adding one nucleon to a 12-nucleon core in which the isospins of the individual particles have combined to give $I = 0$.

The pair (^{14}C, ^{14}O) are formed from the 12-nucleon core by adding two nucleons (two neutrons in the case of ^{14}C, two protons for ^{14}O). These two nuclei have $I_3 = \pm 1$ and must belong to the triplet of states with $I = 1$ which can be formed by adding the isospins of the two nucleons outside the core. The third member of this triplet, with $I_3 = 0$, must be a state of ^{14}N. Another state of ^{14}N will be formed by adding the two nucleons in the state $I = 0$. These two states of ^{14}N must satisfy

$$\langle \psi_0 | e^{-iH_{st}t} | \psi_1 \rangle = 0 \tag{6.2}$$

since they have different values of $\mathbf{I}^2$ and H_{st} commutes with $\mathbf{I}$; thus there will be no transitions between these states on the time scale of the strong interaction. The actual situation is shown in Fig. 6.1. The ground state of ^{14}N, which is stable, has much lower energy than the ground states of ^{14}C and ^{14}O; these are unstable and decay to the ground state of ^{14}N, but by the weak interaction (i.e.

Fig. 6.1.
Isospin multiplets in nuclear physics.

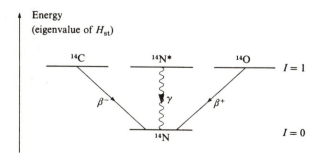

by β-decay), not the strong. There is also a long-lived excited state of ^{14}N, which decays to the ground state by the electromagnetic interaction (γ-decay), not the strong. This excited state is close in energy to the ground states of ^{14}C and ^{14}O, and the differences between the three can be accounted for by the differences in their electrostatic energies. Thus the four states of Fig. 6.1 clearly show how the energy levels of the strong Hamiltonian carry representations of the invariance group SU(2), in accordance with ●3.13.

So far isospin transformations have only been defined on states containing nucleons. By analogy with rotations, they should be defined on all states in the fictitious world in which only the strong force operates, i.e. on all states of hadrons. Then, as with nuclei, the energy levels of hadrons will carry representations of SU(2) labelled by the total isospin I. Since each single-particle state is an eigenstate of the strong Hamiltonian, this means that the hadrons are classified into *isospin multiplets*: a multiplet with isospin I contains $2I+1$ particles which all have the same mass (because of the relativistic equivalence between mass and energy). The particles are not necessarily eigenstates of the weak and electromagnetic Hamiltonians, so this equality of mass in a multiplet is only approximate. Some of these multiplets are listed in Table 6.1. In each multiplet the value of I_3 is related to the electric charge Q by a formula of the form

$$I_3 = Q - \tfrac{1}{2}Y \tag{6.3}$$

where Y is an integer which is characteristic of the multiplet. It is called the **hypercharge** of the multiplet.

The isospin of hadrons can be understood in terms of the quarks which

Table 6.1. *Isospin multiplets*

	I_3	I	Y	J^P
Baryons ($B=0$)				
Nucleons (n, p)	$(-\tfrac{1}{2}, \tfrac{1}{2})$	$\tfrac{1}{2}$	1	$\tfrac{1}{2}^+$
$(\Sigma^-, \Sigma^0, \Sigma^+)$	$(-1, 0, 1)$	1	0	$\tfrac{1}{2}^+$
Λ^0	0	0	0	$\tfrac{1}{2}^+$
(Ξ^-, Ξ^0)	$(-\tfrac{1}{2}, \tfrac{1}{2})$	$\tfrac{1}{2}$	-1	$\tfrac{1}{2}^+$
$(\Delta^-, \Delta^0, \Delta^+, \Delta^{++})$	$(-\tfrac{3}{2}, -\tfrac{1}{2}, \tfrac{1}{2}, \tfrac{3}{2})$	$\tfrac{3}{2}$	1	$\tfrac{3}{2}^+$
Mesons ($B=0$)				
Pions (π^-, π^0, π^+)	$(-1, 0, 1)$	1	0	0^-
Kaons $\begin{cases}(K^0, K^+)\\(K^-, \overline{K^0})\end{cases}$	$(-\tfrac{1}{2}, \tfrac{1}{2})$ $(-\tfrac{1}{2}, \tfrac{1}{2})$	$\tfrac{1}{2}$ $\tfrac{1}{2}$	1 -1	0^- 0^-
η	0	0	0	0^-
(ρ^-, ρ^0, ρ^+)	$(-1, 0, 1)$	1	0	1^-
ω	0	0	0	1^-

(J = spin, P = parity)

make them up, in the same way as the isospin of nuclei is understood in terms of nucleons. The u and d quarks form an isospin doublet, with d having $I_3 = -\frac{1}{2}$ (isospin *down*), u having $I_3 = \frac{1}{2}$ (isospin *up*); all other quarks have $I = 0$. Then all isospin transformations follow from the following basic transformation on u and d states:

$$U|u\rangle = \alpha|u\rangle + \beta|d\rangle$$
$$U|d\rangle = \gamma|u\rangle + \delta|d\rangle \tag{6.4}$$

Each of the multiplets in Table 6.1 is to be regarded as a set of states of a single particle, like the nucleon; this particle then has a complete state space of the form $\mathscr{W} \otimes \mathscr{S} \otimes \mathscr{I}$. The discussion of many-particle states in §4.6 shows that the multiplet particle can be treated as a fermion if the individual members of the multiplet are fermions, and as a boson if they are bosons: the requirement of symmetry or antisymmetry should be applied to the total state, including the isospin state. The baryons are fermions, and the state of two baryons from the same multiplet must be antisymmetric overall (e.g. symmetric in the orbital state, antisymmetric in spin, symmetric in isospin); the mesons are bosons, and the two-particle states must be symmetric overall.

Example 1. The deuteron. The deuteron (the nucleus of deuterium, ^{2}H) is a bound state of a neutron and a proton. Since it is formed from two isospin-$\frac{1}{2}$ particles, its isospin I is either 0 or 1. If it had $I = 1$, it would be a member of a triplet; the other members would be proton–proton and neutron–neutron bound states. Such states do not exist, so the deuteron must have $I = 0$. This is the antisymmetric combination of the two $I = \frac{1}{2}$ states. From our experience with the hydrogen atom and the harmonic oscillator (Fig. 4.5) it seems likely that the lowest-energy bound state will have relative orbital angular momentum $l = 0$ (intuitively, if l is non-zero the particles experience a centrifugal force tending to break up the bound state); then the orbital state will be symmetric. Since nucleons are fermions, the overall state must be antisymmetric, so the spin state must be symmetric; therefore the total spin is $s = 1$. Thus the total intrinsic angular momentum of the deuteron, the sum of $l = 0$ and $s = 1$, is 1; and its parity is $(-1)^l = +1$.

A number of consequences of isospin conservation follow from the fact that the time evolution operator e^{-iHt} commutes with isospin transformations and is therefore an **isoscalar** operator, the isospin analogue of a scalar operator in the theory of angular momentum. We can apply the Wigner–Eckart theorem (4.75) to deduce that if $|I\,I_3\,\alpha\rangle$ is a set of states labelled by isospin and some other quantity α, then

$$\langle I\,I_3\,\alpha|e^{-iHt}|I'I'_3\alpha'\rangle = \delta_{II'}\delta_{I_3I'_3}\langle\alpha\|U_I(t)\|\alpha'\rangle. \tag{6.5}$$

The significance of this equation is twofold: since isospin is conserved, the values of I and I_3 are the same at time t as at $t = 0$; and since the process is invariant under isospin transformations, the probability amplitude is independent of I_3.

Example 2. Nucleon–nucleon scattering. Take $|\alpha\rangle$ to be the spin/orbital state of the two nucleons. If this state is symmetric, the isospin state must be antisymmetric and so the total isospin is $I = 0$; by conservation of isospin, the value of I remains 0 and therefore the spin/orbital state remains symmetric. The reduced matrix element $\langle \alpha \| U_0(t) \| \alpha' \rangle$ in (6.5) then describes the neutron–proton scattering in the symmetric spin/orbital states. If the spin/orbital state is antisymmetric, the isospin state must be symmetric, so $I = 1$; there are then three possible states, corresponding to p–p, n–p and n–n scattering, and (6.5) says that all three scattering processes are governed by the same amplitude $\langle \alpha \| U_1(t) \| \alpha' \rangle$. This shows how the charge independence of nuclear forces is incorporated in isospin invariance.

Example 3. Decay rates. Each Δ-particle decays into a neutron and a pion. We write

$$\Delta \to N + \pi \tag{6.6}$$

which covers the six decays

$$\Delta^{++} \to p + \pi^+, \quad \Delta^+ \to p + \pi^0, \quad \Delta^+ \to n + \pi^+,$$
$$\Delta^0 \to n + \pi^0, \quad \Delta^0 \to p + \pi^-, \quad \Delta^- \to n + \pi^-. \tag{6.7}$$

Consider the two decays of the Δ^+. This has $I = \frac{3}{2}$, $I_3 = \frac{1}{2}$; by conservation of isospin it evolves into the nucleon–pion state with the same values of I and I_3. Adding the isospin of the nucleon and the pion is done by means of Clebsch–Gordan coefficients, which give

$$\left| \tfrac{3}{2} \tfrac{1}{2} \right\rangle = \sqrt{\tfrac{2}{3}} \left| \tfrac{1}{2} \tfrac{1}{2}, 1\,0 \right\rangle + \sqrt{\tfrac{1}{3}} \left| \tfrac{1}{2} - \tfrac{1}{2}, 1\,1 \right\rangle$$
$$= \sqrt{\tfrac{2}{3}} \left| p\,\pi^0 \right\rangle + \sqrt{\tfrac{1}{3}} \left| n\,\pi^+ \right\rangle. \tag{6.8}$$

The probability that this state will be observed as $p + \pi^0$ is $\frac{2}{3}$; the probability of $n + \pi^+$ is $\frac{1}{3}$. Hence the decay $\Delta^+ \to p + \pi^0$ occurs twice as often as $\Delta^+ \to n + \pi^+$:

$$\Gamma(\Delta^+ \to p + \pi^0) = 2\Gamma(\Delta^+ \to n + \pi^+) \tag{6.9}$$

where Γ denotes the rate (probability per unit time) of the decay.

Using (6.5), we can relate the rates of decays of different Δ-particles. Taking the labels α, α' to refer to Δ states and $N\pi$ states, and writing $|\Delta\,m\rangle$ for the Δ state with $I_3 = m$, we have

$$\left. \begin{array}{l} \langle N\pi; \tfrac{3}{2}m | e^{-iHt} | \Delta\,m \rangle = \langle N\pi \| U_{\frac{3}{2}}(t) \| \Delta \rangle \\ \langle N\pi; \tfrac{1}{2}m | e^{-iHt} | \Delta\,m \rangle = 0 \end{array} \right\}. \tag{6.10}$$

Thus all six decays (6.7) are governed by a single function of t, which, according to §3.5, is approximately exponential:

$$\left| \langle N\pi \| U_{\frac{3}{2}}(t) \| \Delta \rangle \right|^2 \simeq e^{-\Gamma t}. \tag{6.11}$$

Here Γ is the total decay rate of any one state $|\Delta m\rangle$; thus, for example,

$$\Gamma(\Delta^{++} \to p + \pi^+) = \Gamma(\Delta^+ \to p + \pi^0) + \Gamma(\Delta^+ \to n + \pi^+). \tag{6.12}$$

Isospin and charge conjugation   The statement that the charge conjugation operator C acts on a state so as to change every particle in the state into its antiparticle does not completely specify it as an operator, since there is an ambiguity of a possible phase factor.

Some of this ambiguity can be removed by reference to isospin transformations.

Let X_m $(m = -I, \ldots, I)$ be particles forming a multiplet with isospin I, and denote their states (for a given spin/orbital state) by $|X\ m\rangle$. Then

$$\mathbf{I}|X\ m\rangle = \sum_n \mathbf{t}^I_{nm}|X\ n\rangle \qquad (6.13)$$

where $\mathbf{t}^I$ is the set of three $(2I + 1) \times (2I + 1)$ matrices representing the hermitian generators of isospin transformations (given by (4.50)–(4.51) with $j = I$). Since all quantum numbers are reversed on going from a particle to its antiparticle, $C|X\ m\rangle$ is an eigenstate of I_3 with eigenvalue $-m$. Suppose the phases are such that the states

$$|\bar{X} - m\rangle = (-1)^m C|X\ m\rangle \qquad (6.14)$$

behave like $|X\ -m\rangle$ with respect to isospin:

$$\mathbf{I}|\bar{X} - m\rangle = \sum_n \mathbf{t}^I_{-n,\ -m}|\bar{X} - n\rangle. \qquad (6.15)$$

From the formulae (4.50)–(4.51) it can be seen that

$$\mathbf{t}^I_{-n,\ -m} = (-1)^{m+n+1}\mathbf{t}^I_{mn} = (-1)^{m+n+1}\overline{\mathbf{t}^I_{nm}} \qquad (6.16)$$

(the second equality because the matrices $\mathbf{t}^I$ are hermitian), so that (6.15) gives

$$\mathbf{I}C|X\ m\rangle = -\sum_n \mathbf{t}^I_{mn}C|X\ n\rangle = -\sum_n \overline{\mathbf{t}^I_{nm}}C|X\ n\rangle. \qquad (6.17)$$

Hence the effect of an isospin transformation R (the isospin version of a rotation through angle θ about axis $\mathbf{n}$) is

$$U(R)C|X\ m\rangle = \exp(i\theta\mathbf{n}\cdot\mathbf{I})C|X\ m\rangle$$

$$= \sum_n [\exp(-i\theta\mathbf{n}\cdot\mathbf{t}^I)]_{nm}C|X\ n\rangle$$

$$= \sum_n \overline{d^I_{nm}(R)}C|X\ n\rangle \qquad (6.18)$$

where $d(R) = \exp(-i\theta\mathbf{n}\cdot\mathbf{t})$ is the $(2I+1) \times (2I+1)$ matrix representing the isospin transformation R. Thus the charge conjugate states $C|X\ m\rangle$ transform by the *complex conjugate* representation of isospin transformations. We will take it to be part of the definition of the charge conjugation operator C that it has this relation to isospin.

The requirement (6.18) does not completely fix the operator C, since it is still possible to multiply all the states $C|X\ m\rangle$ by the same phase factor. If the multiplet contains a particle X^0 which is its own antiparticle, this one remaining phase is determined by the charge conjugation parity η_c of X^0. Since X^0 must be totally neutral, it must have $I_3 = 0$ and electric charge $Q = 0$; hence the multiplet has hypercharge $Y = 0$. The antiparticle of each multiplet particle X_m will be another member of the multiplet, X_{-m} (the multiplet as a whole is **self-conjugate**). From (6.14) we know that $C|X\ m\rangle$ must be a multiple of

$(-1)^m |X \; -m\rangle$, and by taking $m=0$ we see that

$$C|X \; m\rangle = (-1)^m \eta_c |X \; -m\rangle. \tag{6.19}$$

Thus η_c is a property of the multiplet as a whole. It is often discussed in terms of **G-parity**, which is the eigenvalue of the operator

$$G = C e^{i\pi I_2}. \tag{6.20}$$

It can be shown (by the reader: problem 6.8) that (6.19) makes C an isospin analogue of reflection in the (13)-plane; the isospin transformation in (6.20) is the analogue of rotation through π about the y-axis, so that G is an isospin analogue of space inversion. It commutes with isospin transformations, so that every member of a self-conjugate multiplet is an eigenstate of G with eigenvalue $\alpha \eta_c$, where α depends only on the isospin I of the multiplet (see problem 6.9). The internal properties of isospin and G-parity are collected together in a single symbol I^G like the symbol J^P for the space–time properties of spin and parity. Some values of I^G are given in Table 6.2.

Irreducible isospin operators The behaviour of operators under isospin transformations can be discussed in the same way as their behaviour under rotations. An **irreducible operator of isospin type I** is a set of $2I+1$ operators Ω^I_m ($m = -I, \ldots, I$) satisfying

$$[\mathbf{I}, \Omega^I_m] = \sum_n \mathbf{t}^I_{nm} \Omega^I_n \tag{6.21}$$

(cf. (4.73)). An example of such a set of operators is provided by the creation operators $a_{Xm}{}^\dagger$ for a multiplet X_m with isospin I; for if $|\Psi\rangle$ is any state, $a_{Xm}{}^\dagger |\Psi\rangle$ is a state with an added X-particle, and the isospin operators $\mathbf{I}$ act on this as a sum of an operator which acts on the state $|\Psi\rangle$ and an operator which acts on the state of the X-particle:

$$\mathbf{I} a_{Xm}{}^\dagger |\Psi\rangle = a_{Xm}{}^\dagger \mathbf{I} |\Psi\rangle + \left[\sum_n \mathbf{t}^I_{nm} a_{Xm}{}^\dagger \right] |\Psi\rangle. \tag{6.22}$$

Hence

$$[\mathbf{I}, a_{Xm}{}^\dagger] = \sum_n \mathbf{t}^I_{nm} a_{Xn}{}^\dagger. \tag{6.23}$$

The annihilation operators a_{Xm} have different isospin properties, which can be found by taking the hermitian conjugate of (6.23):

$$[\mathbf{I}, a_{Xm}] = -\sum \overline{\mathbf{t}^I_{nm}} \, a_{Xn} = -\sum \mathbf{t}^I_{mn} a_{Xn} \tag{6.24}$$

Table 6.2. *Self-conjugate multiplets*

Multiplet	J^P	I^G	η_c
π	0^-	1^-	$+$
η	0^-	0^+	$+$
ρ	1^-	1^+	$-$
ω	1^-	0^-	$-$

(since the matrices $\mathbf{t}^l$ are hermitian). Comparing with (6.17), we see that the annihilation operators a_{Xm} have the same properties as the creation operators $Ca_{Xm}{}^\dagger C$ which create the antiparticle states $C|X\,m\rangle$. Conversely, the antiparticle annihilation operators $Ca_{Xm}C$ obey the same equation as the creation operators $a_{Xm}{}^\dagger$.

These commutation relations can be conveniently expressed by collecting the creation operators $a_{Xm}{}^\dagger$ into a row vector $a_X{}^\dagger = (a_{X,-I}{}^\dagger, \ldots, a_{XI}{}^\dagger)$ and the annihilation operators into a column vector a_X, and the antiparticle operators $Ca_{Xm}{}^\dagger C$ and $Ca_{Xm}C$ into a column vector $a_{\bar{X}}{}^\dagger$ and a row vector $a_{\bar{X}}$. Then we have

$$[\mathbf{I}, a_X{}^\dagger] = a_X{}^\dagger \mathbf{t}^l, \quad [\mathbf{I}, a_X] = -\mathbf{t}^l a_X$$
$$[\mathbf{I}, a_{\bar{X}}] = a_{\bar{X}} \mathbf{t}^l, \quad [\mathbf{I}, a_{\bar{X}}{}^\dagger] = -\mathbf{t}^l a_{\bar{X}}{}^\dagger \tag{6.25}$$

The quantum fields of the particles X_m are

$$\phi_{Xm} = a_{Xm} + Ca_{Xm}{}^\dagger C. \tag{6.26}$$

We can collect these into a column vector $\phi_X = a_X + a_{\bar{X}}{}^\dagger$ and their hermitian conjugates into a row vector $\phi_X{}^\dagger$; these satisfy

$$[I, \phi_X] = -\mathbf{t}^l \phi_X, \quad [\mathbf{I}, \phi_X{}^\dagger] = \phi_X{}^\dagger \mathbf{t}^l. \tag{6.27}$$

If X is a self-conjugate multiplet the fields are

$$\phi_{Xm} = a_{Xm} + Ca_{Xm}{}^\dagger C = a_{Xm} + (-1)^m \eta_c a_{X-m}{}^\dagger,$$

by (6.19); thus they satisfy the hermiticity condition

$$\phi_{Xm}{}^\dagger = (-1)^m \eta_c \phi_{X-m}. \tag{6.28}$$

Particularly important types of irreducible operator are the isospin analogues of scalar and vector operators. An **isoscalar** operator is one which commutes with all isospin operators; this is another name for an irreducible operator with $I = 0$. An **isovector** operator is a set of three operators (V_1, V_2, V_3) which satisfy

$$[I_i, V_j] = i\varepsilon_{ijk} V_k. \tag{6.29}$$

This is another name (and a different choice of basis) for an irreducible operator of isospin type $I = 1$, for from such an operator $T^1{}_m$ $(m = 0, \pm 1)$ we can form an isovector operator V_i by defining

$$V_1 = (1/\sqrt{2})(T^1{}_{-1} - T^1{}_{+1}), \quad V_2 = (i/\sqrt{2})(T^1{}_{+1} + T^1{}_{-1}), \quad V_3 = T^1{}_0. \tag{6.30}$$

As with ordinary (rotation group) vector operators, from two isovector operators **V** and **W** we can form an isoscalar operator $\mathbf{V} \cdot \mathbf{W}$ and an isovector operator $\mathbf{V} \times \mathbf{W}$.

Let ϕ_X be the column vector of quantum fields of a multiplet X with isospin I, and let

$$\mathbf{V}_X = \phi_X{}^\dagger \mathbf{t}^l \phi_X. \tag{6.31}$$

This is an isovector operator, for

$$[I_i, V_{Xj}] = [I_i, \phi_X^\dagger] t^l{}_j \phi_X + \phi_X^\dagger t^l{}_j [V_i, \phi_X]$$
$$= \phi_X (t^l{}_i t^l{}_j - t^l{}_j t^l{}_i) \phi_X$$
$$= \phi_X^\dagger (i\varepsilon_{ijk} t^l{}_k) \phi_X = i\varepsilon_{ijk} V_{Xk} \qquad (6.32)$$

since the generators $t^l{}_i$ satisfy the SU(2) commutation relations. Note that as a consequence of the components of $\mathbf{t}^l$ being hermitian matrices, the components of $\mathbf{V}_X$ are hermitian operators.

Let $(\alpha^-, \alpha^0, \alpha^+)$ be a self-conjugate isospin triplet with $\eta_c = +1$. According to (6.30), the fields $\phi_{\alpha m}^\dagger$ yield an isovector operator $\mathbf{A}$. Since the multiplet is self-conjugate the fields satisfy (6.28), from which it follows that the components of $\mathbf{A}$ are hermitian. Hence if the fields ϕ_X and ϕ_α commute, the operator

$$H' = \phi_X^\dagger \mathbf{t}^l \phi_X \cdot \mathbf{A} \qquad (6.33)$$

is a hermitian isoscalar, which is therefore qualified to be a part of a Hamiltonian describing an isospin–invariant interaction.

Let us take X_m $(m = \pm\tfrac{1}{2})$ to be the nucleons and α_m $(m = 0, \pm 1)$ to be the pions. Then, adopting the common practice of denoting the quantum field of a particle by the same symbol as the particle itself, the Hamiltonian (6.33) becomes

$$H' = N^\dagger \boldsymbol{\tau} N \cdot \boldsymbol{\pi} \quad \text{where } N = \begin{pmatrix} p \\ n \end{pmatrix}$$

$$= (p^\dagger \, n^\dagger) \begin{pmatrix} \pi^0 & -\sqrt{2}\pi^+ \\ \sqrt{2}\pi^- & -\pi^0 \end{pmatrix} \begin{pmatrix} p \\ n \end{pmatrix}$$

$$= p^\dagger p \pi^0 - \sqrt{2} p^\dagger n \pi^+ + \sqrt{2} n^\dagger p \pi^- - n^\dagger n \pi^0. \qquad (6.34)$$

This is the **Yukawa–Kemmer** interaction. It represents a theory of the strong force between two nucleons in which the field quanta are pions, and which has the feature of isospin invariance.

Isospin in electromagnetic and weak interactions   Since isospin invariance proclaims the equivalence of states with different electric charge, it is flagrantly violated by electromagnetic interactions. Likewise, the decay of the neutron involves a change in the value of I_3 (the leptons having no isospin), so the weak interactions also do not conserve isospin. However, this only means that the electromagnetic and weak Hamiltonians have non-zero commutators with the isospin operators; it is possible to apply isospin to these interactions by determining the exact form of the commutators.

The formula (6.3) shows that electric charge is the sum of I_3, which is a component of the isovector $\mathbf{I}$, and a quantity $-\tfrac{1}{2}Y$ which has the same value for all members of an isospin multiplet and therefore commutes with all isospin transformations, i.e. it is an isoscalar. The electromagnetic Hamiltonian H_{em} has a similar structure:

$$H_{\text{em}} = e J_{\text{em}} \phi_\gamma = e(J^0 + J^1{}_0) \phi_\gamma \qquad (6.35)$$

where ϕ_γ is the photon field and the **electromagnetic current** J_{em} is the sum of an isoscalar J^0 and the $I_3 = 0$ member of an isospin triplet $\{J^1{}_m: m = 0, \pm 1\}$.

Example 4. Radiative decay. The ρ-mesons (ρ^-, ρ^0, ρ^+) are an isospin triplet. One of their rarer decay modes is to emit a photon and become a pion. Since the electromagnetic Hamiltonian contains the small parameter e, the decays can be treated by first-order perturbation theory; this gives the amplitude for $\rho^+ \to \pi^+ + \gamma$ (ignoring kinematical factors) as

$$M_+ = \langle \pi^+ \gamma | H_{em} | \rho^+ \rangle = e \langle \pi^+ | J_{em} | \rho^+ \rangle. \tag{6.36}$$

Labelling the particle states as $|\rho\ I\ I_3\rangle = |\rho\ 1\ m\rangle$ and $|\pi\ 1\ m\rangle$, we have from the Wigner–Eckart theorem (p. 148)

$$\langle \pi\ 1\ m | J^0 | \rho\ 1\ m\rangle = \langle 1\ m | 0\ 0,\ 1\ m\rangle \langle \pi \| J^0 \| \rho\rangle = \langle \pi \| J^0 \| \rho\rangle$$

and

$$\langle \pi\ 1\ m | J^1{}_0 | \rho\ 1\ m\rangle = \langle 1\ m | 1\ 0,\ 1\ m\rangle \langle \pi \| J^1 \| \rho\rangle.$$

Hence, using the Clebsch–Gordan coefficients in Appendix III,

$$\begin{aligned}
M_+ &= \langle \pi^+ | J_{em} | \rho^+ \rangle = \langle \pi\ 1\ 1 | (J^0 + J^1{}_0) | \pi\ 1\ 1\rangle \\
&= \langle \pi \| J^0 \| \rho\rangle + \sqrt{\tfrac{1}{2}} \langle \pi \| J^1 \| \rho\rangle, \\
M_0 &= \langle \pi^0 | J_{em} | \rho^0 \rangle = \langle \pi \| J^0 \| \rho\rangle
\end{aligned}$$

and

$$M_1 = \langle \pi^- | J_{em} | \rho^- \rangle = \langle \pi \| J^0 \| \rho\rangle - \sqrt{\tfrac{1}{2}} \langle \pi \| J^1 \| \rho\rangle.$$

Thus the three amplitudes are related by

$$M_+ + M_- = 2M_0. \tag{6.37}$$

The *weak* interaction, at the level of hadrons and leptons, can be described by an effective Hamiltonian

$$H_w = g_w{}' J_{ch}{}^\dagger J_{ch} \tag{6.38}$$

where the **charged weak current** J_{ch} is the sum of a number of terms which include the **leptonic current**

$$J_{lep} = e^\dagger v_e + \mu^\dagger v_\mu + \tau^\dagger v_\tau \tag{6.39}$$

and also a term $J^1{}_-$ which belongs to the same isospin triplet as the electromagnetic operator $J^1{}_0$. The third member of this triplet, $J^1{}_+$, occurs in $J_{ch}{}^\dagger$. The terms $J_{lep}{}^\dagger J^1{}_-$ and $J^1{}_+ J_{lep}$ in the weak Hamiltonian are responsible for nuclear β-decay processes like those on the left and right of Fig. 6.1; the electromagnetic Hamiltonian $J^1{}_0 \phi_\gamma$ is responsible for the γ-decay in the centre of that figure. Thus the three processes shown there are related to the three components of a single isovector, and are governed by a single reduced matrix element (see problem 6.11). This is a pointer towards the unified electroweak theory, which will be described in §6.7.

It is remarkable that all weak processes, whatever particles they involve (whether hadrons, leptons or both), can be described in terms of the single coupling constant $g_w{}'$. This fact is called the **universality** of the weak interactions.

6.2. Strangeness   The property of strangeness was discovered in the new particles that are created in the collisions between cosmic rays and terrestrial matter. Cosmic rays themselves are mainly ordinary nuclei which originate outside the solar system and arrive here with enormous energies; when they collide with other nuclei this energy makes it possible to produce a numerous burst of particles. A number of the particles listed in Table 6.1 were first observed in these cosmic ray bursts, in particular the Λ^0, the $\Sigma^\pm$ and the Ξ doublet among the baryons, and the two K doublets among the mesons. They are unstable, decaying as follows:

$$\Lambda^0 \to N + \pi, \quad \Sigma^\pm \to N + \pi, \quad \Xi \to \Lambda + \pi; \tag{6.40}$$

$$K \to 2\pi, \qquad K \to 3\pi, \qquad K \to \pi + l + \bar{\nu}_l, \tag{6.41}$$

where N denotes a nucleon and l a lepton ($e^\pm$ or $\mu^\pm$). Their lifetimes are of the order of 10^{-10} to 10^{-8} s, which is time enough for them to leave a visible track (centimetres long) in a photographic emulsion or, if neutral, a visible gap between the point of production and the point of decay. This time scale is characteristic of the weak interactions.

On the other hand, large numbers of these particles appear in cosmic ray collisions: there is a high probability that they will be produced. This means that the Hamiltonian governing the process has large matrix elements; in fact the rate of production is consistent with its being governed by the *strong* Hamiltonian. This seems to conflict with the fact that their decays are governed by the weak interactions. Most of the decays involve strongly interacting particles; one would therefore expect them to be governed by the strong interaction (the particles involved in the original production being available as a virtual intermediate state), and to exhibit the typical strong-interaction time scale of 10^{-23} s. There is something strange about these particles.

The resolution of this puzzle was proposed by Pais, Gell-Mann and Nishijima in 1952. They suggested that the reason that the decays did not take place by the strong interaction was that they involved a change in a quantity which was conserved by the strong force. This quantity belongs to the strange new particles but not to the other particles (nucleons, pions and leptons); the decay of a strange particle can therefore only take place by the weak force, which does not conserve strangeness. The *production* of strange particles can take place by the strong force if strangeness can take positive and negative values, like electric charge and the other quantum numbers, and if particles are produced in pairs with total strangeness 0. Observation confirmed that strange particles are produced in pairs and confirmed the idea of strangeness.

The assignment of strangeness to particles is shown in Table 6.3. The baryons all have strangeness of the same sign (conventionally chosen as negative; their antiparticles have positive strangeness), while one of the K-meson doublets has strangeness $+1$ and the other has strangeness -1. At moderate energies the production process will yield a strange baryon together with a positively strange K-meson, e.g.

$$p + p \to n + \Sigma^+ + K^+. \tag{6.42}$$

To produce a negatively strange K-meson requires more energy, since it must be accompanied by a particle (either another K-meson, or an antibaryon) which cannot be obtained by converting a nucleon. Thus the K^- is rarer than the K^+; in this sense K^+ and K^0 go together with nucleons and other baryons to count as 'matter', while K^- and $\bar{K}^0$ are antimatter.

In the decays (6.40)–(6.41), in which the particles lose their strangeness, the total strangeness changes by just one unit at a time. Thus the Ξ-particles, with strangeness $S = -2$, must decay twice before becoming a non-strange particle. For this reason Ξ is sometimes pronounced 'cascade'. An example of such a two-stage decay is shown for the antiparticle of the Ξ^- in Fig. 6.2.

Strangeness is a property of an isospin multiplet as a whole: all members of the multiplet have the same strangeness. In this respect it is like baryon

Table 6.3. *Strangeness*

Particle	Σ	Λ	Ξ	K^0, K^+	$\bar{K}^0, K^-$	N	π
Strangeness	-1	-1	-2	$+1$	-1	0	0

Fig. 6.2. Two-stage decay of $\overline{\Xi}^+ (= \overline{\overline{\Xi}^-})$ (photo: CERN).

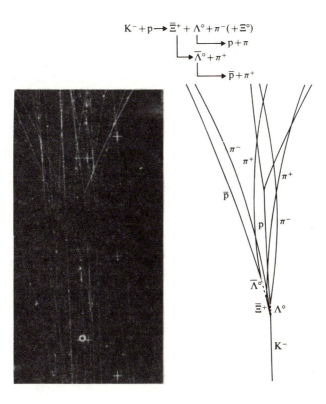

number. There is a relation between these two properties and the multiplet property of hypercharge: a comparison of Tables 6.1 and 6.3 shows that

$$Y = B + S. \tag{6.43}$$

As was described in Chapter 1, strangeness is now understood as the characteristic property of a third quark s which is heavier than the u and d quarks. The Σ and Λ particles each contain one s quark, as do K^- and $\bar{K}^0$; the Ξ particles contain two s quarks, while K^0 and K^+ contain the antiquark $\bar{s}$.

The decays of strange particles can all be understood in terms of the decay of the strange quark:

$$s \rightarrow u + W^- \quad \text{or} \quad \bar{s} \rightarrow \bar{u} + W^+, \tag{6.44}$$

followed by the creation of a pair of particles from the W. If the particles created are a quark and an antiquark, the result is a set of quarks and antiquarks which then rearrange themselves into the final particles. It is found that, to a good degree of accuracy, the effect of this whole process is to change the total isospin by $\frac{1}{2}$: the process is described by an effective Hamiltonian which transforms as one member of an isospin doublet. Thus in the decays $\Lambda \rightarrow N + \pi$ the final state is the combination of $|n\pi^0\rangle$ and $|p\pi^-\rangle$ with total isospin $\frac{1}{2}$, which is $\sqrt{\frac{1}{3}}|n\pi^0\rangle + \sqrt{\frac{2}{3}}|p\pi^-\rangle$ (Clebsch–Gordan coefficients); hence

$$\Gamma(\Lambda^0 \rightarrow p + \pi^-):\Gamma(\Lambda^0 \rightarrow n + \pi^0) = 2:1. \tag{6.45}$$

The actual ratio is 64:36.

This $\Delta I = \frac{1}{2}$ **rule** appears to be hopelessly wrong in the decay $K^+ \rightarrow \pi^+ + \pi^0$, for since all the particles are spinless the two-pion state, with zero relative orbital angular momentum, is symmetric in its spin/orbital state and therefore must be symmetric in its isospin state. This means that its total isospin is 0 or 2; but $I_3 = 1$, so $I = 2$. Since the K-meson has $I = \frac{1}{2}$, the isospin carried by the Hamiltonian must be $\Delta I = \frac{3}{2}$ or $\frac{5}{2}$, and the amount of $\Delta I = \frac{1}{2}$ is strictly zero. However, the rule successfully passes the test of this exception, for this decay is considerably slower than the decay $K^0 \rightarrow 2\pi$ which does obey the $\Delta I = \frac{1}{2}$ rule, the ratio of the rates being 1:138. This indicates that the part of the Hamiltonian which has isospin $\frac{1}{2}$ is very much larger than the other parts, which is the sense in which the $\Delta I = \frac{1}{2}$ rule is to be understood.

The **semileptonic** decays of strange particles, i.e. the last decay of (6.41) and rare decays like

$$\Lambda^0 \rightarrow p + e^- + \bar{\nu}_e, \tag{6.46}$$

can be understood in the same way as β-decay, in terms of the current–current Hamiltonian (6.38) (the purely hadronic decays do not fit simply into this scheme). They can be incorporated by adding a strangeness-changing term

$$J^{\frac{1}{2}}_- = \Lambda^\dagger p + (\pi^0)^\dagger K^+ + \cdots \tag{6.47}$$

to the weak current. As the notation indicates, $J^{\frac{1}{2}}_-$ transforms as a member of a doublet under isospin transformations. Like the leptonic term (6.39), this current involves a change in electric charge; it also involves a change in strangeness of the same amount. Thus in all decays like (6.46), the change in charge and strangeness of the hadrons are equal:

$$\Delta S = \Delta Q. \tag{6.48}$$

One further modification must be made to this current–current Hamiltonian. The force responsible for (6.46) and similar strangeness-changing decays is weaker than that of β-decay by a factor of about 20; thus the Hamiltonians governing these decays must be of the form $g_w' J^1_- J_{lep}$ and $g_w'' J^{\frac{1}{2}}_- J_{lep}$ with $g_w' \neq g_w''$. This spoils the universality of the weak interactions. An insight into this can be obtained by writing

$$g_w' = g_w \cos \theta_C, \quad g_w'' = g_w \sin \theta_C \tag{6.49}$$

and taking the full current to be

$$J_w = J^1_- \cos \theta_C + J^{\frac{1}{2}}_- \sin \theta_C + J_{lep}. \tag{6.50}$$

The hadronic current is then a superposition of two parts J^1_- and $J^{\frac{1}{2}}_-$ which are regarded as orthogonal. The angle θ_C is called the **Cabibbo angle**; its value is about $\sin^{-1}(0.23)$. The Hamiltonian $J_w J_w^\dagger$ constructed from this current differs from (6.38) in its action on non-strange particles: the various parts of this new Hamiltonian have coupling constants $g_w \cos^2 \theta_C$, $g_w \cos \theta_C$ and g_w instead of having a single universal coupling constant. Since $\cos \theta_C \simeq 1$, the differences are small; experiments confirm that (6.50) is the true weak current.

Neutral K-mesons Strangeness is not the only quantity that is conserved in the production of strange particles but not in their decay; the same applies to parity. This is illustrated by K-mesons. The production of kaons, and all scattering processes involving kaons and other hadrons, are consistent with parity conservation if the intrinsic parity of the kaon is taken to be negative; this is also what would be expected from its quark composition, by the same argument as for pions (see p. 153). But in the decay $K \to 2\pi$ the final state has parity $(-1)^l$, where l is the relative orbital angular momentum, since all pions have the same intrinsic parity; and since pions and kaons are all spinless, $l = 0$ and so the parity is $+1$. Thus the weak interactions do not conserve parity and therefore are not invariant under mirror reflections.

Let us consider the possibility that the weak interactions are invariant under the combined operation CP, so that there is symmetry between particles and their mirror-image antiparticles. This has particular consequences for the neutral kaons K^0 and $\bar{K}^0$, which we will denote collectively by $(K)^0$.

The only difference between K^0 and its antiparticle $\bar{K}^0$ is the value of strangeness. Since this is not conserved, it does not commute with the total Hamiltonian and it is not necessary for the eigenstates of S in the $(K)^0$ system to be eigenstates of the Hamiltonian; K^0 and $\bar{K}^0$ need not be stationary states. If the system is invariant under CP, the stationary states must be eigenstates of CP (more precisely, there must be a complete set of simultaneous eigenstates of

H and CP). If we consider the states of a $(K)^0$ at rest, the effect of CP on the K^0 state is to take it to a $\bar{K}^0$ state, and we can define the states so that there is no phase factor:

$$CP|K^0\rangle = |\bar{K}^0\rangle, \quad CP|\bar{K}^0\rangle = |K^0\rangle. \tag{6.51}$$

Then the eigenstates of CP are

$$|K_S^0\rangle = \sqrt{\tfrac{1}{2}}(|K^0\rangle + |\bar{K}^0\rangle) \quad \text{and} \quad |K_L^0\rangle = \sqrt{\tfrac{1}{2}}(|K^0\rangle - |\bar{K}^0\rangle) \tag{6.52}$$

with eigenvalues $+1$ and -1 respectively.

Now consider the decays $(K)^0 \to 2\pi$. Any state of two pions with zero total charge must be an eigenstate of CP with eigenvalue $+1$, for the effect of CP is to interchange the particles (P doing the job in the orbital state and C in the charge state), and pions are bosons. Hence if CP is conserved, only the $+$ eigenstate K_S^0 is free to decay into two pions; the other eigenstate must decay into three pions or a pion and two leptons. The phase space factor is much larger for the 2π decay than for the 3π decay, so the 2π decay has a faster rate. This rate is also faster than that of the other three-body decay $K \to \pi + l + \bar{\nu}_l$. It follows that the $+$ eigenstate K_S^0 has a shorter lifetime than the $-$ eigenstate K_L^0.

These conclusions are confirmed experimentally. There are two observed neutral kaons, the short-lived K_S^0 which decays into two pions with a lifetime of $\tau_S = \Gamma_S^{-1} = 9 \times 10^{-11}$ s, and the long-lived K_L^0 whose lifetime is $\tau_L = \Gamma_L^{-1} = 5 \times 10^{-8}$ s.

The existence of the superpositions (6.52) offers a clear demonstration of the basic principles of quantum mechanics. When the particles are first produced (by a strangeness-conserving strong interaction), they are in an eigenstate of strangeness; as explained above, a collision process at moderate energy will produce the positively strange K^0. This can be distinguished from $\bar{K}^0$ by its interactions with ordinary matter; the K^0 undergoes only elastic or charge–exchange scattering

$$K^0 + n \to K^0 + n, \quad K^0 + p \to K^+ + n, \tag{6.53}$$

while the $\bar{K}^0$ can be absorbed:

$$\bar{K}^0 + p \to \Lambda^0 + \pi^+ \tag{6.54}$$

(it behaves like antimatter). Roughly speaking, matter is transparent to K^0 but opaque to $\bar{K}^0$.

The K^0 is a superposition of K_S^0 and K_L^0 with equal coefficients, so it has equal probabilities of decaying quickly into 2π or slowly into 3π. Thus if a beam of K^0 particles is prepared and left for several short lifetimes, about half† of the particles will remain and they will all be K_L^0, decaying only into 3π. This is a superposition of K^0 and $\bar{K}^0$ with equal coefficients, so if the beam is passed through a slab of matter half of the particles will behave like $\bar{K}^0$ and be absorbed (whereas none of them would have done immediately after

---

† To be precise, $\tfrac{1}{2}[\exp(-\Gamma_L t) + \exp(-\Gamma_S t)]$; $\Gamma_L t$ is small and $\Gamma_S t$ is large.

production). The particles that emerge from the slab will be K^0, and again half of them will decay into 2π. The K_S^0 particles have been regenerated from the K_L^0 beam.

This behaviour is strikingly analogous to the behaviour of polarised light passing through crossed polaroids (see Fig. 6.3). K^0 and $\bar{K}^0$ are analogous to light polarised in the north–south and east–west directions, while K_L^0 and K_S^0 are analogous to light polarised NE–SW and NW–SE. A slab of matter acts like a polaroid with its axis pointing north, while the analogue of passing light through a polaroid oriented NE is the operation of waiting several short lifetimes. A $(K)^0$ exhibiting 2π-decay is like light passing through a polaroid oriented NW. Now the fact that a slab of matter will regenerate K_S^0 particles, causing a resurgence of 2π decays, corresponds to the fact that no light can pass NE-oriented and NW-oriented polaroids placed together, but some light can pass if a N-oriented polaroid is placed between them. Leaving the $(K)^0$ system to evolve acts like a measurement of *CP*, forcing it into an eigenstate $|K_L^0\rangle$ or $|K_S^0\rangle$; scattering it off nuclei acts like a measurement of strangeness, forcing it into an eigenstate $|K^0\rangle$ or $|\bar{K}^0\rangle$.

Let us look more closely at the time development of the K^0 after its production. If the particle is at rest the eigenvalues of the Hamiltonian are $m_L c^2$ and $m_S c^2$ where m_L and m_S are the masses of the K_L^0 and K_S^0 particles; hence, taking into account the decays of the particles (and taking $\hbar = c = 1$), the time

Fig. 6.3.
Polarised light and neutral K-mesons: (*a*) polarised light; (*b*) neutral K-mesons.

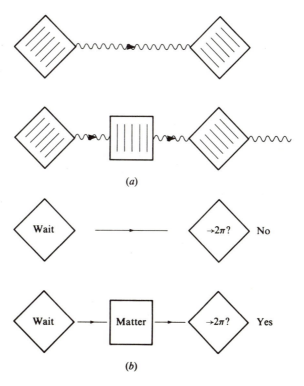

(*a*)

(*b*)

development of K^0 can be represented as

$$|K^0\rangle = \sqrt{\tfrac{1}{2}}(|K_L^0\rangle + |K_S^0\rangle)$$
$$\rightarrow |K(t)\rangle = \sqrt{\tfrac{1}{2}}[\exp(-\tfrac{1}{2}\Gamma_L t - im_L t)]|K_L^0\rangle$$
$$+ \exp(-\tfrac{1}{2}\Gamma_S t - im_S t)|K_S^0\rangle. \tag{6.55}$$

The probability that the particle will behave as K^0 after a time t is therefore

$$|\langle K^0|K(t)\rangle|^2 = \tfrac{1}{4}[e^{-\Gamma_L t} + e^{-\Gamma_S t} + 2e^{-\tfrac{1}{2}(\Gamma_L + \Gamma_S)t}\cos(\Delta mt)] \tag{6.56}$$

where $\Delta m = m_L - m_S$. Thus the system oscillates with frequency Δm. The period of these oscillations is of the order of the short lifetime τ_S, and is easily measurable; it yields a mass difference Δm with the astonishingly small value of 10^{-11} MeV. This makes the $(K)^0$ system a sensitive detector for delicate effects, including even gravitational effects.

CP non-conservation   It is ironic that the K_L^0 particle, whose existence was deduced from *CP* invariance, was the agent of the discovery that *CP* is not conserved. In 1964 it was found that the K_L^0 does decay into 2π. Thus either K_L^0 is not an eigenstate of *CP*, in which case *CP* does not commute with the Hamiltonian in the $(K)^0$ system, or it is, in which case *CP* is not conserved in the decay $K_L^0 \rightarrow 2\pi$. In either case *CP* invariance breaks down. The effect is small, and is not understood.

6.3. **The eightfold way**   Isospin transformations can all be built up from the fundamental transformations (6.4), which change the u and d quarks into combinations of each other. The transformations act on any hadron by acting on all the quarks in the hadron; quarks other than u and d are left unchanged. Clearly this can be extended so as to include the other quarks. To start with, we will just include the strange quark s; this requires transformations like

$$\left.\begin{array}{l} |u\rangle \rightarrow \alpha|u\rangle + \beta|d\rangle + \gamma|s\rangle \\ |d\rangle \rightarrow \delta|u\rangle + \varepsilon|d\rangle + \zeta|s\rangle \\ |s\rangle \rightarrow \eta|u\rangle + \theta|d\rangle + \kappa|s\rangle \end{array}\right\}, \tag{6.57}$$

where the matrix of coefficients belongs to SU(3). These give rise to a group of transformations of hadron states which are called simply SU(3) transformations.

If the strong force had the same effect on all three quarks, u, d and s, these SU(3) transformations would commute with the Hamiltonian and particles would form equal-mass multiplets carrying representations of the group SU(3). This cannot be quite true, since the s quark is significantly more massive than the u and d quarks, and at least the free-particle part of the Hamiltonian will not be invariant under SU(3) transformations. But this does not affect the existence of the multiplets, it only introduces differences in mass between members of a multiplet.

In order to find the form of these multiplets we will examine the structure and representations of the group SU(3). As in the case of the rotation group, this is best done by examining the Lie algebra of the group. The Lie algebra of SU(3) consists of all antihermitian 3×3 matrices with zero trace (see p. 108). We multiply these by i to obtain hermitian generators. The set of all hermitian 3×3 matrices is a nine-dimensional *real* (not complex) vector space (three real numbers are needed to specify the diagonal entries, six to specify the three independent complex entries off the diagonal). The condition of tracelessness removes one dimension, so SU(3) has eight independent hermitian generators. A standard set of generators is the following:

$$\lambda_1 = \begin{bmatrix} 0 & 1 & 0 \\ 1 & 0 & 0 \\ 0 & 0 & 0 \end{bmatrix}, \quad \lambda_2 = \begin{bmatrix} 0 & -i & 0 \\ i & 0 & 0 \\ 0 & 0 & 0 \end{bmatrix}, \quad \lambda_3 = \begin{bmatrix} 1 & 0 & 0 \\ 0 & -1 & 0 \\ 0 & 0 & 0 \end{bmatrix},$$

$$\lambda_4 = \begin{bmatrix} 0 & 0 & 1 \\ 0 & 0 & 0 \\ 1 & 0 & 0 \end{bmatrix}, \quad \lambda_5 = \begin{bmatrix} 0 & 0 & -i \\ 0 & 0 & 0 \\ i & 0 & 0 \end{bmatrix}, \quad \lambda_6 = \begin{bmatrix} 0 & 0 & 0 \\ 0 & 0 & 1 \\ 0 & 1 & 0 \end{bmatrix},$$

$$\lambda_7 = \begin{bmatrix} 0 & 0 & 0 \\ 0 & 0 & -i \\ 0 & i & 0 \end{bmatrix}, \quad \lambda_8 = \frac{1}{\sqrt{3}} \begin{bmatrix} 1 & 0 & 0 \\ 0 & 1 & 0 \\ 0 & 0 & -2 \end{bmatrix}. \tag{6.58}$$

These are called **Gell-Mann's λ-matrices**. They have the property

$$\mathrm{tr}\,(\lambda_i \lambda_j) = 2\delta_{ij}. \tag{6.59}$$

Their commutation relations are written

$$[\lambda_i, \lambda_j] = 2i f_{ijk} \lambda_k \tag{6.60}$$

(here, as in all equations in this section, suffices i, j, k take values from 1 to 8 and the summation convention is used for repeated indices). The constants are the structure constants of SU(3) (in this basis); clearly $f_{ijk} = -f_{jik}$, and as a consequence of (6.59) we have

$$f_{ijk} = -\tfrac{1}{4}i\,\mathrm{tr}\,(\lambda_i \lambda_j \lambda_k - \lambda_j \lambda_i \lambda_k) = -\tfrac{1}{4}i\,\mathrm{tr}\,(\lambda_k \lambda_i \lambda_j - \lambda_i \lambda_k \lambda_j) = f_{kij}. \tag{6.61}$$

Thus the array f_{ijk} is totally antisymmetric. It does the same job for SU(3) as ε_{ijk} does for SU(2).

The generators (6.58) are chosen so as to bring out the structure of the group SU(3). The matrices λ_1, λ_2 and λ_3, consisting of the Pauli σ-matrices bordered by 0s, are the generators of a subgroup which is isomorphic to SU(2); referring to (6.57) we see that these generators act in the subspace spanned by $|u\rangle$ and $|d\rangle$, so this is the isospin (or **I-spin**) subgroup. The matrices λ_6 and λ_7, together with

$$\mu_3 = \begin{bmatrix} 0 & 0 & 0 \\ 0 & 1 & 0 \\ 0 & 0 & -1 \end{bmatrix} = -\frac{1}{2}\lambda_3 + \frac{\sqrt{3}}{2}\lambda_8, \tag{6.62}$$

generate another SU(2) subgroup which acts in the (d, s) subspace; this is called the **U-spin** subgroup. A third SU(2) subgroup, acting in the (u, s) subspace, is generated by λ_4, λ_5 and

$$v_3 = \begin{bmatrix} 1 & 0 & 0 \\ 0 & 0 & 0 \\ 0 & 0 & -1 \end{bmatrix} = \frac{1}{2}\lambda_3 + \frac{\sqrt{3}}{2}\lambda_8; \tag{6.63}$$

this is called the **V-spin** subgroup.

The physical SU(3) operations are represented by a group of unitary operators on the state space of all hadronic states. Their hermitian generators are a set of eight hermitian operators $F_1, \ldots, F_8$ which bear the same relation to the Gell-Mann matrices as the isospin operators I_1, I_2, I_3 do to the Pauli matrices; they have the commutation relations

$$[F_i, F_j] = if_{ijk}F_k. \tag{6.64}$$

For $a = 1, 2, 3$ we have $F_a = I_a$; the other two SU(2) subgroups have generators (U_1, U_2, U_3) corresponding to $(\lambda_5, \lambda_6, \mu_3)$, and (V_1, V_2, V_3) corresponding to $(\lambda_4, -\lambda_5, -v_3)$. Thus we have

$$I_3 = F_3, \quad U_3 = -\tfrac{1}{2}F_3 + \tfrac{1}{2}\sqrt{3}F_8, \quad V_3 = -\tfrac{1}{2}F_3 - \tfrac{1}{2}\sqrt{3}F_8. \tag{6.65}$$

The raising and lowering operators in these subgroups are

$$I_\pm = F_1 \pm iF_2, \quad U_\pm = F_6 \pm iF_7, \quad V_\pm = F_4 \mp iF_5. \tag{6.66}$$

We will now see how their roles of raising and lowering operator fit together in the wider context of SU(3), and at the same time obtain a more meaningful form of the commutation relations (6.64).

Among the matrices (6.58) two which commute are the diagonal matrices λ_3 and λ_8; none of the others commute with both of them. Hence F_3 and F_8 form a complete set of commuting operators in any representation of SU(3), and we can label states by their simultaneous eigenvalues for these operators. Now $I_\pm$ act as raising and lowering operators for F_3, and commute with F_8; hence their effect on a simultaneous eigenstate $|r, s\rangle$ is

$$I_\pm |r, s\rangle = |r \pm 1, s\rangle. \tag{6.67}$$

Let us regard (r, s) as the components of a two-dimensional vector $\mathbf{l}$ and write the simultaneous eigenstate $|r, s\rangle$ as $|\mathbf{l}\rangle$; then

$$I_\pm |\mathbf{l}\rangle = |\mathbf{l} \pm \mathbf{i}\rangle \tag{6.68}$$

where $\mathbf{i} = (1, 0)$. The effects of $U_\pm$ are similar. They act as raising and lowering operators for $U_3 = -\tfrac{1}{2}F_3 + \tfrac{1}{2}\sqrt{3}F_8$ and commute with the orthogonal combination $\tfrac{1}{2}\sqrt{3}F_3 + \tfrac{1}{2}F_8$; $|\mathbf{l}\rangle$ is an eigenstate of U_3 with eigenvalue $\mathbf{u} \cdot \mathbf{l}$ where $\mathbf{u} = (-\tfrac{1}{2}, \tfrac{1}{2}\sqrt{3})$, so the effect of $U_\pm$ is

$$U_\pm |\mathbf{l}\rangle = |\mathbf{l} \pm \mathbf{u}\rangle. \tag{6.69}$$

Similarly,

$$V_\pm |\mathbf{l}\rangle = |\mathbf{l} \pm \mathbf{v}\rangle \tag{6.70}$$

where $\mathbf{v} = (-\tfrac{1}{2}, -\tfrac{1}{2}\sqrt{3})$. The vector $\mathbf{l}$ of simultaneous eigenvalues of F_3 and F_8 is called a **weight**, and the simultaneous eigenvector $|\mathbf{l}\rangle$ is called a **weight vector**.

Because $I_\pm$, $U_\pm$, $V_\pm$ shift the weights around in a plane, we will call them **shift operators**.

The Lie algebra of SU(3) (i.e. the commutators of $F_1, \ldots, F_8$) can now be described as follows. It is characterised by the six special vectors $\pm\mathbf{i}$, $\pm\mathbf{u}$, $\pm\mathbf{v}$ in the plane $\mathbb{R}^2$ (see Fig. 6.4), which are called the **roots** of SU(3). For each root $\mathbf{a}$ there is a shift operator $E(\mathbf{a})$ (e.g. $E(-\mathbf{u}) = U_-$); we define $E(\mathbf{a}) = 0$ if $\mathbf{a}$ is not one of the six roots. Let $\mathbf{H} = (F_3, F_8)$ and let $\mathbf{a}$ and $\mathbf{b}$ be any two-component vectors; then the commutators are

$$[\mathbf{a} \cdot \mathbf{H}, \mathbf{b} \cdot \mathbf{H}] = 0, \tag{6.71a}$$

$$[\mathbf{H}, E(\mathbf{a})] = \mathbf{a} E(\mathbf{a}), \tag{6.71b}$$

$$[E(\mathbf{a}), E(-\mathbf{a})] = 2\mathbf{a} \cdot \mathbf{H}, \tag{6.71c}$$

$$[E(\mathbf{a}), E(\mathbf{b})] = E(\mathbf{a} + \mathbf{b}). \tag{6.71d}$$

(a) is the statement that F_3 and F_8 commute; (b) shows the $E(\mathbf{a})$ as shift operators; (c) shows the SU(2) algebra in which $E(\pm\mathbf{a})$ play the role of $J_\pm$. The form of (d) is a consequence of the Jacobi identity:

$$[\mathbf{H}, [E(\mathbf{a}), E(\mathbf{b})]] = [[\mathbf{H}, E(\mathbf{a})], E(\mathbf{b})] + [E(\mathbf{a}), [\mathbf{H}, E(\mathbf{b})]]$$
$$= (\mathbf{a} + \mathbf{b})[E(\mathbf{a}), E(\mathbf{b})] \tag{6.72}$$

which, by comparison with (b), shows that $[E(\mathbf{a}), E(\mathbf{b})]$ must be a multiple of $E(\mathbf{a} + \mathbf{b})$. Note that

$$I_3 = \mathbf{i} \cdot \mathbf{H}, \quad U_3 = \mathbf{u} \cdot \mathbf{H}, \quad V_3 = \mathbf{v} \cdot \mathbf{H}. \tag{6.73}$$

A representation of SU(3) can be described by giving the pairs of simultaneous eigenvalues of F_3 and F_8 in the representation; thus it corresponds to a diagram consisting of a number of points in a plane with Cartesian axes labelled F_3 and F_8, each point representing a simultaneous eigenvector. This is called the **weight diagram** of the representation. It must have the property that neighbouring points are connected by one of the root vectors of Fig. 6.4; also, because of the symmetry between $\mathbf{i}$, $\mathbf{u}$ and $\mathbf{v}$ in the structure of the Lie algebra (which comes from the symmetry between the three SU(2) subgroups in the group) the weight diagram must be symmetrical under rotations through $120°$ about the origin. We will not undertake a general

Fig. 6.4.
The roots of SU(3).

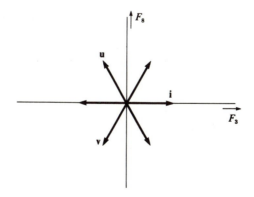

description of all possible representations of SU(3), but will just describe a few which are particularly important for particle physics.

The **fundamental** representation is the three-dimensional representation in which each element of SU(3) is represented by itself as a matrix; the representation space consists of complex 3×1 column vectors. F_3 and F_8 are represented by the diagonal matrices λ_3 and λ_8, so their eigenvalues are given by the diagonal entries. This gives the weight diagram of Fig. 6.5(*a*). The points of the diagram are labelled by the quarks which form the corresponding eigenstates in the representation (6.57).

The **conjugate** representation is the three-dimensional representation in which each element U of SU(3) is represented by the complex conjugate matrix $\bar{U}$. The hermitian generators of this representation are of the form

$$X' = i \frac{d}{ds} \overline{U(s)}\Big|_{s=0} = -\overline{\left[i \frac{d}{ds} U(s)\right]} = -\bar{X} \tag{6.74}$$

where X is the hermitian generator of the fundamental representation corresponding to the sequence $U(s)$ of group elements. Thus the eigenvalues are the negatives of those in the fundamental representation, and the weight diagram is obtained by inverting that of the fundamental representation through the origin (Fig. 6.5(*b*)). The eigenvalues thus obtained are the quantum numbers of the antiparticles of the quarks of Fig. 6.5(*a*).

If the fundamental representation ρ is regarded as acting on a three-dimensional state space, like the quark space with basis $|u\rangle, |d\rangle, |s\rangle$ with which we opened this section, then the conjugate representation $\bar{\rho}$ acts on the space of bras as follows (see problem 6.19):

$$\bar{\rho}(U)\langle\psi| = \langle\psi|U^\dagger. \tag{6.75}$$

The **adjoint** representation acts on the space of all traceless 3×3 matrices A according to

$$A \to UAU^{-1} = \rho(U)A. \tag{6.76}$$

The hermitian generators of this representation are given by

$$A \to i \frac{d}{ds} [U(s)AU(s)^{-1}]_{s=0} = [X, A] \tag{6.77}$$

where X is as in (6.74). This operation, of taking the commutator with X, is

Fig. 6.5.
Weight diagrams for SU(3):
(*a*) fundamental; (*b*) conjugate;
(*c*) adjoint.

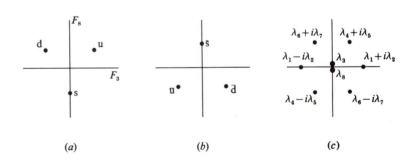

(*a*)   (*b*)   (*c*)

denoted by ad$X$. Thus the generators F_3 and F_8 for the adjoint representation are

$$F_3 = \mathrm{ad}\lambda_3, \quad F_8 = \mathrm{ad}\lambda_8. \tag{6.78}$$

The weights of the representation are the pairs of eigenvalues of these. Now the commutator (6.71d) shows that in any representation the effect of ad$\mathbf{H}$ on the shift operator $E(\mathbf{a})$ is to multiply it by the root vector $\mathbf{a}$. Taking $\mathbf{H} = (\lambda_3, \lambda_8)$, this shows that each root vector is a pair of eigenvalues for (adλ_3, adλ_8). Also $(0,0)$ occurs twice as a pair of eigenvalues, the eigenvectors being λ_3 and λ_8 themselves. Thus the weights of the adjoint representation are the eight points shown in Fig. 6.5(c).

The adjoint representation can also be described using the form (6.64) of the Lie algebra, by saying that the representation space is eight-dimensional with a basis $|v_i\rangle$ $(i = 1,\ldots,8)$ and the hermitian generators X_i act according to

$$X_i |v_j\rangle = i f_{ijk} |v_k\rangle. \tag{6.79}$$

These three representations are also called the **triplet**, **antitriplet** and **octet** representations and denoted by **3**, **$\bar{3}$** and **8**. The **singlet** representation **1** is the trivial one-dimensional representation in which all elements of SU(3) are represented by the identity operator, and all hermitian generators are zero.

The triplet representation is the first of a series of representations with triangular weight diagrams; the next two representations, the sextet and the decuplet, are shown in Fig. 6.6. These representations can be defined as follows. The representation Δ_n whose weight diagram is a triangle with sides n times as long as the triangle of the fundamental representation, Fig. 6.5(a), acts on symmetrised products of n vectors taken from the fundamental representation space $\mathscr{V}$; i.e. the representation space of Δ_n is $\mathscr{V} \vee \mathscr{V} \vee \cdots \vee \mathscr{V}$ (n times). The weight vectors of Δ_n are of the form $S(|\mathbf{l}_1\rangle \cdots |\mathbf{l}_n\rangle)$ where $\mathbf{l}_1, \ldots, \mathbf{l}_n$ are weights of the fundamental representation; there is one such weight vector for every unordered choice of n weights $\mathbf{l}_i$, and its weight is $\mathbf{l}_1 + \cdots + \mathbf{l}_n$. They can all be obtained by starting with $|\mathbf{l}_0\rangle |\mathbf{l}_0\rangle \cdots |\mathbf{l}_0\rangle$, where $\mathbf{l}_0$ is the top right-hand weight of Fig. 6.5(a), and changing up to n of the weights $\mathbf{l}_0$ into one of the other two; the total weight is obtained correspondingly by

Fig. 6.6.
The sextet and decuplet representations.

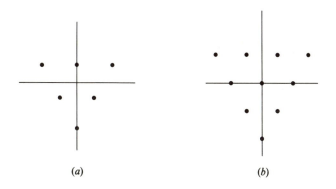

(a)　　　　　　　(b)

starting with nl_0 and moving up to n times along one of the arrows $-\mathbf{i}$ or $\mathbf{v}$. This gives the triangular weight diagram.

By starting with the conjugate representation we can obtain a similar series of triangular representations, the triangles now, like that of Fig. 6.5(*b*), being the right way up.

We have already seen the patterns of Figs. 6.5(*c*) and 6.6(*b*) in Figs. 1.4 and 1.5, which show how hadronic particles do indeed fall into multiplets carrying representations of SU(3). The lightest baryons (which have zero charm, beauty or truth and decay only by the weak force) form an octet like Fig. 6.5(*c*), while the next lightest baryons, which (with one exception, to which we will return) are unstable to decay by the strong force, form a decuplet like Fig. 6.6(*b*). The mesons also fall into SU(3) multiplets, all octets and singlets. These multiplets are collected together in Fig. 6.7; their decays are summarised in Table 6.4.

Quark structure   Let us see how the SU(3) multiplets of mesons and baryons can be described in terms of quarks. The quarks u, d and s form an SU(3) triplet, as in Fig. 6.5(*a*). Let $\mathcal{Q}$ be the three-dimensional state space of these quarks, and let $\bar{\mathcal{Q}}$ be the state space of their antiparticles, so that $\mathcal{Q}$ and $\bar{\mathcal{Q}}$ are the representation spaces for $\mathbf{3}$ and $\bar{\mathbf{3}}$. The states of a quark and an antiquark form the two-particle state space $\mathcal{Q} \otimes \bar{\mathcal{Q}}$. In this space the eigenvalues of F_3 and F_8 are obtained by adding the eigenvalues for the individual particles; so each weight of $\mathcal{Q} \otimes \bar{\mathcal{Q}}$ is the vector sum of a weight of $\mathcal{Q}$ and a weight of $\bar{\mathcal{Q}}$. Hence the weight diagram of $\mathcal{Q} \otimes \bar{\mathcal{Q}}$ is the union of three copies of the weight diagram of $\bar{\mathcal{Q}}$, each centred on one of the points in the weight diagram of $\mathcal{Q}$. This yields the octet diagram with one extra point (Fig. 6.8), which suggests that $\bar{\mathcal{Q}} \otimes \bar{\mathcal{Q}}$ splits into an octet and a

Fig. 6.7.
SU(3) multiplets: (*a*) baryons $J^P = \frac{1}{2}^+$, lifetimes $\sim 10^{-10}$ sec;
(*b*) baryons $J^P = \frac{3}{2}^+$, lifetimes $\sim 10^{-23}$ sec;
(*c*) mesons $J^P = 0^-$, lifetimes 10^{-8} to 10^{-10} sec;
(*d*) mesons $J^P = 1^-$, lifetimes $\sim 10^{-22}$ sec;
(*e*) meson $J^P = 0^-$, lifetime 10^{-20} sec;
(*f*) meson $J^P = 1^-$, lifetime 10^{-22} sec.

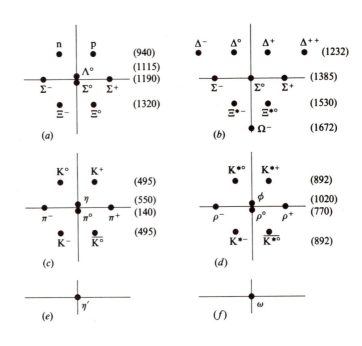

singlet:

$$3 \otimes \bar{3} = 8 \oplus 1. \tag{6.80}$$

Algebraically, this decomposition can be understood as follows. If $\mathscr{Q}$ is a space of kets, $\bar{\mathscr{Q}}$ can be regarded as the space of bras. Then $\mathscr{Q} \otimes \bar{\mathscr{Q}}$ (which is isomorphic to $\bar{\mathscr{Q}} \otimes \mathscr{Q}$) is spanned by products $|\phi\rangle\langle\psi|$, i.e. operators on $\mathscr{Q}$. They transform under SU(3) transformations, according to (6.75), by

$$|\phi\rangle\langle\psi| \rightarrow U|\phi\rangle\langle\psi|U^{\dagger} = U|\phi\rangle\langle\psi|U^{-1} \tag{6.81}$$

(since U is unitary), i.e. in accordance with (6.78). The octet representation space, consisting of traceless operators, is a subspace of this $\bar{\mathscr{Q}} \otimes \mathscr{Q}$; the orthogonal one-dimensional subspace is the space of multiples of the identity operator, which is invariant under the transformations (6.78) and so constitutes the singlet representation.

SU(3) invariance now fixes the quark composition of the three neutral, non-strange mesons π^0, η and η'. If the η' is an SU(3) singlet, it corresponds to the

Table 6.4. *Principal decays of the particles in Fig. 6.7*

Weak decays (lifetimes 10^{-10} to 10^{-8} s)[a]

	Baryon octet	0^- meson octet
$\Delta S = 0$	$n \rightarrow p + e^- + \bar{\nu}_e$	$\pi^- \rightarrow \mu^- + \bar{\nu}_\mu, \quad \pi^+ \rightarrow \mu^+ + \nu_\mu$
$\Delta S = \pm 1$	$\Lambda^0 \rightarrow N + \pi$ $\Sigma^\pm \rightarrow N + \pi$ $\Xi \rightarrow \Lambda + \pi$	$K^- \rightarrow \mu^- + \bar{\nu}_\mu, \quad K^+ \rightarrow \mu^+ + \nu_\mu$ $K^\pm \rightarrow 2\pi, 3\pi$ $K_S^0 \rightarrow 2\pi, \quad K_L^0 \rightarrow 3\pi$

Baryon decuplet: $\Omega^- \rightarrow \Xi + \pi, \Omega^- \rightarrow \Lambda^0 + K^-$

Electromagnetic decays (lifetimes $\sim 10^{-20}$ s)[b]

Baryon octet	0^- meson octet	0^- meson singlet
$\Sigma^0 \rightarrow \Lambda^0 + \gamma$	$\pi^0 \rightarrow 2\gamma$ $\eta \rightarrow 2\gamma$ $\eta \rightarrow 3\pi$	$\eta' \rightarrow \rho^0 + \gamma$ $\eta' \rightarrow \eta + 2\pi$

Strong decays (lifetimes $\sim 10^{-23}$ s)[b]

Baryon decuplet	1^- meson octet	1^- meson singlet
$\Delta \rightarrow N + \pi$ $\Sigma^* \rightarrow \Sigma + \pi, \quad \Lambda + \pi$ $\Xi^* \rightarrow \Xi + \pi$	$\rho \rightarrow 2\pi$ $\phi \rightarrow K + \bar{K}, \quad 3\pi$ $K^* \rightarrow K + \pi$	$\omega \rightarrow 3\pi$

[a] The neutron lifetime is exceptionally long because of a very small phase space factor.
[b] The question of how particles with such short lifetimes are observed will be taken up in §6.4.

identity operator

$$\mathbf{1} = |u\rangle\langle u| + |d\rangle\langle d| + |s\rangle\langle s| \tag{6.82}$$

on $\mathcal{Q}$. Since each bra vector transforms like the corresponding antiquark state, this gives

$$|\eta'\rangle = (1/\sqrt{3})(|u\bar{u}\rangle + |d\bar{d}\rangle + |s\bar{s}\rangle) \tag{6.83}$$

as the normalised quark–antiquark singlet state. The octet particles π^0 and η must correspond to traceless operators on $\mathcal{Q}$ and therefore to quark–antiquark states of the form

$$\sum_{\alpha,\beta=1}^{3} c_{\alpha\beta} |q_\alpha\rangle |\bar{q}_\beta\rangle \quad \text{with} \sum_\alpha c_{\alpha\alpha} = 1, \tag{6.84}$$

writing $(u, d, s) = (q_1, q_2, q_3)$. The π^0 has isospin 1, so it must be composed of the isospin-$\frac{1}{2}$ particles $(u, \bar{u}, d, \bar{d})$; the neutral $I = 1$ combination of these is

$$|\pi^0\rangle = \sqrt{\tfrac{1}{2}}(|u\bar{u}\rangle - |d\bar{d}\rangle) \tag{6.85}$$

since from (6.14) the antiquark isospin doublet is $(i|\bar{u}\rangle, -i|\bar{d}\rangle)$ and we can ignore the phase factor i. Now the η must be orthogonal to both (6.83) and (6.85):

$$|\eta\rangle = \sqrt{\tfrac{1}{6}}(|u\bar{u}\rangle + |d\bar{d}\rangle - 2|s\bar{s}\rangle). \tag{6.86}$$

The full set of nine quark–antiquark combinations is called a **nonet**. Fig. 6.7 shows two such nonets, one with $J^P = 0^-$ (which can be identified as the set of quark–antiquark states with $l = 0$ and $s = 0$, since the quark and antiquark have opposite intrinsic parities), and one with $J^P = 1^-$ $(l = 0, s = 1)$. In each nonet there are two isospin singlet states, the SU(3) singlet like (6.83) and the SU(3) octet state like (6.86); the states observed as particles can be combinations of these if SU(3) invariance is not respected. It appears [Perkins 1982, §5.4] that in the 0^- nonet the observed particles η and η' are close to the SU(3) octet and singlet states, but in the 1^- nonet this is not true and the particles are better represented as

$$|\phi\rangle = |s\bar{s}\rangle, \quad |\omega\rangle = \sqrt{\tfrac{1}{2}}(|u\rangle|\bar{u}\rangle + |d\rangle|\bar{d}\rangle). \tag{6.87}$$

The baryons are three-quark states. To understand these, first consider the two-quark state space $\mathcal{Q} \otimes \mathcal{Q}$. The procedure of Fig. 6.8 leads to the weight diagram of Fig. 6.9(a). We know that this must contain the sextet representation, since this acts on the symmetric subspace $\mathcal{Q} \vee \mathcal{Q}$ of $\mathcal{Q} \otimes \mathcal{Q}$; if this is separated out, as in Fig. 6.9(b), what is left is the antitriplet representation:

$$\mathbf{3} \otimes \mathbf{3} = \mathbf{6} \oplus \bar{\mathbf{3}}. \tag{6.88}$$

To obtain the weight diagram for three-quark states, we must superimpose the quark weight diagram on both diagrams for $\mathbf{6}$ and $\bar{\mathbf{3}}$, in the manner of Figs. 6.8 and 6.9. We know that this must contain the decuplet, which was defined to be the symmetric three-particle state space $\mathcal{Q} \vee \mathcal{Q} \vee \mathcal{Q}$; this occurs in $\mathbf{6} \otimes \mathbf{3}$, which also contains an octet. Thus we have

$$3 \otimes 3 \otimes 3 = (6 \oplus \bar{3}) \otimes 3 = (6 \otimes 3) \oplus (\bar{3} \otimes 3) = 10 \oplus 8 \oplus 8 \oplus 1. \quad (6.89)$$

So the three-quark states include a decuplet and an octet, which occur as multiplets of baryons.

As expected, these SU(3) multiplets do not contain particles with equal mass. In the baryon decuplet, for example, the mass of each isospin multiplet is greater than the one above by about 145 MeV. Since the particles in each isospin multiplet contain one more s quark than those in the one above, this can be simply understood as being due to the s quark's having a greater mass than the u and d quarks. This also gives a qualitative understanding of the mass differences in the baryon and meson octets. A more quantitative understanding can be obtained by assuming definite SU(3) transformation properties for the Hamiltonian (see problem 6.22).

The particle at the bottom of the baryon decuplet, the Ω^-, is composed of three s quarks and has strangeness $S = -3$. A state with this strangeness cannot be made up from any set of particles with total mass less than that of the Ω^-; hence the decay of this particle must involve a change of strangeness and

Fig. 6.8.
$3 \otimes \bar{3} = 8 \oplus 1$.

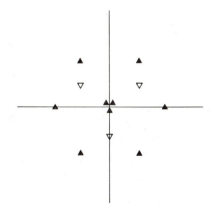

Fig. 6.9.
$3 \otimes 3 = 6 \oplus \bar{3}$.

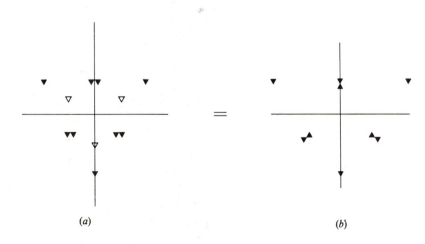

(a)   (b)

so it must be a weak process. Thus the lifetime of the Ω^- is of the order of the weak interaction times, about 10^{-8} s. At the time when the theory of SU(3) symmetry was developed the Ω^- was not known but the other particles in the decuplet were. The discovery in 1964 of this long-lived particle with such a relatively high mass confirmed the ideas of SU(3) symmetry, in a striking parallel to the confirmation of Mendeleev's periodic table by the discovery of missing elements. Fig. 6.10 is a bubble-chamber photograph of the production and decay of an Ω^-.

Operators with definite SU(3) transformation properties can be formed from the fields of particle multiplets in the same way as for isospin. Let the 3×3 matrix Y be an element of SU(3), and let $U(Y)$ be the corresponding unitary operator on the full state space; then if ϕ_α is the set of fields of the quark triplet, regarded as a column vector Φ,

$$\left.\begin{aligned} U(Y)\Phi U(Y)^{-1} &= Y\Phi \\[4pt] \text{and} \qquad U(Y)\Phi^\dagger U(Y)^{-1} &= \Phi^\dagger Y^\dagger \end{aligned}\right\}, \tag{6.90}$$

of which the infinitesimal form is

$$[F_i, \Phi] = \lambda_i \Phi \quad \text{and} \quad [F_i, \Phi^\dagger] = -\Phi^\dagger \lambda_i. \tag{6.91}$$

Fig. 6.10.
Production and decay of an
Ω^- (photo: CERN).

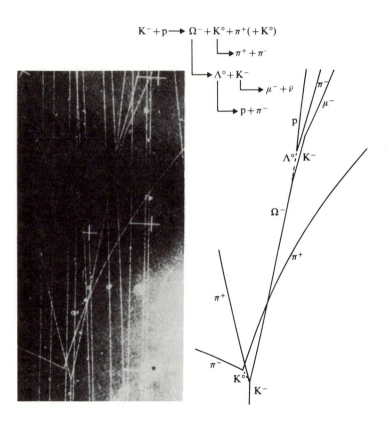

$K^- + p \longrightarrow \Omega^- + K^\circ + \pi^+ (+K^\circ)$
$\qquad\qquad\quad \longrightarrow \pi^+ + \pi^-$
$\qquad\qquad \longrightarrow \Lambda^\circ + K^-$
$\qquad\qquad\qquad\quad \longrightarrow \mu^- + \bar\nu$
$\qquad\qquad\qquad \longrightarrow p + \pi^-$

The combinations

$$V_i = \Phi^\dagger \lambda_i \Phi \qquad (6.92)$$

form an octet operator, i.e.

$$[F_i, V_j] = i f_{ijk} V_k. \qquad (6.93)$$

More generally, if $\phi_1, \ldots, \phi_N$ are the fields of any SU(3) multiplet of particles regarded as a column vector Φ, and t_i are the $N \times N$ matrices which are the hermitian generators of SU(3) in this representation, then $\Phi^\dagger t_i \Phi$ is an octet operator. This is the SU(3) counterpart of the SU(2) vector operator (6.31).

If V_i and W_i are both octet operators in the sense of (6.93), then $V_i W_i$ is an SU(3) scalar and $f_{ijk} V_j W_k$ is another octet operator. These correspond to the scalar product and vector product of two SU(2) vector operators.

An SU(3) version of the Yukawa–Kemmer Hamiltonian (6.34) can be constructed with a self-conjugate octet of particles, i.e. an octet in which the particles at diametrically opposite points of the weight diagram are antiparticles of each other. An example of a self-conjugate octet is the meson octet of Fig. 6.7(c). Such an octet yields a set of hermitian fields α_i which form an octet operator, defined by comparing the weight diagram with Fig. 6.5(c) (for example, from Fig. 6.7(c) we have $\phi_{K^+} = \alpha_4 + i\alpha_5$). The particles of a self-conjugate octet can be taken as field quanta for an SU(3)-symmetric force acting on any SU(3) multiplet; the corresponding Yukawa–Kemmer Hamiltonian is

$$H' = \Phi^\dagger t_i \Phi \cdot \alpha_i \qquad (6.94)$$

where Φ is the column vector of fields for the multiplet.

SU(4, 5 and 6) The notion of symmetry between quarks can of course be extended successively to the quarks c, b and t. This leads in turn to the symmetry groups SU(4), SU(5) and SU(6), which, in the same way as SU(2) and SU(3), can be used to classify particles in multiplets and to obtain relations between decay rates. The representations of SU(n) are described by ($n-1$)-dimensional weight diagrams; as an example, Fig. 6.11 shows the weight diagram for the adjoint representation of SU(4). This has 15 points in a three-dimensional figure (a cuboctahedron). There should be a multiplet of 0^- mesons corresponding to this figure, comprising an SU(3) octet, a singlet, a triplet and an antitriplet. These all contain a quark and an antiquark, as shown in Fig. 6.11. The axes in this figure label I_3, strangeness and charm. The relation between I_3 and electric charge is modified in the presence of charm, and modified again by further flavours; the general relation can be expressed by the formula (6.3) where the hypercharge is

$$Y = B + S + C - B' + T \qquad (6.95)$$

(B = baryon number, S = strangeness, C = charm, B' = beauty, T = truth).

As the mass differences between successive quarks get steadily larger, the symmetry between them becomes less and less real, and it becomes more meaningful to classify particles in simple terms of quark content rather than by SU(n) multiplets. The difference is a matter of the degree of superposition of different quark structures, and can be illustrated by referring to the totally neutral mesons π^0, η and η'. From the three quarks u, d, s and their antiquarks one might naively expect to form three totally neutral mesons u$\bar{\text{u}}$, d$\bar{\text{d}}$, s$\bar{\text{s}}$. In fact, because of the very good symmetry between u and d, the observed particle π^0 is an equal superposition of u$\bar{\text{u}}$ and d$\bar{\text{d}}$, and as shown in (6.83) and (6.86) the other two particles are also superpositions. However, because SU(3) is not an exact symmetry the observed particles are not actually given by these expressions but by combinations of them, so that η is closer to s$\bar{\text{s}}$ and η' to u$\bar{\text{u}}$ + d$\bar{\text{d}}$. In the 1^- octet this is even more true, and the observed ϕ-meson is almost exactly s$\bar{\text{s}}$. With the heavier quarks this trend continues, so that the totally neutral mesons formed from them can be considered as c$\bar{\text{c}}$, b$\bar{\text{b}}$ and t$\bar{\text{t}}$.

6.4. Hadron spectroscopy

Once it is known that hadrons (baryons and mesons) are composed of quarks, the object of the study of hadrons becomes to understand the forces between quarks. There are precedents for this in the study of the higher-level composite systems, molecules, atoms and nuclei, which we will now briefly review.

In quantum mechanics forces are described by a Hamiltonian, whose most characteristic feature is its spectrum, the set of energy levels of the system. In an atom or molecule these energy levels are directly accessible to experiment, since the differences between them give the frequencies of the radiation emitted by the atom. Thus in atomic and molecular spectroscopy the basic information-giving event is the decay of an excited state with the emission of a photon.

Fig. 6.11
The adjoint representation of SU(4): thick lines outline SU(3) multiplets.

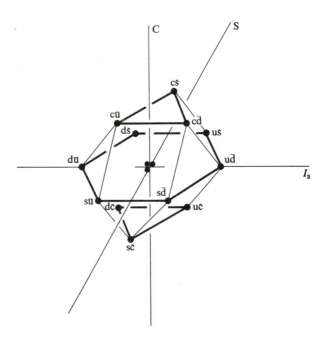

Another piece of information that is given by this event (or rather, by a large number of similar events) is the lifetime of the excited state. This is not measured directly (it is typically of the order of 10^{-14} s), but deduced from a second property of the emitted radiation, the **width** of the spectral line. The radiation in a given line of the spectrum of a substance (i.e. that associated with a given pair of energy levels) is not all emitted at precisely the same frequency, but is distributed over a narrow range of frequencies, so that the line observed in a spectrometer is not infinitely thin but has a finite width. This is to be expected from the general discussion of decay in §3.5; the unstable excited state, not being a stationary state, is a superposition of energy eigenstates with coefficients $\rho(E)$ which, for the case of exponential decay, satisfy the Breit–Wigner formula

$$|\rho(E)|^2 = \frac{\Gamma/2\pi}{\frac{1}{4}\Gamma^2 + (E - E_0)^2}. \tag{6.96}$$

After the photon has been emitted, the decay interaction can be neglected and the energy can be taken to be the sum of the energies of the atom and the photon. Since the energy of the atom is fixed, the total energy is measured by measuring the frequency of the photon, which will show a distribution like (6.96).

For an atom to emit a photon from an excited state it must have been excited in the first place. One way in particular in which this can happen is for the atom to absorb a photon. In this case the whole process can be regarded as a collision in which the photon is scattered by the atom, since it will not in general be emitted in the same direction as it was absorbed from. If the atom returns to the state it was in before the collision, the emergent photon will have the same energy as the incident one; this is called **elastic** scattering. Time-dependent perturbation theory shows that the scattering will only take place if the eigenvalues of H_0 are nearly the same before and after, so that conservation of energy can be applied by simply adding the energies of the atom and the photon (as one would expect: when they are far apart, the potential energy of interaction is negligible). Thus the absorbed photon has the same energy as the emitted one, and it is distributed as in (6.96). More generally, the atom will decay to another excited state and the overall process will be **inelastic** scattering:

$$X + \gamma \to X' + \gamma. \tag{6.97}$$

It is still true that the energies of both absorbed and emitted photons are distributed in the Breit–Wigner form (6.96).

The upshot is that if the amount of scattering is plotted against the energy of the incident photons, the graph shows a number of peaks at the energies of the excited states of the atom, as in Fig. 6.12. In the case of atoms and molecules, the scattering of photons when a beam of light is passed through a material causes a depletion of the beam in the incident direction; peaks like those of Fig.

6.12 show up as dark lines when the different frequencies of light are spread out in a spectrometer. This is called an **absorption spectrum**.

In general, for any scattering process in which a beam of one type of particle is scattered off a target of another type of particle, we can plot the amount of scattering against the total energy of one beam particle plus one target particle. The 'amount of scattering' is defined as

$$\sigma = \frac{\text{number of particles scattered per unit time}}{\text{number of target particles} \times \text{flux of beam}} \tag{6.98}$$

where the **flux** of the beam is the number of particles crossing unit area in unit time. This σ has the dimensions of area, and is called the **scattering cross-section**, because if each target particle presented a cross-sectional area which a beam particle must hit if it is to be scattered, (6.98) would give that area. (If the direction of scattering is specified, and (6.98) is taken to mean the number of particles scattered per unit solid angle at that direction, the result is the **differential cross-section** $d\sigma/d\Omega$; the **total cross-section** σ is the integral of this over all directions.)

Eq. (6.98) is also used to define the cross-section for a process in which the particles may change their identity, i.e.

$$A + B \rightarrow C + D + E + \cdots. \tag{6.99}$$

The top line of (6.98) must then be understood to mean the number of beam particles which initiate the particular process being considered.

A peak in the graph of scattering cross-section against energy is called a **resonance**; if it can be approximated by (6.96) near some energy E_0, Γ is called the **width** of the resonance. We can now generalise from our discussion of atomic and molecular spectra to draw the following moral:

> A resonance in a scattering cross-section, at energy E_0 and with width Γ, is an indication of an excited state of the target with energy E_0 and lifetime Γ^{-1}.

In the inelastic process (6.97) there are two excited states involved: a first one which is produced by the absorption of the incident photon, and a second one

Fig. 6.12.
Resonances.

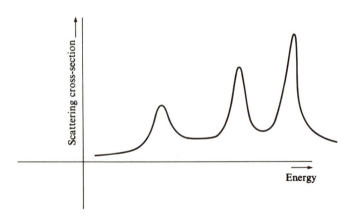

formed by the decay of the first. This second state will also decay, so that the end result is

$$X + \gamma \to X + 2\gamma, \tag{6.100}$$

the full description of the intermediate processes being

$$X + \gamma \to X'' \to X' + \gamma$$
$$| $$
$$\to X + \gamma. \tag{6.101}$$

If both decays occur so quickly that only (6.100) is observed, then both excited states must be treated as resonances. The first is a resonance in the total cross-section, as has already been described (in this context it might be more appropriate to replace the word 'scattered' by 'absorbed' in the definition (6.98) of the cross-section); this is called **formation** of the resonance. The second excited state X', however, can only be detected by looking at the emitted photons: if the energy of each photon is added to that of the final ground-state atom, it will be found that there are a large number of such $X\gamma$ pairs with energy close to that of the excited state X', and their energy is distributed about the X' energy according to the resonance formula (6.96). This is called **production** of the X' resonance.

These considerations are relevant not only to emission and absorption spectroscopy, which we have been describing as the study of photon–atom scattering, but also to other forms of atomic scattering. An atom can be put into an excited state by a collision with another atom or ion or an electron, so the scattering of beams of ions or electrons off atoms can give information about excited states. Also, an excited state can sometimes decay by the emission of an electron; such a state is called an **autoionising** state. An autoionising state of an atom X, which decays into the ion X^+ plus an electron, will show up as a resonance in the scattering of electrons from the ion X^+; this resonance represents not an excited state of the target but a bound state of the target and the beam particle combined together.

Thus there are many scattering processes of the form (6.99) which can give information about the excited states of atoms and molecules. The same is true in nuclear physics: the excited states of a nucleus are determined by means of scattering experiments in which the nucleus is bombarded with α-particles, protons or neutrons.

To apply these ideas to particle physics we will have to take account of the fact that particle collisions occur at high energy; the recoil of the target will be significant, and relativistic mechanics must be used. In a process like (6.99), if the two initial particles form a state X it will be characterised not by its energy (which depends on its velocity, i.e. the centre-of-mass velocity of A and B), but by its rest-mass m_X given by

$$m_X{}^2 c^4 = E_X{}^2 - \mathbf{p}_X{}^2 c^2$$
$$= m_{AB}{}^2 c^4 = (E_A + E_B)^2 - (\mathbf{p}_A + \mathbf{p}_B)^2 c^2, \tag{6.102}$$

by conservation of energy and momentum. It is this quantity m_{AB}, the centre-of-mass energy or **invariant mass** of A and B, which is significant in scattering experiments: resonance formation is shown by a peak in the scattering cross-section as a function of m_{AB}. Similarly, production of a resonance which then decays into C + D is indicated by a peak in the number of CD pairs as a function of m_{CD}.

All the 'particles' mentioned in this chapter which were ascribed lifetimes of the order of 10^{-23} s, characteristic of the strong interactions, are observed only as resonances. Thus the isospin-$\frac{3}{2}$ Δ multiplet is a set of resonances formed in pion–nucleon scattering; the other unstable particles in the baryon decuplet, the Σ^* triplet and the Ξ^* doublet, are produced as $\Sigma\pi$ and $\Xi\pi$ resonances and seen in the final states of antikaon–nucleon scattering:

$$\bar{K} + N \rightarrow \Sigma^* + \pi$$
$$| $$
$$\rightarrow \Sigma + \pi, \tag{6.103}$$
$$\bar{K} + N \rightarrow \Xi^* + K + \pi$$
$$|$$
$$\rightarrow \Xi + \pi. \tag{6.104}$$

These particles can only be seen as production resonances and not in formation experiments, because their masses are too low for them to be formed in $\bar{K}$–N scattering.

There are many baryon resonances with masses in the region from 1 GeV to 3 GeV, with spins up to $\frac{15}{2}$ and possibly greater. They all have isospin and strangeness the same as one of the particles we have already met, N, Σ, Λ, Ξ or Δ. These symbols, together with the mass, are used as names for the resonances: for example, $\Sigma(1670)$ denotes a triplet of resonances with isospin 1, strangeness -1 and mass 1670 MeV. The most prominent baryon resonances are shown on a plot of spin against mass in Fig. 6.13.

Fig. 6.13.
Baryon resonances.

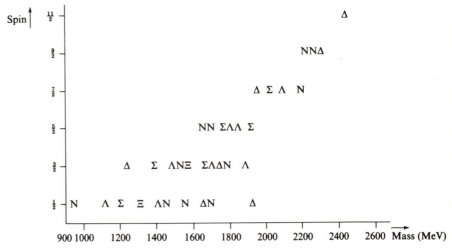

There is a similar profusion of meson resonances, including the nonet with $J^P = 1^-$ which we encountered in §6.3. Their isospin and strangeness are restricted to the values exhibited by π, K, $\bar{\text{K}}$ and η. They are mainly observed in production processes such as

$$\pi + \text{N} \rightarrow \text{N} + \rho$$
$$\big|$$
$$\rightarrow 2\pi. \tag{6.105}$$

This applies also to the η, the $I = 0$ member of the 0^- octet whose other members live long enough to form visible bubble-chamber tracks. However, there is a class of meson resonances, namely totally neutral resonances with $J^P = 1^-$, which are also observed in formation processes in electron–positron scattering. Thus the ρ^0, ω and ϕ mesons appear as resonances in the cross-sections for

$$\text{e}^- + \text{e}^+ \rightarrow \pi^- + \pi^+ \quad \text{or} \quad \pi^- + \pi^+ + \pi^0. \tag{6.106}$$

These facts about resonances all go to confirm that hadrons are composed of quarks, each baryon being a bound state of three quarks and each meson a bound state of a quark and an antiquark. The isospin and strangeness values of N, Σ, Λ, Ξ, Δ and Ω are just those that can be obtained by putting together three quarks chosen from u, d and s, while the isospin and strangeness of π, K, $\bar{\text{K}}$ and η are just those of the quark–antiquark combinations of these three. Thus the original particles with these names seem to be ground states of the 3q or q$\bar{\text{q}}$ systems, and the resonances are excited states which decay to the ground state. The set of masses of the resonances with given charge and strangeness form a spectrum, like the set of energies of the excited states of an atom, which should give information about the forces between the relevant quarks; thus Fig. 6.13 should be regarded as being the same sort of diagram as Fig. 4.6.

These excited states usually decay by emitting a pion or kaon; this is not analogous to the decay of an excited state of an atom, with the emission of a photon, but involves the creation of a quark and its antiquark. This is illustrated in Fig. 6.14 for the decays of the Δ^{++} and the ϕ. The quark–antiquark pairs in these processes are created by virtual gluons, the quanta of the interquark force, which play the same role inside hadrons as photons do inside atoms.

Fig. 6.14(c) shows a method of decay for a meson resonance which is much rarer than the type of decay shown in Fig. 6.14(b). The **Zweig rule** states that any process involving an intermediate state containing only gluons is suppressed, i.e. has a very low rate compared with processes like Figs. 6.14(a) and (b), in which there are quark lines which join the initial state to the final state.

Charm, beauty and truth The formation of neutral meson resonances in electron–positron annihilation can be understood by means of the Feynman diagram Fig. 6.15. The quark and the antiquark in a totally neutral meson are antiparticles of each other, and so they can annihilate to give a (virtual) photon, or be created from a photon, in one of the basic events of the electromagnetic force. The probability of this pair creation will be enhanced in a resonance-like way at energies at which the quark and antiquark form a bound state. Such a bound state must have the same angular momentum and parity as the photon, namely $J^P = 1^-$. It can decay to produce hadrons (with the aid of further quark–antiquark pair creation, as in Fig. 6.14), or it can annihilate to give another virtual photon, which then creates either an electron–positron pair or a $\mu^- \mu^+$ pair. Thus the formation of neutral 1^- meson resonances shows up as a simultaneous peak in the cross-sections for $e^- + e^+ \to$ hadrons (like (6.106)), $e^- + e^+ \to \mu^- + \mu^+$, and the elastic scattering $e^- + e^+ \to e^- + e^+$.

Fig. 6.14.
Quark diagrams for resonance decays: (*c*) is suppressed by the Zweig rule.

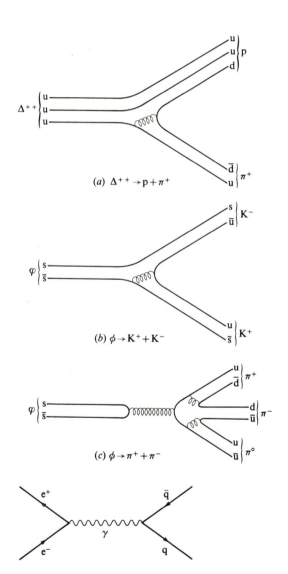

(*a*) $\Delta^{++} \to p + \pi^+$

(*b*) $\phi \to K^+ + K^-$

(*c*) $\phi \to \pi^+ + \pi^-$

Fig. 6.15.
Formation of a neutral 1^- meson resonance.

This process provided one of the first pieces of evidence for the existence of the charmed quark. In 1974 an extremely sharp resonance in e^-e^+ scattering, with a mass of 3100 MeV, was observed at SLAC (Stanford, California). It was observed simultaneously in a production experiment at Brookhaven, New York, where collisions of high-energy protons with beryllium nuclei produced e^-e^+ pairs whose invariant mass showed a sharp peak at 3100 MeV. This double discovery led to a double christening for the resonance, which is still known as the J/ψ.

The width of the J/ψ is 0.06 MeV, which is very much smaller than the widths of the other resonances and corresponds to a lifetime of about 10^{-20} s, characteristic of electromagnetic rather than strong interactions. If it was composed of u, d and s quarks it would decay into pions and kaons and its width would be of the order of 10 to 100 MeV like those of the other resonances. This is reminiscent of the strangely long lifetimes of particles containing the s quark, and can be explained in a somewhat similar way by supposing that the J/ψ is a bound state of the fourth quark c and its antiquark c̄, which carry the new quantum number of **charm**. In the case of strange particles an s quark and an s̄ antiquark are created by strong interactions and are prevented from being destroyed by them because they move apart in two hadrons. In the case of the J/ψ the annihilation of the c and the c̄ is prevented not by physical separation but by the Zweig rule, which would allow the J/ψ to decay only as in Fig. 6.16(a), for example. This will be impossible if all mesons

Fig. 6.16.
J/ψ decay: in accordance with standard practice, gluon lines are not shown.

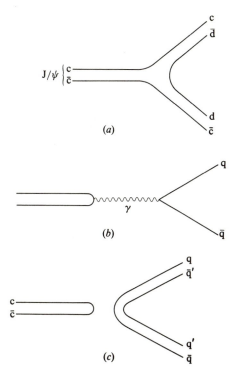

with non-zero charm (like c$\bar{\text{d}}$ and $\bar{\text{c}}$d) have masses greater than half the J/ψ mass. This was subsequently found to be the case. Thus the J/ψ can only decay as in Fig. 6.16(b), which is an electromagnetic process, and Fig. 6.16(c), which violates the Zweig rule and so has rate similar to that of the electromagnetic process.

The existence of charmed quarks was confirmed by the discovery of the D-mesons. These comprise two isospin doublets (D^0, D$^+$) and (D$^-$, $\bar{\text{D}}^0$) and are the charmed analogues of the K-mesons; their quark composition is (c$\bar{\text{u}}$, c$\bar{\text{d}}$) and (d$\bar{\text{c}}$, u$\bar{\text{c}}$). They have lifetimes of the order of 10^{-12} s (like kaons, they can only decay weakly). There are also charmed baryons Λ_{c}^+ (an isospin singlet containing c, u and d quarks) and Σ_{c} (an isospin triplet), and a strange charmed meson F$^+$ = c$\bar{\text{s}}$.

The composition of the J/ψ as a particle–antiparticle bound state is confirmed by some further spectroscopy. There are several resonances in the neighbourhood of the J/ψ (some formed as resonances in e$^-$e$^+$ scattering, some produced in the decay of the first type) which can be understood as states of the c$\bar{\text{c}}$ system. Instead of classifying these by their spin and parity (J^P) it is more instructive to try to deduce the relative orbital angular momentum L and total spin S of the quark and antiquark; values of these which are consistent with the J^P values of the resonances are presented in Fig. 6.17(a) on a plot of L against the mass of the resonance. For comparison, Fig. 6.17(b) shows the stationary states of positronium (this is a refined version of Fig. 4.5(a), the diagram of stationary states of the hydrogen atom, in which the effects of special relativity and the magnetic properties of the electron and positron have been taken into account; these cause a separation between the degenerate eigenvalues of the nonrelativistic, spin-independent Hamiltonian of §4.4). The resemblance between the two is so close that there can be little doubt that the resonances are states of a particle–antiparticle system like positronium. Because of this resemblance, the c$\bar{\text{c}}$ system is called **charmonium**. The spacings between the energy levels of charmonium are greater (by factors of 10^7) than those in positronium, showing that the force responsible for binding the c and $\bar{\text{c}}$

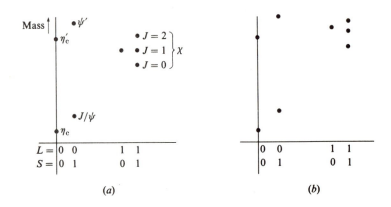

Fig. 6.17.
(a) The ψ family;
(b) positronium.

together is not the electromagnetic force but a very much stronger force of similar form.

The discovery of the b quark followed a similar course to that of the c quark. A narrow resonance was observed in the invariant mass of e^-e^+ pairs produced in collisions between protons and uranium nuclei; it has a mass of 9460 MeV and a width of 0.04 MeV. This particle, called the Υ (capital upsilon) is interpreted as a bound state of the fifth quark b and its antiparticle $\bar{b}$. It has been investigated by means of electron–positron scattering, which again shows a spectrum of resonances very like the spectrum of positronium. The b$\bar{b}$ system is called **beautonium** or (more commonly) **bottomonium**. A pair of particles with non-zero value of the associated quantum number B′ (beauty) has been observed; these are $(B^0, B^+) = (d\bar{b}, u\bar{b})$ and $(B^-, \bar{B}^0) = (b\bar{u}, b\bar{d})$.

The sequence of 1^- resonances ρ^0, ω (which are linear combinations of u$\bar{u}$ and d$\bar{d}$ states), ϕ ($=$ s$\bar{s}$), J/ψ and Υ gives a consistent method for estimating quark masses. Assuming that these states are close to the limit of zero binding energy, we have

$$m_u \simeq m_d \simeq \tfrac{1}{2}m_{\rho,\omega} \simeq \;\; 390 \text{ MeV}$$
$$m_s \simeq \tfrac{1}{2}m_\phi \simeq \;\; 500 \text{ MeV}$$
$$m_c \simeq \tfrac{1}{2}m_{J/\psi} \simeq 1600 \text{ MeV} \tag{6.107}$$
$$m_b \simeq \tfrac{1}{2}m_\Upsilon \simeq 5000 \text{ MeV}$$

The sixth quark t was observed in a different way from the others; it was produced not by strong but by weak interactions, in the decay of the W boson:

$$W \to t + \bar{b}. \tag{6.108}$$

This will be discussed in §6.6. The mass of the t quark is not known with certainty, but it seems likely that

$$m_t \simeq 40 \text{ GeV}. \tag{6.109}$$

6.5. The colour force
Colour

There are a number of empirical reasons for believing that each quark has an additional property beyond those we have already considered, which has three possible values, so that the state space for quarks of a given flavour (u, d, s, c, b or t) is not the usual spin/orbital state space $\mathscr{W} \otimes \mathscr{S}$ but $\mathscr{W} \otimes \mathscr{S} \otimes \mathscr{C}$ where $\mathscr{C}$ is three-dimensional. We will describe two pieces of evidence for this further degree of freedom, which is called **colour**.

The first comes from baryon spectroscopy. The spins, parities and isospins (and SU(3) multiplets) of baryon resonances formed from u, d and s quarks correspond very well to the *totally symmetric* states of three particles, each having spin $\tfrac{1}{2}$ and belonging to an SU(3) triplet, occupying the states of a central potential like the harmonic oscillator or the hydrogen atom (the relevant feature of which is that the states fall into orbital angular momentum multiplets). For example, the Δ multiplet, with spin $\tfrac{3}{2}$ and isospin $\tfrac{3}{2}$, is the symmetric combination of three spin-$\tfrac{1}{2}$ isospin-$\tfrac{3}{2}$ particles all having orbital angular momentum $l=0$ (this extends to the $\tfrac{3}{2}^+$ decuplet, which is the symmetric combination of three SU(3) triplets); the Δ multiplets with spins $\frac{1}{2}, \frac{3}{2}$, $\frac{5}{2}, \frac{7}{2}$ with masses around 1900 MeV (see Fig. 6.13) can be obtained as symmetric combinations in which two particles have $l = 1$ and one has $l = 0$. But since quarks have spin $\frac{1}{2}$ they should be fermions and should only exist in *antisymmetric* combinations. Thus there must be another degree of freedom to provide the antisymmetry; if quark states have a factor in the colour space $\mathscr{C}$, many-quark states can be symmetric in space, spin and isospin and antisymmetric in colour.

The second piece of evidence comes from the production of hadrons in electron–positron annihilation. We have seen how resonances occur in this process through the formation of quark–antiquark bound states as in Fig. 6.15. If the energy is not equal to that of a bound state the photon in this diagram can still produce a quark–antiquark pair, which will then separate and combine with other quarks and antiquarks formed from the vacuum as in Fig. 6.14, to produce a final state of many hadrons. Away from resonances the virtual photon is equally likely to produce any particle–antiparticle pair of a given charge, provided it has sufficient energy. In general, the probability that it will create a particular particle–antiparticle pair is proportional to the square of the charge on the particle (the charge occurs as a factor in the term of the Hamiltonian describing the photon–particle–antiparticle vertex, and therefore in the amplitude for the process; this must be squared to give the probability). Thus at a given energy E we can compare the total probability of quark–antiquark creation, leading to a final state of hadrons, with the probability of $\mu^- \mu^+$ creation, which does not lead to hadrons:

$$R(E) = \frac{\Gamma(e^- e^+ \rightarrow \text{hadrons})}{\Gamma(e^- e^+ \rightarrow \mu^- \mu^+)} = \sum_i q_i^2 \tag{6.110}$$

where q_i is the charge of quark number i and the sum is over all quarks with mass below $\frac{1}{2}E$.

The experimentally determined form of the function $R(E)$ is shown in Fig. 6.18. It shows a steplike rise each time the energy reaches a value at which another quark–antiquark pair can be produced; these **thresholds** are marked by resonances showing the formation of a bound state at energies slightly lower (because of binding energy) than that at which the quark and antiquark can separate. This qualitatively confirms the picture that the elementary constituents of hadrons, in increasing order of mass, are u, d, s, c and b (the t threshold has not yet been reached). But the value of R is not what this picture would give; instead of being $(\frac{2}{3})^2 + (-\frac{1}{3})^2 = \frac{5}{9}$ in the low-energy region where only u and d quarks are involved, it is more like $\frac{5}{3}$, and it continues to be too large by a factor of about 3. This suggests that every q^2 in the sum (6.110) occurs three times, i.e. that there are three different quarks for each of u, d, ...

Although we have evidence that the three different colour states of quarks exist, we have no idea what the difference between them is; it does not correspond to any detectable difference between the particles we observe. This has two very important consequences. First, it means that the laws of physics are unchanged if the different colours are changed round: thus it implies the existence of another symmetry group in nature. Unlike the isospin and SU(3) symmetries (**flavour** symmetries) which declared the approximate equivalence of particles which were in fact detectably different, this colour symmetry is *exact*. All the states in the three-dimensional colour space are equivalent to each other under this symmetry, so the symmetry operations consist of all transformations of these states which preserve the basic quantum-mechanical relations of linear dependence and inner product; as with the flavour symmetries, this leads us to the group of all unitary transformations of the colour space. Separating off the multiples of the identity (which refer to the symmetry of all states under multiplication by phase factors, and are not specific to colour symmetry), we find another symmetry group with the mathematical structure of the group SU(3).

Since colour differences between states are not observable, all physical states must be unaffected by colour transformations. This does not mean that the transformations are meaningless, for they act on single-quark states and isolated quarks are never found in nature; all physical states are combinations of several quarks. The effect of colour transformations on such combinations is obtained by combining the representations of the colour SU(3) group, in the same way as was discussed for the flavour SU(3) group in §6.3. This must result in the representation in which all colour transformations act as the identity operator. Thus the second consequence of the unobservability of colour is that *all physical states belong to singlet representations of the colour SU(3) group.*

This rule explains why there can be quark–antiquark combinations (mesons) and three-quark combinations (baryons) but no two-quark

Fig. 6.18.
The ratio
$$R = \frac{\Gamma(e^+e^- \rightarrow \text{hadrons})}{\Gamma(e^+e^- \rightarrow \mu^+\mu^-)}.$$

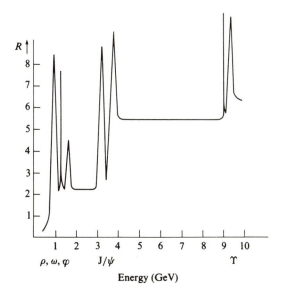

Energy (GeV)

combinations. Each quark belongs to the triplet representation of colour SU(3), each antiquark to the antitriplet; the combination of these two contains a singlet (see (6.80)). The combination of two triplets does not contain a singlet (eq. (6.88)), but the combination of three triplets does (eq. (6.89)). Moreover, this singlet is the antisymmetric combination of the three triplets, as it must be if colour is to do the job of restoring antisymmetry to the quarks in a baryon.

Gluons   The eight hermitian generators of the colour SU(3) group represent the observables which are conserved as a result of the symmetry. They are called **colour charges**. In this case 'observable' is a misnomer, since different values of these quantities distinguish different colour states, and this difference is *not* observable. Nevertheless, the colour charges are of great physical significance, for they act as the source of the strong force between quarks in the same way as the electric charge is the source of the electric field.

Each of the eight colour charges is associated with a boson, the quantum of the field generated by the charge, in the same way as electric charge is associated with the photon (this is explained more fully in §7.4). These bosons are called **gluons**; they form an octet representation of the colour SU(3) group, which is self-conjugate (the antiparticle of a gluon is another gluon in the octet). As discussed on p. 255, such an octet is described by a set of hermitian fields γ_i, and a Hamiltonian describing the colour-symmetric interaction between gluons and quarks is

$$H' = \alpha_s \Phi^\dagger \lambda_i \Phi \cdot \gamma_i \tag{6.111}$$

where α_s is a coupling constant giving the strength of the interaction and $\Phi^\dagger = (\phi_R{}^\dagger, \phi_B{}^\dagger, \phi_Y{}^\dagger)$ is the set of fields for a colour triplet of quarks (say red, blue and yellow). This refers to a given flavour of quark (e.g. u quarks only); the Hamiltonian for the strong interactions of all quarks is

$$H' = \alpha_s C_i^q \gamma_i \quad \text{where } C_i^q = \sum_f \Phi_f{}^\dagger \lambda_i \Phi_f, \tag{6.112}$$

the sum in C_i^q being over all flavours $f = u, d, s, c, b, t$. This interaction Hamiltonian has an exact SU(6) symmetry under flavour transformations (the coupling constant α_s being the same for each flavour); unlike the colour symmetry, this is not an exact symmetry of the full Hamiltonian since it is broken by the different masses in the quark free-particle terms.

The operators C_i^q of (6.112) are the components of the **quark colour current**; each of them can be written as

$$C_i^q = \sum_f (a_i a_i^\dagger + \bar{a}_i^\dagger a_i^\dagger + a_i \bar{a}_i + \bar{a}_i \bar{a}_i^\dagger) \tag{6.113}$$

where a_i and $a_i^\dagger$ are the annihilation and creation operators for the colour state of a quark which is an eigenstate of the ith colour charge, $\bar{a}_i$ and $\bar{a}_i^\dagger$ refer to the antiquark, and the sum is over all flavours. Thus C_i^q gives rise to lines in a Feynman diagram along which the ith colour charge flows in the same way as electric charge flows along the electron–positron lines of electromagnetic Feynman diagrams; the ith gluon is attached to these lines in the same way as the photon is attached to electron–positron lines.

Since the gluons also carry colour charge, they must themselves experience the colour force and therefore must contribute to the total colour current. Their contribution will be

$$C_i^g = \Gamma^\dagger t_i \Gamma \qquad (6.114)$$

where Γ is the eight-component column vector containing the gluon fields γ_i, and t_i is the 8×8 matrix representing the ith hermitian generator of SU(3). This is given by (6.93), which, together with the fact that the gluon fields γ_i are hermitian, gives

$$C_i^g = i f_{ijk} \gamma_j \gamma_k. \qquad (6.115)$$

If each gluon had only one state the gluon colour current C_i^g would vanish, since f_{ijk} is antisymmetric. But in fact gluons are spin-1 particles, and the different spin states give a non-zero contribution to (6.115). This then gives an extra interaction term which can be written as

$$H_g' = \alpha_s C_i^g \gamma_i = \alpha_s f_{ijk} \gamma_i \cdot (\gamma_j \times \gamma_k) \qquad (6.116)$$

where the spin-1 gluon fields are regarded as vector operators under the rotation group. This vector character of the gluon fields plays a crucial role in making the field theory of the colour force a gauge theory; this is explained in Chapter 7.

The term (6.116) gives rise to three-gluon vertices in Feynman diagrams: gluons, having colour charge, can themselves emit and absorb gluons. It can be seen that this term depends on the structure constants f_{ijk} being non-zero, which is equivalent to the group SU(3) being non-abelian. The gauge theory also requires a four-gluon term

$$H_g'' = \alpha_s^2 C_i^g C_i^g. \qquad (6.117)$$

If gluons are attracted to each other by a strong force, as the three-gluon and four-gluon vertices suggest, then it seems possible that they might form bound states. Since the product of two SU(3) octets contains a singlet, a bound state of two gluons could exist as an observable physical particle. Such a hypothetical particle is called a **glueball**. There are one or two meson resonances which might be candidates for identification as a glueball, but there is no incontrovertible evidence that glueballs exist.

Asymptotic freedom and confinement
The dynamical consequences of a gauge field theory cannot be treated properly in this book; we will simply summarise the main conclusions. The effect of complicated Feynman diagrams on the rate of a particular process is to reproduce simple, low-order diagrams but with a changed coupling constant α_s which depends on the momenta of the particles involved; in particular, for two-body scattering $A + B \rightarrow C + D$, α_s depends on the

momentum transfer

$$q^2 = (p_A - p_A')^2 \qquad (6.118)$$

where p_A are the energy-momentum 4-vectors of A before and after the collision, and the square denotes the Lorentz-invariant square of a 4-vector (see Appendix I). This can also be regarded as a dependence on the distance between the particles, with high values of q^2 corresponding to small distances and vice versa. The fact that quantum chromodynamics is a gauge field theory has significant consequences for this variation of α_s.

It has been *proved* that α_s becomes small at large values of q^2, so that quarks behave like free particles. This fact, called **asymptotic freedom**, explains the results of high-energy electron–proton scattering experiments (the hadronic analogue of Rutherford scattering off nuclei), in which electrons which are deflected through large angles and therefore transfer high momenta to the quarks (so that q^2 is large) scatter elastically† off the quarks as if they were free point particles. More precisely, the quarks behave like particles bound in a weak potential (with small coupling constant) like the electrons in an atom.

It is *conjectured* that α_s is large at small values of q^2, corresponding to large distances. This feature, called **confinement**, would explain why isolated quarks have never been observed and why all physical states are colour singlets. Together with confinement of quarks goes confinement of gluons; gluons, being coloured objects, cannot escape indefinitely far from their quark sources (even though they are massless and, like photons, are associated with an apparently infinite-range force). Thus the lines of force of the gluon field must all begin and end on quarks, forming a tube of force as in Fig. 6.19(*a*). This gives rise to a constant force between quarks, so that the amount of energy required to separate them is proportional to the separation, and beyond a certain

† But note that this occurs in 'deep *inelastic* scattering'. This is because the energy which is transferred to the quark is then used to disrupt the hadron it belongs to and to create more hadrons, as explained below.

Fig. 6.19.
(*a*) Confinement of colour lines of force; (*b*) jet formation.

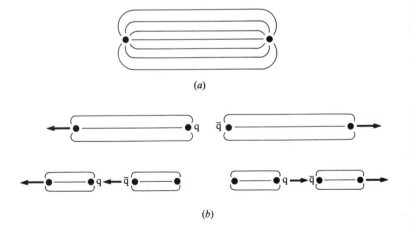

(*a*)

(*b*)

separation is sufficient to create a quark–antiquark pair from the vacuum.

If a quark and an antiquark are in a state with high energy, as for example when they are created from a photon in electron–positron collision experiments, then in their centre-of-mass frame of reference they will be moving away from each other with high momentum. This stretches the tube of force between them until another quark–antiquark pair is created, as in Fig. 6.19(*b*); this is repeated until eventually a crowd of hadrons has been produced. If each quark–antiquark pair is created as soon as there is sufficient energy in the tube of force, they will have small relative velocity; in particular, the new quark and antiquark in Fig. 6.19(*b*) have low transverse velocity relative to the original tube of force, and so the two new tubes have momenta in the same directions as the original two particles. As these in turn get stretched and produce new particles, the momenta will continue to be in the two original directions. On the basis of this intuitive classical reasoning, it is conjectured that the hadrons produced by this process will appear in two **jets** of particles: in each jet the momenta of the particles are all in approximately the same direction, which is the direction of motion of the original quark or antiquark. A jet of hadrons can also be expected to form if a high-energy gluon is radiated by a quark.

The hadrons produced in electron–positron collisions do indeed appear in the form of jets (see Fig. 6.20). Events with both two and three jets have been observed.

The ideas of asymptotic freedom and confinement are confirmed by the spectroscopy of the heavy quark systems of charmonium and beautonium. The excited states of these systems are consistent with a potential of the form

$$V(r) = \frac{\alpha}{r} + \beta r \qquad (6.119)$$

with $\alpha \simeq \frac{1}{2}$. The smallness of the coupling constant on the Coulomb-like term α/r shows the effect of asymptotic freedom; the linear term will give rise to confinement.

The Zweig rule The 1^- mesons ϕ, J/ψ and Υ are each a bound state of a quark and its antiquark. They could decay by the mutual annihilation of this pair, but this decay has a very slow rate. The structure of quantum chromodynamics, in conjunction with the rule that all physical states are colour singlets, provides an explanation for this.

Colour conservation prevents the quark and antiquark annihilating to form a single gluon, since they are in a colour singlet state and the gluon is an octet state. If they annihilate to form two gluons, these must be in the SU(3) singlet state formed from the product of two octets, which is a symmetric combination (as an SU(3) scalar, it is given by the scalar product $V_i W_i$). Hence the two-gluon state has charge conjugation parity $+$, for the effect of charge conjugation in this totally neutral state is to interchange the two gluons. But the mesons in

Fig. 6.20.
Jets produced in a proton–antiproton collision (photo: CERN).

question are produced from photons and therefore have odd charge conjugation parity. Thus they cannot form two gluons, but must form at least three. There are therefore at least three $q\bar{q}g$ vertices in the Feynman diagram for the process, and the amplitude has a factor of $\alpha_s{}^3$. Since α_s is small at the distance involved, the probability for the process is considerably reduced.

6.6. **The electroweak force**
Non-conservation of parity

Before discussing the unified theory of the weak and electromagnetic forces we need to review some characteristic properties of the weak force. The first of these, which we noted in discussing neutral K-mesons, is that the weak force does not conserve parity. The clearest manifestation of this is that neutrinos, which engage in no other interactions except gravity, are always left-handed (have helicity $-\frac{1}{2}$); thus the parity operator P, which would change a left-handed neutrino into a right-handed one travelling in the opposite direction, is not defined on the state space of a neutrino. (Antineutrinos are always right-handed, so the combined operation CP is defined.)

The failure of reflection symmetry in weak interactions was shown experimentally by C. S. Wu in 1957, by observing the electrons emitted in the β-decay

$$^{60}\text{Co} \rightarrow {}^{60}\text{Ni} + \text{e}^- + \bar{\nu}_e. \tag{6.120}$$

The ^{60}Co nuclei, which have spin 5, were placed in a strong magnetic field **B**. The effect of this is to add a term $-\mu\mathbf{J}\cdot\mathbf{B}$ to the Hamiltonian of the nucleus (the spin $\mathbf{J}$ giving the nucleus a magnetic moment $\mu\mathbf{J}$); thus if the magnetic field is along the z-axis the ground state of the nucleus is the eigenstate of J_z with eigenvalue 5. (Like a compass needle, the spin of the nucleus points along the magnetic field.) It was found that the electrons in the decay (6.120) had an angular distribution

$$N(\theta) = 1 - \frac{v}{c}\cos\theta \tag{6.121}$$

where $N(\theta)\,d\theta$ is the number of electrons with velocity v emitted between angles θ and $\theta + d\theta$ to the magnetic field. Thus more electrons are emitted in the opposite direction to the nuclear spin than in the same direction. Fig. 6.21 shows that this situation is not symmetric under reflection in a mirror placed parallel to the nuclear spin.

The asymmetry shown in this experiment can be explained qualitatively by the hypothesis that the electrons all have negative helicity. The ^{60}Ni nucleus has spin 4, so the ^{60}Co nucleus has lost angular momentum $\Delta\mathbf{J}$ with magnitude at least 1 (i.e. $\Delta\mathbf{J}^2 = j(j+1)$ with $j \geqslant 1$) and z-component $\delta J_z \geqslant 1$; this must be made up from the spins of the electron and the antineutrino, their orbital angular momentum and that of the ^{60}Ni nucleus. Let us ignore orbital angular momentum for the moment, and consider electrons emitted in the z-direction ($\theta = 0$ or π). Electrons emitted upwards ($\theta = 0$) have spin component $s_z = -\frac{1}{2}$ (by hypothesis), so they cannot make up the required ΔJ_z; this can only be done by a downward electron ($s_z = \frac{1}{2}$) accompanied by an upward antineutrino ($s_z = \frac{1}{2}$ since antineutrinos are righthanded). Thus if only spin is taken into account, electrons can be emitted in the direction $\theta = \pi$ but not in the direction $\theta = 0$. The effects of orbital angular momentum will not distinguish between the two directions, so there will be a net surplus of electrons in the direction $\theta = \pi$, as is shown by (6.121).

It is not consistent with special relativity to postulate that all electrons produced in β-decay are left-handed, for if an electron is left-handed in one frame of reference it is right-handed in a frame of reference moving faster than the electron in the same direction (see Fig. 6.22). Instead, one must postulate that an electron produced in β-decay will be in a spin state

$$|L\rangle = \sqrt{(\tfrac{1}{2}(1 - v/c))}|+\rangle + \sqrt{(\tfrac{1}{2}(1 + v/c))}|-\rangle \tag{6.122}$$

where $|\pm\rangle$ are states with helicity $\pm\frac{1}{2}$, and v is the velocity of the electron. It is shown in §7.2 that this is a relativistically invariant statement. An electron in the state $|L\rangle$ is more likely to be left-handed than right-handed, and this likeliness becomes certainty (as for a neutrino) as $v \to c$.

The orthogonal state to $|L\rangle$ is

$$|R\rangle = \sqrt{(\tfrac{1}{2}(1 + v/c))}|+\rangle - \sqrt{(\tfrac{1}{2}(1 - v/c))}|-\rangle. \tag{6.123}$$

This is the appropriate state for the antiparticle of a particle in the state $|L\rangle$. All positrons produced by weak interactions are in the state $|R\rangle$.

Since the state $|L\rangle$ becomes the negative-helicity state $|-\rangle$ when $v = c$, we see that both electrons and neutrinos are in the state $|L\rangle$ when produced by weak

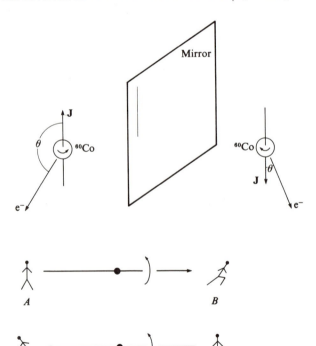

Fig. 6.21.
Parity violation.

Fig. 6.22.
Helicity of an electron: the electron is left-handed in the frame of reference in which A is at rest, but right-handed in the frame in which B is at rest.

interactions, while their antiparticles are in the state $|R\rangle$. This statement is true of all the basic fermions (leptons and quarks), and shows a general left-handedness of the weak force. However, it must be modified to take account of angular momentum conservation: the final state of fermions in $|L\rangle$ and $|R\rangle$ states must be projected onto the subspace with the same angular momentum as the initial state.

Example: pion decay

Charged pions decay predominantly into a muon and the appropriate neutrino. By the universality of the weak interactions, there should be an equal amplitude for the decay of the pion into an electron or a muon. Assuming that this is so, we will calculate the ratio of the rates of the decays

$$\pi^- \rightarrow \mu^- + \bar{\nu}_\mu \quad \text{and} \quad \pi^- \rightarrow e^- + \bar{\nu}_e. \tag{6.124}$$

The decay Hamiltonian W acts on the initial π^- to produce an equal superposition of electron and muon states:

$$W|\pi^-\rangle = g\Pi|\mu^-, L\rangle|\bar{\nu}_\mu, R\rangle + g\Pi|e^-, L\rangle|\bar{\nu}_e, R\rangle \tag{6.125}$$

where g is a coupling constant and Π is the projection onto the initial angular momentum state: in this case, since the pion is spinless, and taking it to be at rest, Π is the projection onto the state with zero angular momentum. By conservation of momentum the lepton and the antineutrino have opposite momentum; since the antineutrino has helicity $+\frac{1}{2}$, the lepton must have helicity $-\frac{1}{2}$ to give zero component of the total angular momentum in the direction of motion. Thus Π projects the lepton state $|L\rangle$ onto the positive-helicity state $|+\rangle$; from (6.122), this gives

$$\langle l^- \bar{\nu}_l | W | \pi^- \rangle = g\sqrt{(\tfrac{1}{2}(1 - v_l/c)} \tag{6.126}$$

where $l = e$ or μ, and v_l is the velocity of the lepton l. As in §3.5, the relativistic kinematics of the decay give the lepton's energy and momentum as

$$E_l = \frac{m_\pi^2 + m_l^2}{2m_\pi}c^2, \quad p = \frac{m_\pi^2 - m_l^2}{2m_\pi}c, \tag{6.127}$$

so that

$$\frac{v_l}{c} = \frac{pc}{E} = \frac{m_\pi^2 - m_l^2}{m_\pi^2 + m_l^2}. \tag{6.128}$$

From (3.200) (taking A to be the lepton and B the antineutrino, so that $E_A = E_l$ and $E_B = pc$), the phase-space factor is

$$\rho_l = \frac{p^2 E_l}{m_\pi c^3} = \frac{(m_\pi^2 - m_l^2)^2(m_\pi^2 + m_l^2)c}{8m_\pi^4}. \tag{6.129}$$

Hence the ratio of the rates of the decays is

$$\frac{\Gamma(\pi^- \rightarrow e^- \bar{\nu}_e)}{\Gamma(\pi^- \rightarrow \mu^- \bar{\nu}_\mu)} = \frac{\rho_e |\langle e^- \bar{\nu}_e | W | \pi^- \rangle|^2}{\rho_\mu |\langle \mu^- \bar{\nu}_\mu | W | \pi^- \rangle|^2} = \frac{m_e^2(m_\pi^2 - m_e^2)^2}{m_\mu^2(m_\pi^2 - m_\mu^2)^2} \simeq 10^{-4}. \tag{6.130}$$

To summarise: the electronic decay is suppressed because it can only take place with a right-handed electron and the weak force prefers to produce a left-handed electron; this preference is stronger for the lighter electron than for the heavier muon. This consideration outweighs the fact that the electronic decay is more favourable energetically than the muonic.

The Salam–Weinberg
Hamiltonian

The examples just discussed show that an electron in the state $|L\rangle$ and an electron neutrino (in its sole helicity state) form a pair under the weak force; they are produced in the form of a particle–antiparticle pair in β-decay and pion decay. The third helicity state of the electron-type leptons, namely the $|R\rangle$ state of an electron, stands alone, unaffected by the weak force. The Salam–Weinberg theory of weak interactions introduces an SU(2) group of transformations called **weak isospin** transformations which mix the first two states and leave the third as it is. Thus we have a weak isospin doublet (e_L, v_e) and a singlet e_R. The other leptons are classified similarly, forming doublets (μ_L, v_μ) and (τ_L, v_τ) and singlets μ_R and τ_R.

The quarks also form left-handed doublets and right-handed singlets under weak isospin transformations. The situation here is a little more complicated, since the u quark is coupled both to the d quark (as in β-decay, where one of the d quarks in the neutron becomes a u quark) and to the s quark (as in the semileptonic Λ^0 decay (6.46), where the s quark in the Λ^0 becomes a u). Since the coefficients of the strangeness-conserving and strangeness-changing parts of the hadronic weak current are given by the Cabibbo angle θ_C, the weak isospin doublet containing $|u, L\rangle$ is taken to include the superposition

$$|d', L\rangle = |d, L\rangle \cos \theta_C + |s, L\rangle \sin \theta_C. \tag{6.131}$$

The orthogonal combination

$$|s', L\rangle = |s, L\rangle \cos \theta_C - |d, L\rangle \sin \theta_C \tag{6.132}$$

is taken to form a doublet together with the charmed quark state $|c, L\rangle$. The third quark doublet contains the left-handed states of the t and b quarks. The b component is in fact a superposition and contains an admixture of d and s, and correspondingly the d' and s' states should contain some b. These terms appear to be small and for simplicity we will ignore them, but they must be present since otherwise the b quark would be stable.

For each of these particles except the neutrinos we have two possible states and therefore there will be two annihilation and creation operators; from these we can form two field operators which we will denote by the name of the particle with the suffix L or R. Thus for the electron we have the fields

$$\left. \begin{aligned} e_L &= a_{e^-, L} + (a_{e^+, R})^\dagger \\ e_R &= a_{e^-, R} + (a_{e^+, L})^\dagger \end{aligned} \right\}. \tag{6.133}$$

Now we arrange these fields in column vectors according to their weak isospin properties, as in Table 6.5. In each multiplet the electric charge is given by

$$Q = i_3 + \tfrac{1}{2} y \tag{6.134}$$

where i_3 is a generator of the weak isospin group and y is a property of the multiplet called the **weak hypercharge**. The values of y are also shown in Table 6.5.

Let Ψ be one of these two-component column vectors of fields. Then we can form a weak isovector $\Psi^\dagger \tau \Psi$ where τ denotes the vector of Pauli matrices. The total weak current is the sum of all these operators:

$$\mathbf{J}_w = \sum_{l=e,\mu,\tau} \Psi_l^\dagger \tau \Psi_l + \sum_{q=u,c,t} \sum_{i=1}^{3} \Psi_{qi}^\dagger \tau \Psi_{qi} \tag{6.135}$$

(the second term includes a sum over quark colours i). The Salam–Weinberg theory postulates a set of bosons $W^\pm$, W^0 which form a weak isospin triplet, so that their fields can be considered as a weak isovector operator $\mathbf{W}$ (cf. 6.30)), and which go together with the current $\mathbf{J}_w$ in an interaction Hamiltonian of Yukawa–Kemmer type. The theory also includes a single boson B^0 which forms a weak isospin singlet, and which couples with the weak isoscalar operator

$$J_0 = \sum_f y \Psi_f^\dagger \Psi_f + \sum_{f'} y \phi_{f'}^\dagger \phi_{f'}. \tag{6.136}$$

where the first sum extends over all the doublet (left-handed) fields Ψ, the second over all the singlet (right-handed) fields ϕ, and y is the weak hypercharge. Thus the full electroweak interaction Hamiltonian is

$$H_{ew} = g \mathbf{J}_w \cdot \mathbf{W} + g' J_0 B \tag{6.137}$$

where g and g' are independent coupling constants.

The four bosons in this theory correspond to the four generators of a group

$$G_{ew} = SU(2)_w \times U(1)_y \tag{6.138}$$

in which $SU(2)_w$ is the group of weak isospin transformations and $U(1)_y$ is a one-parameter group whose single generator is the weak hypercharge operator. As a group, $U(1)_y$ is isomorphic to the group of complex numbers of unit modulus; the element $e^{i\theta}$ is represented by an operator $U(\theta)$ which acts on a state with weak hypercharge y by multiplying it by $e^{in\theta}$ where $n = 3y$. Classifying states by their weak isospin and weak hypercharge is equivalent to putting them into representations of G_{ew}.

Table 6.5. *Weak isospin multiplets*

$y = -1$:	$\Psi_e = \begin{pmatrix} v_e \\ e_L \end{pmatrix}$	$\Psi_\mu = \begin{pmatrix} v_\mu \\ \mu_L \end{pmatrix}$	$\Psi_\tau = \begin{pmatrix} v_\tau \\ \tau_L \end{pmatrix}$
$y = -2$:	e_R	μ_R	τ_R
$y = \frac{1}{3}$:	$\Psi_u = \begin{pmatrix} u \\ d' \end{pmatrix}_L$	$\Psi_c = \begin{pmatrix} c \\ s' \end{pmatrix}_L$	$\Psi_t = \begin{pmatrix} t \\ b \end{pmatrix}_L$
$y = \frac{4}{3}$:	u_R	c_R	t_R
$y = -\frac{2}{3}$:	d'_R	s'_R	b_R

$$(d' = d \cos \theta_C + s \sin \theta_C, \; s' = -d \sin \theta_C + s \cos \theta_C)$$

We will now see how the Hamiltonian (6.137) incorporates both weak and electromagnetic effects. Each term involving $W^\pm$ yields a Feynman vertex in which one member of a weak isodoublet enters and the other one leaves, with the emission or absorption of a $W^\pm$. These vertices account for all the weak processes we have considered so far: some examples are shown in Fig. 6.23. Note that the doublets containing d' and s' give rise to processes involving d and s in which the amplitude is multiplied by a factor of $\cos\theta_C$ or $\sin\theta_C$ for each vertex. Thus, for example, the ratio of the rates of the decays of the charmed meson D^0 shown in Figs. 6.23(g) and (h) is

$$\frac{\Gamma(D^0 \to K^+\pi^-)}{\Gamma(D^0 \to K^-\pi^+)} = \frac{|\sin^2\theta_C|^2}{|\cos^2\theta_C|^2} = \tan^4\theta_C \simeq 3 \times 10^{-3}. \qquad (6.139)$$

Note that if the bosons $W^\pm$ are very massive, then at low energy the part of the Hamiltonian containing them, namely

$$H_{ch} = g(J_{ch}W_- + J_{ch}{}^\dagger W_+) \qquad (6.140)$$

(where $J_{ch} = J_{w+}$, $J_{ch}{}^\dagger = J_{w-}$; 'ch' stands for 'charged') can, according to (4.194),

Fig. 6.23.

Weak processes:

(a) $n \to p + e^- + \bar{v}_e$

(b) $\Lambda^0 \to p + e^- + \bar{v}_e$

(c) $\Lambda^0 \to p + \pi^-$

(d) $K^- \to \pi^- + \pi^0$

(e) $\pi^+ \to \mu^+ + v_\mu$

(f) $\Lambda_c^+ \to \Lambda^0 + K^+ + \pi^0$

(g) $D^0 \to K^- + \pi^+$

(h) $D^0 \to K^+ + \pi^-$.

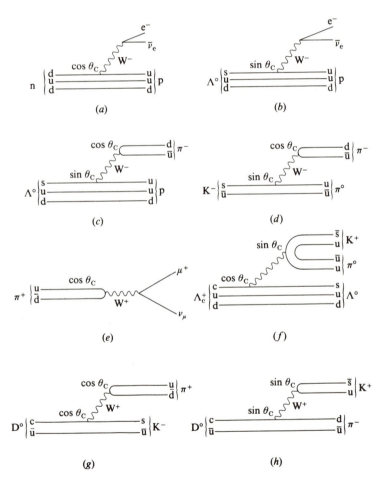

be replaced by the effective Hamiltonian

$$H_{\text{eff}} = \frac{g^2}{m_W} J_{\text{ch}}^{\dagger} J_{\text{ch}}, \tag{6.141}$$

describing second-order processes like those of Fig. 6.23. This is the Hamiltonian of (6.38).

The remaining terms in the electroweak Hamiltonian (6.137) are those containing the neutral bosons W^0 and B^0:

$$H_{\text{neut}} = g J_3 W_3 + g' J_0 B. \tag{6.142}$$

Now for any isodoublet $\Psi = (\phi_1, \phi_2)^T$ we have

$$\tfrac{1}{2}\Psi^{\dagger}\tau_3\Psi = \tfrac{1}{2}\phi_1^{\dagger}\phi_1 - \tfrac{1}{2}\phi_2^{\dagger}\phi_2 = \sum_k i_3 \phi_k^{\dagger}\phi_k. \tag{6.143}$$

Hence

$$J_3 = \sum_f \Psi_f^{\dagger}\tau_3\Psi_f = \sum_k 2i_3 \phi_k^{\dagger}\phi_k \tag{6.144}$$

where the second sum is taken over all the individual fields in the doublets, and might as well include the singlet fields since these have $i_3 = 0$. From (6.136) we can write J_0 also as a sum over individual fields, so H_{neut} can be written as

$$H_{\text{neut}} = \sum_k \phi_k^{\dagger}\phi_k(2gi_3 W^0 + g' y B) \tag{6.145}$$

(since the third component of the isovector W is the W^0 field).

The fields in (6.145) are those of eigenstates of weak isospin, and therefore create the Cabibbo-rotated states d′, s′ rather than the physical particles d and s. However, because these have the same values of i_3 and y, and because they are orthogonal combinations of d and s, the fields occur as

$$d'^{\dagger}d' + s'^{\dagger}s' = d^{\dagger}d + s^{\dagger}s. \tag{6.146}$$

Thus the fields can after all be taken as referring to the physical particles.

Write

$$W^0 = Z \cos \theta_W + A \sin \theta_W,$$
$$B = -Z \sin \theta_W + A \cos \theta_W, \tag{6.147}$$

where

$$\tan \theta_W = \frac{g'}{g}; \tag{6.148}$$

then

$$H_{\text{neut}} = \sum_k e(i_3 + \tfrac{1}{2}y)\phi_k^{\dagger}\phi_k A + \sum_k e(2i_3 \cot \theta_W - \tfrac{1}{2}y \tan \theta_W)\phi_k^{\dagger}\phi_k Z \tag{6.149}$$

where

$$e = 2g \sin \theta_W. \tag{6.150}$$

Since $i_3 + \tfrac{1}{2}y$ is the electric charge in units of the charge on the electron, the first term in (6.149) is the electromagnetic Hamiltonian if e is the charge of the electron and A is the photon field. The second term describes new interactions, mediated by the neutral particle Z, in which the identity of particles does not change; for example, this term gives rise to a force on neutrinos which will cause elastic scattering of neutrinos off nuclei. This is called a **weak neutral current** force, and was thought to be non-existent before the Salam–Weinberg theory was proposed.

If the d' field was not balanced by the s' field, the Hamiltonian would contain a term

$$d'^{\dagger}d'Z = [d^{\dagger}d \cos^2 \theta_C + (d^{\dagger}s + s^{\dagger}d) \cos \theta_C \sin \theta_C + s^{\dagger}s \sin^2 \theta_C]Z, \quad (6.151)$$

which would cause processes in which quarks did change their identity: an s quark could change into a d quark and emit a Z. Such a strangeness-changing neutral current has long been known to be absent: if it existed, for example, neutral kaons could decay into $\mu^+ + \mu^-$ by a similar process to pion decay (see Fig. 6.24). It was the need to cancel the $d^{\dagger}s$ terms in (6.151), by means of an equation like (6.146), that led Glashow, Iliopoulos and Maiani to the hypothesis that the c quark must exist as a weak isospin partner of s', some four years before it was discovered experimentally. The cancellation of the strangeness-changing neutral current so achieved is known as the **GIM mechanism**.

Like quantum electrodynamics, the Salam–Weinberg theory is a gauge theory; the significance of this will be explained in Chapter 7. Unlike the gluons of quantum chromodynamics, the bosons $W^{\pm}$, Z and γ of quantum flavourdynamics are not all massless; it will be shown that this is related to the fact that weak isospin is not an exact symmetry like colour.

Experimental confirmation of the Salam–Weinberg theory

Events in which a muon neutrino interacted with a proton without being changed into a muon were observed in 1974. An example of such an event is

$$\nu_{\mu} + p \rightarrow n + \pi^+ + \nu_{\mu}. \quad (6.152)$$

This is a neutral-current event, of the sort which is required by the existence of the Z boson.

The W and Z bosons were discovered in 1983 at CERN, Geneva. They were produced in high-energy proton–antiproton collisions, in which a quark and an antiquark produce a boson according to the Feynman diagrams of Fig. 6.25, and were detected by their subsequent decay into leptons:

$$W^+ \rightarrow e^+ + \nu_e, \quad W^- \rightarrow e^- + \bar{\nu}_e, \quad Z^0 \rightarrow e^+ + e^- \quad (6.153)$$

Fig. 6.24

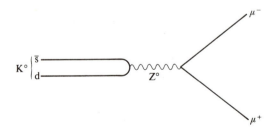

and similar processes with muons. The characteristic feature which signals the presence of the W and Z bosons is the extraordinarily high energy of the electrons and positrons. The mass of the W is 81 GeV (162 000 times that of the electron), and that of the Z is 93 GeV, and all this mass is converted into energy of the leptons. The W and the Z also decay into a quark and an antiquark, producing two jets of hadrons.

The angle θ_W is called the **Weinberg angle**. Its value is determined by the coupling constant g, which can be obtained from the low-energy effective weak coupling constant once the mass of the W is known. The result is

$$\sin \theta_W \simeq 0.48. \tag{6.154}$$

6.7. **Further speculations**
Grand unification

The success of the Salam–Weinberg theory in reducing the number of unrelated fundamental interactions from three to two encourages attempts to continue the process and develop a theory which combines the electroweak force with the colour and gravitational forces. Gravity is a special case, because of the unique relation to the geometry of space-time which is accorded it by general relativity; this means that it is still problematic whether it can even be combined with quantum mechanics satisfactorily. It seems, therefore, that the most promising order of attack is to try to unify the electroweak and colour forces first.

The bosons of the colour and electroweak forces correspond to the generators of the group

$$G_{ccw} = SU(3)_c \times SU(2)_w \times U(1)_y \tag{6.155}$$

(colour × weak isospin × weak hypercharge). There are three independent coupling constants because of the three commuting subgroups in this group. A unified theory, with only one coupling constant, could be obtained by embedding G_{cew} in an algebraically simple group G_{UT} (one with no factor subgroups) whose generators would correspond to the bosons of the unified theory and whose representations should classify the known particle states. The simplest possibility, which was proposed by Georgi and Glashow, is

$$G_{UT} = SU(5) \tag{6.156}$$

with the subgroups as indicated below, where the 5×5 matrix belonging to SU(5) is partitioned as $(3+2) \times (3+2)$:

$$SU(3)_c \ni \begin{bmatrix} U & 0 \\ 0 & 1_2 \end{bmatrix}, \quad SU(2)_w \ni \begin{bmatrix} 1_3 & 0 \\ 0 & U \end{bmatrix}, \quad U(1)_y \ni \begin{bmatrix} e^{2i\theta}1_3 & 0 \\ 0 & e^{-3i\theta}1_2 \end{bmatrix}. \tag{6.157}$$

In each family there are 15 pairs of fermion states, each consisting of a left-handed particle state and the right-handed state of its antiparticle; there is a

Fig. 6.25 Production of W and Z bosons.

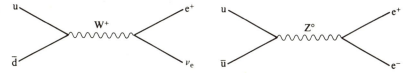

field operator for each pair. In the first family the left-handed states, arranged in representations of G_{cew}, are as follows:

$$(u_r, u_y, u_b, d_r, d_y, d_b)_L = \mathbf{3} \times \mathbf{2} \ (n = 1);$$
$$(\bar{u}_r, \bar{u}_y, \bar{u}_b)_L = \mathbf{3} \times \mathbf{1} \ (n = -4);$$
$$(\bar{d}_r, \bar{d}_y, \bar{d}_b)_L = \mathbf{3} \times \mathbf{1} \ (n = 2);$$
$$(v_e, e^-)_L = \mathbf{1} \times \mathbf{2} \ (n = -3);$$
$$(e^+)_R = \mathbf{1} \times \mathbf{1} \ (n = 6) \tag{6.158}$$

where the Roman subscripts label the colour states of quarks (red, yellow and blue, say); bold numerals denote the representations of $SU(3)_c \times SU(2)_w$ (for example, the six states in the first multiplet each belong to a triplet of $SU(3)_c$ and a doublet of $SU(2)_w$); and the representation of $U(1)_y$ is denoted by the value of $n = 3y$. (Compare Table 6.5.) Now $SU(5)$ has a five-dimensional representation (consisting of 5-component vectors u_α) and a 10-dimensional representation (consisting of antisymmetric second-rank tensors $t_{\alpha\beta}$) which break up into representations of the $SU(3) \times SU(2) \times U(1)$ subgroup as follows:

$$\mathbf{5} = \mathbf{3} \times \mathbf{1} \ (n = 2) \oplus \mathbf{1} \times \mathbf{2} \ (n = -3), \tag{6.159}$$

$$\mathbf{10} = \mathbf{3} \times \mathbf{2} \ (n = 1) \oplus \mathbf{3} \times \mathbf{1} \ (n = -4) \oplus \mathbf{1} \times \mathbf{1} \ (n = 6). \tag{6.160}$$

Thus between them these representations can accommodate all the left-handed fermion states in a family.

Because the weak hypercharge subgroup $U(1)_y$ occurs as a subgroup of a simple group, and not as an abelian subgroup commuting with everything else, the group representation theory requires its eigenvalues to be quantised (in very much the same way as the representation theory of the rotation group, based on the angular momentum commutation relations, forces J_z to be quantised). Thus this theory would explain the quantisation of electric charge. In fact, (6.157) shows that the hypercharge generator has the same eigenvalue for members of a colour multiplet, and the sum of its eigenvalues in any $SU(5)$ multiplet must be nought; the theory therefore explains why the electric charges on colour triplets (quarks) are multiples of a third of the charges on colour singlets (leptons).

$SU(5)$ has $5^2 - 1 = 24$ generators, so a theory based on this group will have 24 bosons carrying the force between the fermions. The photon, $W^\pm$, Z^0 and gluons are 12 of these. The remaining 12 correspond to 5×5 hermitian matrices with entries in the off-diagonal blocks in the partition of (6.157). These bosons X, Y couple quarks to leptons, in the same way as the $W_\pm$ bosons couple the u to the d and s quarks; thus the theory contains Feynman diagram vertices like those of Fig. 6.26. Just as the W-mediated weak interactions do not conserve strangeness, the X-mediated processes will not conserve baryon number. They should lead to the decay of the proton into leptons. The lifetime of the proton is expected to be about 10^{33} years (for comparison, the age of the universe is about 10^{10} years). Experiments looking for proton decay are under way at present; so far there are no reliable indications of it.

The fact that SU(5) is a simple group means that there can only be one coupling constant in a theory based on this group. Thus the two coupling constants g and g' (equivalently, e and θ_W) in electroweak theory, and the coupling constant α_s of quantum chromodynamics, should be related to each other by purely group-theoretical factors. The calculation is complicated by the fact that in the full theory the coupling constants change with distance, in the way that was sketched for the colour force in §6.5. After allowing for this, it is possible to calculate the Weinberg angle θ_W from the grand unified SU(5) theory. The result agrees reasonably well with the empirical value.

The masses of the extra bosons X and Y can also be estimated from the dependence on distance of the electroweak and colour coupling constants. Calculations in these theories show that the coupling constants become equal at a distance of about 10^{-30} m. This corresponds to the enormous mass of 10^{15} GeV. It is unlikely that this will ever be accessible to experiment; so if the grand unified theory is correct, there will be no very interesting experiments to be done between the present energies of a few hundred GeV, the scale of the electroweak force, and this grand-unification energy of 10^{15} GeV.

Supersymmetry   In the search for a unified description of particles the importance of internal symmetry groups, from the SU(2) of isospin to the SU(5) of grand unification, is obvious. However, there are fundamental limitations on the extent to which particles can be unified by being grouped together according to representations of a symmetry group. All the particle multiplets we have seen contain particles with the same spin. Now the spin states of a particle form a representation of the rotation group R, so if we have a multiplet of particles forming a representation $\mathscr{U}$ of an internal symmetry group G, and if they all have the same spin s, then their spin states form a space $\mathscr{U} \otimes \mathscr{D}_s$ which carries a representation of the product group $G \times R$. To get multiplets which contain particles with different spin it would be necessary to consider representations of a group which contained the rotation group as a subgroup in a less trivial way than this (see problem 6.31).

More generally, the spin/orbital states of a particle in a theory which incorporates special relativity must carry a representation of a group which includes rotations, translations and Lorentz transformations. This group is called the **Poincaré group** P. An irreducible representation of P is determined

Fig. 6.26.
Feynman diagram vertices in grand unified theories.

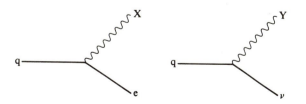

by two parameters m and s acts on the state space of a particle with mass m and spin s. The **Coleman–Mandula theorem** states that, given certain physically reasonable assumptions, any symmetry group which contains the Poincaré group as a subgroup must be of the form $G \times P$. This means that it is impossible to have symmetries which relate particles with different spins.

The idea of supersymmetry is to circumvent this result by extending the concept of symmetry. We find the representations of a symmetry group by finding the representations of its Lie algebra, which is a mathematical structure defined by commutators. In a supersymmetry this is replaced by a **Lie superalgebra**, which is a mathematical structure defined by commutators and anticommutators. The precise definition is that a Lie superalgebra L is a direct sum of two vector spaces, $L = L_0 \oplus L_1$, with an antisymmetric bilinear map $[X, Y]$ from $L_0 \times L_0$ to L_0 which makes L_0 a Lie algebra, a bilinear map $[X, Y]$ from $L_0 \times L_1$ to L_1, and a symmetric bilinear map $\{X, Y\}$ from $L_1 \times L_1$ to L_0, which satisfy

$$\left.\begin{aligned}
[[X, Y], Z] &= [X, [Y, Z]] - [Y, [X, Z]] \quad (X, Y \in L_0; \ Z \in L_1) \\
[X, \{Y, Z\}] &= \{[X, Y], Z\} + \{Y, [X, Z]\} \quad (X \in L_0; \ Y, Z \in L_1) \\
[\{X, Y\}, Z] &+ [\{Y, Z\}, X] + [\{Z, X\}, Y] = 0 \quad (X, Y, Z \in L_1)
\end{aligned}\right\}.$$

$$(6.161)$$

L_0 is called the **even** part of the superalgebra, L_1 the **odd** part.

A **representation** of a Lie superalgebra L is an assignment of operators $\rho(X)$ to the elements X of the superalgebra in such a way that

$$\left.\begin{aligned}
\rho([X, Y]) &= [\rho(X), \rho(Y)] \quad (X \in L_0; \ Y \in L_0 \text{ or } L_1) \\
\rho(\{X, Y\}) &= \{\rho(X), \rho(Y)\} \quad (X, Y \in L_1)
\end{aligned}\right\},$$

$$(6.162)$$

where the square brackets and curly brackets on the right-hand sides denote the usual commutator and anticommutator of operators.

With Lie superalgebras it is possible to find examples which contain the Lie algebra of the Poincaré group in a non-trivial way, and which have irreducible representations containing particles with different spins. Here we will give only a non-relativistic version to show how the idea works; for the relativistic version see problem 7.4.

Let L_0 be the Lie algebra spanned by the angular momentum J_i, position x_i, momentum p_i and energy H of a single free particle, together with the identity operator. The Lie brackets in L_0 are given by the basic commutation relations of ●3.10 together with

$$[H, J_i] = [H, p_i] = 0, \quad [H, x_i] = p_i/m \tag{6.163}$$

where m is the mass of the particle. Let L_1 be a four-dimensional space with basis elements $Q_\alpha, Q_\alpha^\dagger$ $(\alpha = 1, 2)$, and define the brackets

$$\left.\begin{aligned}
[J_i, Q_\alpha] &= -\tfrac{1}{2}(\sigma_i)_{\alpha\beta} Q_\beta, \quad [J_i, Q_\alpha^\dagger] = \tfrac{1}{2}(\sigma_i)_{\beta\alpha} Q_\beta^\dagger \\
[H \text{ or } p_i &\text{ or } x_i, Q_\alpha \text{ or } Q_\alpha^\dagger] = 0 \\
\{Q_\alpha, Q_\beta\} &= \{Q_\alpha^\dagger, Q_\beta^\dagger\} = 0, \quad \{Q_\alpha, Q_\beta^\dagger\} = m\delta_{\alpha\beta}
\end{aligned}\right\},$$

$$(6.164)$$

where σ_i are the Pauli matrices and the summation convention applies to Greek (spinor) indices.

The Lie algebra L_0 has irreducible representations corresponding to single particles, with representation spaces of the form $\mathscr{W} \otimes \mathscr{D}_s$ where $\mathscr{W}$ is a space of wave functions and $\mathscr{D}_s$ is a $(2s + 1)$-dimensional spin space. The odd part of the superalgebra obeys anticommutation relations like those of the annihilation and creation operators of a two-state fermion. For a given eigenvalue of H these have a representation containing four states: a vacuum, two one-particle states, and an antisymmetric two-particle state. A representation ρ of the full superalgebra, obeying the 'unitarity' condition $\rho(Q_\alpha{}^\dagger) = \rho(Q_\alpha)^\dagger$, can be obtained by taking a representation of L_0 and replacing every eigenstate of H by these four states. If the representation of L_0 described a particle with spin s, the representation of the superalgebra describes four particles with spins $s, s, s - \frac{1}{2}$ and $s + \frac{1}{2}$ (when $s = 0$ it describes three particles with spins $0, 0$ and $\frac{1}{2}$).

Other superalgebras can be constructed to include internal symmetries. It is characteristic that their representations include particles with spins differing by $\frac{1}{2}$. Thus the idea of supersymmetry offers the hope of a unified description of fermions and bosons. It also, because of its relation to space-time transformations, offers the hope of a quantum theory of general relativity and therefore of unifying gravity with the other forces. However, the experimental prognostications are not good. Supersymmetry requires that the spin-2 graviton should belong to a multiplet which also contains a massless spin-$\frac{3}{2}$ fermion called the **gravitino**, and that the gluons and the W and Z bosons should have spin-$\frac{1}{2}$ partners called **gluinos**, **Wino** and **Zino**. Also, the quarks and leptons should be accompanied by spin-0 particles called **squarks** and **sleptons**. None of these supersymmetric partners can be identified with any particles yet observed.

Substructure   The pattern of families of quarks and leptons naturally prompts speculation that they themselves are composite objects. However, there are no dynamical indications of any internal structure to these particles. The compositeness of a particle like the proton means that its electric charge is distributed over an extended region, so that its electrodynamical behaviour differs from that of a point charge: it is not described by the form of quantum electrodynamics appropriate to point particles (see §7.3). The leptons, however, behave electrodynamically like point particles to a very high degree of accuracy. The experimental limits on their deviations from pointlike behaviour imply that their spatial extension is less than 10^{-22} m, or alternatively that the energy needed to liberate their constituents is greater than 10^6 GeV.

For a review of composite models and experimental limits on them see Lyons 1983.

Further reading   For fuller accounts of the material in this chapter the textbooks by Perkins (1982) and Halzen & Martin (1984) are recommended. A less systematic account, intended to communicate the 'oral tradition' of particle physics, is Gottfried & Weisskopf 1984.

A fuller treatment of isospin and other symmetries can be found in Gibson & Pollard 1976. For hadron spectroscopy see Leader & Predazzi 1982, chapters 8–12. The reader who would like to understand the details of electroweak theory without first learning quantum field theory is referred to Aitchison & Hey 1982. For supersymmetry see Wess & Bagger 1983.

Non-technical accounts of the physics which here has been treated sketchily, or omitted, can be found in the following *Scientific American* articles: Bloom & Feldman 1982 (on hadron spectroscopy), Glashow 1975, Schwitters 1977, Lederman 1978 and Mistry, Poling & Thorndike 1983 (on charm and beauty), Ishikawa 1982 and Quigg 1985 (on the colour force), Nambu 1976 and Rebbi 1983 (on quark confinement), Jacob & Landshoff 1980 (on proton–proton scattering and jets), Weinberg 1974 (on the electroweak force), Perl & Kirk 1978 (on heavy leptons), Cline, Rubbia & van der Meer 1982 (on the experiments on W and Z particles), Georgi 1981 and Weinberg 1981 (on grand unification), Freedman & van Nieuwenhuizen 1978 (on supergravity), and Harari 1983 (on substructure). Some of these are collected in Kaufmann 1980. See also the Nobel addresses by Weinberg, Salam and Glashow (all 1980) and the *New Scientist* articles collected in Sutton 1985.

Problems on Chapter 6

1. The ρ^0, which belongs to the isospin triplet (ρ^-, ρ^0, ρ^+), decays into two pions by a strong interaction. What are the charges on the pions? Show that the spin of the ρ^0 is an odd integer and its parity is negative.

2. A neutral particle X^0 is a member of an isospin triplet and decays into two ρ-mesons. If isospin is conserved in the decay, show that the spin of the X^0 is at least 1.

3. The ω-meson is an isospin singlet with spin 1. Explain why it decays predominantly into three (rather than two) pions.

4. Show that in the isospin-conserving, parity-conserving process $n + \bar{p} \to \pi^- + \pi^0$ the relative angular momentum of the two pions is odd, while that of the neutron and the antiproton is even. Deduce that the neutron and the antiproton have parallel spins.

    Discuss the processes $p + \bar{p} \to \pi^+ + \pi^-$ and $p + \bar{p} \to \pi^0 + \pi^0$.

5. The N(1470) is a pair of particles (n^*, p^*) with the same quantum numbers as the neutron and the proton; they decay into Δ-baryons and π-mesons by an interaction which conserves isospin. Find the ratio of the rates of the decays $p^* \to \Delta^{++} + \pi^-$, $p^* \to \Delta^+ + \pi^0$ and $p^* \to \Delta^0 + \pi^+$.

6. $(\Delta_1^-, \Delta_1^0, \Delta_1^+, \Delta_1^{++})$ and $(\Delta_2^-, \Delta_2^0, \Delta_2^+, \Delta_2^{++})$ are two isospin multiplets of baryons; the latter decays into the former by an isospin-conserving process $\Delta_2 \to \Delta_1 + \pi$. Find the ratio of the rates of the decays $\Delta_2^{++} \to \Delta_1^+ + \pi^+$, $\Delta_2^+ \to \Delta_1^+ + \pi^0$ and $\Delta_2^+ \to \Delta_1^0 + \pi^+$.

7. Express the amplitudes for the isospin-conserving processes $\pi^+ + n \to \pi^0 + p$, $\pi^+ + n \to \pi^+ + n$ and $\pi^+ + p \to \pi^+ + p$ in terms of an $I = \frac{1}{2}$ amplitude and an $I = \frac{3}{2}$ amplitude, and hence find a relation between them.

8. Find CI_iC where C is the charge conjugation operator, and deduce that in a self-conjugate isospin multiplet C acts as reflection in the (13)-plane.

9. Show that the eigenvalue of G-parity for a self-conjugate multiplet is $(-1)^I \eta_c$ where I is the isospin of the multiplet and η_c is the charge conjugation eigenvalue of its neutral member.

10. The magnetic moment of an elementary particle is the expectation value of an operator which has the same isospin properties as the electromagnetic Hamiltonian. Show that the magnetic moments $\mu(\Delta)$ of the Δ multiplet are related by

$$\mu(\Delta^0) = 2\mu(\Delta^+) - \mu(\Delta^{++}) = \tfrac{1}{2}\mu(\Delta^-) + \tfrac{1}{2}\mu(\Delta^+).$$

11. Find the ratios of the lifetimes of ^{14}C, $^{14}\text{N*}$ and ^{14}O, decaying as in Fig. 6.1. (Use first-order perturbation theory.)

12. Suppose there was a neutral particle X^0 with spin 1, which decayed into two neutrons by a strong interaction. Show that it would have negative parity and that there would be two other particles with the same properties as X^0 except for electric charge. How would these particles decay?

13. There are particles with spins as high as $\frac{11}{2}$. Why are there no particles with isospin greater than $\frac{3}{2}$? ('Particle' here means something with baryon number 0 or ± 1.)

14. Assuming that the $\Delta I = \frac{1}{2}$ rule applies in the decays $\Sigma^+ \to p + \pi^0$, $\Sigma^+ \to n + \pi^+$ and $\Sigma^- \to n + \pi^-$, find a linear relation between the first-order decay amplitudes.

15. Use the $\Delta I = \frac{1}{2}$ rule to calculate the ratio of the lifetimes of Ξ^- and Ξ^0.

16. Use the $\Delta I = \frac{1}{2}$ rule to explain the following decay rates:

$$\Gamma(K^+ \to \pi^0 \pi^+) = 1.711 \times 10^7 \text{ s}^{-1},$$

$$\Gamma(K_s^0 \to \pi^+ \pi^-) = 7.689 \times 10^9 \text{ s}^{-1},$$

$$\Gamma(K_s^0 \to \pi^0 \pi^0) = 3.942 \times 10^9 \text{ s}^{-1}.$$

17. Give an example of a process which you would expect to produce a K^- meson.

18. Show that parity is conserved in $K \to 3\pi$ decays.

19. Show that the conjugate representation $\bar{\rho}$ of SU(3), defined by (6.75), has a matrix which is the complex conjugate of that of the fundamental representation ρ.

20. Write down the action of the generators on the weight vectors in a triangular representation of SU(3) and its conjugate representation.

21. Express electric charge and hypercharge in terms of the SU(3) operators $\mathbf{i} \cdot \mathbf{H}$, $\mathbf{u} \cdot \mathbf{H}$ and $\mathbf{v} \cdot \mathbf{H}$. Deduce that electric charge is constant in U-spin multiplets.

22. Suppose the strong Hamiltonian has the form $H_{st} = H_0 + H_8$ where H_0 commutes with SU(3) transformations and H_8 is the eighth component of an octet operator. Show that H_{st} is the sum of a U-spin scalar and the third

component of a U-spin vector. Deduce that masses (identified as expectation values of H_{st}) are equally spaced in a U-spin multiplet. Hence (i) reproduce Gell-Mann's prediction of the mass of the Ω^-, given the masses of the other particles in the decuplet; (ii) prove the **Gell-Mann/Okubo** formula

$$\tfrac{1}{2}(m_N + m_\Xi) = \tfrac{1}{4}(m_\Sigma + 3m_\Lambda)$$

for the masses of the octet baryons. (You will need to identify the combination $|\Sigma^0{}_U\rangle$ of $|\Sigma^0\rangle$ and $|\Lambda^0\rangle$ which is the $U_3 = 0$ component of a U-spin triplet.)

Compare with the data (Appendix II). How well does the Gell-Mann/Okubo formula work for the meson octet? Try it with m^2 instead of m.

23. Write out the SU(3) Yukawa–Kemmer Hamiltonian (6.94) in terms of isospin multiplets.

24. Suggest a way of measuring the Σ^0 lifetime.

25. The quantum numbers of strangeness, charm, beauty and truth can be understood as 's quark number', 'c quark number', etc. What are u quark number and d quark number?

26. Consider the model of particle interactions in which each particle has just one state, with interaction Hamiltonian $V = W + W^\dagger$ where $W = (\phi_A{}^\dagger \phi_B + \phi_C{}^\dagger \phi_D)\phi_\alpha$. Show that in second-order perturbation theory the amplitude for $A + B \to C + D$ becomes infinite if $E_A + E_B = E_\alpha$. Show also that if E_α is replaced by $E_\alpha + i\Gamma$ the transition probability has the Breit–Wigner form as a function of $E = E_A + E_B$. [This indicates how resonance behaviour is associated with a certain sort of Feynman diagram.]

27. Define 'left-handed'.

28. Prove (6.129).

29. For each of the following sets of decays, state the quark composition of the particles involved and draw a diagram to show the events underlying the decays in terms of quarks and leptons. Ignoring kinematical factors, find the ratio of the rates of the decays in each set in terms of the Cabibbo angle.

(i) $n \to p + e^- + \bar{\nu}_e$ (ii) $\Sigma^- \to \Lambda^0 + e^- + \bar{\nu}_e$

$\quad \Lambda^0 \to p + e^- + \bar{\nu}_e \qquad \Xi^- \to \Lambda^0 + e^- + \bar{\nu}_e$

(iv) $\Lambda_c{}^+ \to \Lambda^0 + e^+ + \nu_e$ (v) $K^- \to \pi^0 + e^- + \bar{\nu}_e$ (vi) $D^0 \to K^- + e^+ + \bar{\nu}_e$

$\quad \Lambda_c{}^+ \to n + e^+ + \nu_e \qquad \pi^- \to \pi^0 + e^- + \bar{\nu}_e \qquad D^0 \to \pi^- + e^+ + \bar{\nu}_e$

(viii) $D^0 \to K^- + \pi^0 + \pi^+$ (ix) $F^+ \to \eta^0 + \pi^+$

$\qquad D^0 \to K^- + K^0 + \pi^+ \qquad F^+ \to \eta^0 + K^+$

$\qquad D^0 \to \pi^- + \pi^0 + \pi^+ \qquad F^+ \to K^0 + \pi^+$

$\qquad D^0 \to \pi^- + K^0 + \pi^+ \qquad F^+ \to K^0 + K^+.$

30. Find the ratio of the rates of the decays in each of the following sets of decays. (Each ratio is the product of a Cabibbo factor and a phase-space factor. Use $\sin \theta_C = 0.23$ and the particle masses given in Appendix II.)

(i) $\pi^- \to \mu^- + \bar{\nu}_\mu$ (ii) $D^0 \to K^- + \pi^+$

$\quad K^- \to \mu^- + \bar{\nu}_\mu \qquad D^0 \to K^- + K^+$

$\qquad\qquad\qquad\qquad D^0 \to \pi^- + \pi^+$

$\qquad\qquad\qquad\qquad D^0 \to \pi^- + K^+$

31. Prove (6.159)–(6.160).

32. Let $G \times R$ denote the direct product of a group G and the rotation group R. Show that every irreducible representation of $G \times R$ is of the form $\mathscr{D}_G \otimes \mathscr{D}_R$ where $\mathscr{D}_G$ is an irreducible representation of G and $\mathscr{D}_R$ is an irreducible representation of R.

7

Quantum fields

The theoretical framework needed for the full development of the ideas of Chapter 6 is that of quantum field theory. This chapter is intended to take the reader onto the threshold of that theory. It follows directly on from Chapter 4, being concerned with the general theoretical structure, and is independent of Chapters 5 and 6 until §7.4, when the concept of isospin (§6.1) will be needed. In §7.5 and §7.6 the quantum field theory aspects of the colour force and the electroweak force (quantum chromodynamics and quantum flavourdynamics) are described in outline.

This chapter requires a fuller knowledge of special relativity than the other chapters of the book. In particular, it will be assumed that the reader is acquainted with the 4-vector formalism (see Appendix I for a summary). Knowledge of the electromagnetic field will also be assumed.

7.1. Field operators

In §4.6 the notion of a reduced quantum field was introduced. We will now see the full concept from which this reduced notion was derived. We will arrive at this concept from two different directions: first, as the quantum counterpart of the classical concept of a field, and secondly by developing the ideas of §4.6 on a system of an indefinite number of particles.

(a) The electromagnetic field

The classical theory of the electromagnetic field concerns two vector fields, the electric field $\mathbf{E}(\mathbf{r}, t)$ and the magnetic field $\mathbf{B}(\mathbf{r}, t)$, which (in the absence of dielectric or magnetic material) satisfy **Maxwell's equations**:

$$\nabla \cdot \mathbf{E} = \frac{\rho}{\varepsilon_0}, \tag{7.1}$$

$$\nabla \cdot \mathbf{B} = 0, \tag{7.2}$$

$$\nabla \times \mathbf{E} = -\frac{1}{c}\frac{\partial \mathbf{B}}{\partial t}, \tag{7.3}$$

$$\nabla \times \mathbf{B} = \frac{1}{\mu_0}\mathbf{j} + \frac{1}{c}\frac{\partial \mathbf{E}}{\partial t}, \tag{7.4}$$

where ρ and $\mathbf{j}$ are the charge density and current density of the electrically charged matter which generates the fields, ε_0 and μ_0 are dimensional constants, and $c = (\varepsilon_0 \mu_0)^{-\frac{1}{2}}$ is the speed of light. We will suppose that units of electric charge, length and time have been chosen so that $\varepsilon_0 = \mu_0 = 1$ (and therefore $c = 1$); we also continue to take $\hbar = 1$.

Maxwell's equations imply the existence of a scalar field $\phi(\mathbf{r}, t)$ and a vector field $\mathbf{A}(\mathbf{r}, t)$ which determine $\mathbf{E}$ and $\mathbf{B}$ according to

$$\mathbf{E} = -\nabla\phi - \frac{\partial \mathbf{A}}{\partial t}, \tag{7.5}$$

$$\mathbf{B} = \nabla \times \mathbf{A}, \tag{7.6}$$

and which satisfy

$$\frac{1}{c}\frac{\partial \phi}{\partial t} + \nabla \cdot \mathbf{A} = 0, \tag{7.7}$$

$$\frac{\partial^2 \phi}{\partial t^2} - \nabla^2 \phi = \rho, \tag{7.8}$$

$$\frac{\partial^2 \mathbf{A}}{\partial t^2} - \nabla^2 \mathbf{A} = \mathbf{j}. \tag{7.9}$$

The fields ϕ and $\mathbf{A}$ are called the **electromagnetic potentials**.

We will now consider the electromagnetic field as a dynamical system. Since the fields $\mathbf{E}$ and $\mathbf{B}$ are determined by the potentials ϕ and $\mathbf{A}$, we can fix our attention on the latter. The value of ϕ at some given point $\mathbf{r}$ is a varying quantity whose variation in time is given by the second-order differential equation (7.8); in this respect it resembles a single coordinate q_i of a classical mechanical system. The same goes for each component of $\mathbf{A}(\mathbf{r})$. Thus the set of all values $\{\phi(\mathbf{r}), \mathbf{A}(\mathbf{r}): \mathbf{r} \in \mathbb{R}^3\}$ at an instant t can be regarded in the same light as the set of coordinates $(q_1, \ldots, q_n)$ which define the configuration of a mechanical system; the point of space $\mathbf{r}$ plays the role of a label like the index i on the coordinate q_i. When we pass to the corresponding quantum-mechanical system q_i becomes an operator $\hat{q}_i$ labelled by the index i; thus the quantum mechanics of the electromagnetic field will involve operators $\hat{\phi}(\mathbf{r})$, $\hat{\mathbf{A}}(\mathbf{r})$ labelled by $\mathbf{r}$. These operator-valued functions of position are **quantum fields**.

Now let us suppose that $\rho = 0$ and $\mathbf{j} = \mathbf{0}$, so that we are considering the electromagnetic field by itself. The 'equation of motion' (7.8) for $\phi(\mathbf{r})$ involves the values of ϕ at nearby points to $\mathbf{r}$, since these are involved in $\nabla^2 \phi$. We can obtain decoupled equations, each referring to just one varying quantity, by taking the Fourier transform of ϕ:

$$\phi(\mathbf{r}, t) = \frac{1}{(2\pi)^{\frac{3}{2}}} \int \beta(\mathbf{k}, t) e^{i\mathbf{k} \cdot \mathbf{r}} d^3\mathbf{k}. \tag{7.10}$$

At each time the quantities $\beta(\mathbf{k}, t)$ completely specify the field values $\phi(\mathbf{r}, t)$: considering $\beta(\mathbf{k})$ instead of $\phi(\mathbf{r})$ is like changing coordinates in the mechanical system. However, the different $\beta(\mathbf{k})$ are not independent, for the fact that ϕ is real $(\phi = \bar{\phi})$ gives the relation

$$\overline{\beta(\mathbf{k})} = \beta(-\mathbf{k}). \tag{7.11}$$

Applying (7.8), with $\rho = 0$, and taking the inverse Fourier transform gives

$$\frac{\partial^2 \beta}{\partial t^2} + \mathbf{k}^2 \beta(\mathbf{k}) = 0. \tag{7.12}$$

This shows that $\beta(\mathbf{k})$ obeys the same equation of motion as a simple harmonic oscillator with angular frequency $|\mathbf{k}|$.

In the same way we can perform a Fourier transform of the vector potential **A**:

$$\mathbf{A}(\mathbf{r}, t) = \frac{1}{(2\pi)^{\frac{3}{2}}} \int \boldsymbol{\alpha}(\mathbf{k}, t) e^{i\mathbf{k}\cdot\mathbf{r}} \, d^3\mathbf{k} \tag{7.13}$$

with

$$\overline{\boldsymbol{\alpha}(\mathbf{k}, t)} = \boldsymbol{\alpha}(-\mathbf{k}, t); \tag{7.14}$$

the equation of motion (7.9) then shows that each component of $\boldsymbol{\alpha}(\mathbf{k})$ oscillates with angular frequency $|\mathbf{k}|$.

Thus as a dynamical system the electromagnetic field behaves like a set of independent oscillators. The corresponding quantum-mechanical state space will be the tensor product of a set of harmonic-oscillator state spaces, one for each of the independent oscillators among $\alpha_i(\mathbf{k})$ and $\beta(\mathbf{k})$. We know from ●4.12 that this state space is isomorphic to that of a variable number of bosons; these bosons can be identified as photons. Thus the existence of photons (and the fact that they are bosons) is a consequence of the quantum mechanics of the electromagnetic field.

The potentials ϕ and **A** are not uniquely determined by the physically significant quantities **E** and **B**. In the case of a radiation field (no charge or current density) they can be chosen so that

$$\phi = \nabla \cdot \mathbf{A} = 0, \tag{7.15}$$

i.e.

$$\beta(\mathbf{k}) = \mathbf{k} \cdot \boldsymbol{\alpha}(\mathbf{k}) = 0. \tag{7.16}$$

This choice of potentials gives a simple expression for the total energy in the field, which electromagnetic theory shows to be

$$H = \frac{1}{2} \int (\mathbf{E}^2 + \mathbf{B}^2) \, dV. \tag{7.17}$$

From (7.5)–(7.6) we find that the Fourier analyses of **E** and **B** are

$$\mathbf{E} = \frac{-1}{(2\pi)^{\frac{3}{2}}} \int \dot{\boldsymbol{\alpha}}(\mathbf{k}, t) e^{i\mathbf{k}\cdot\mathbf{r}} d^3\mathbf{k}, \quad \mathbf{B} = \frac{-i}{(2\pi)^{\frac{3}{2}}} \int \mathbf{k} \times \boldsymbol{\alpha}(\mathbf{k}, t) e^{i\mathbf{k}\cdot\mathbf{r}} d^3\mathbf{k}. \tag{7.18}$$

where the dot denotes differentiation with respect to t. Hence, using Plancherel's theorem (see problem 2.13),

$$H = \frac{1}{2} \int (|\dot{\boldsymbol{\alpha}}|^2 + |\mathbf{k} \times \boldsymbol{\alpha}|^2) \, d^3\mathbf{k} \tag{7.19}$$

(writing $|\mathbf{v}|^2 = \mathbf{v} \cdot \bar{\mathbf{v}}$ if $\mathbf{v}$ is a vector with complex components). Let α_1 and α_2 be two components of $\boldsymbol{\alpha}(\mathbf{k})$ along two directions perpendicular to $\mathbf{k}$; then, in view of (7.15), (7.18) becomes

$$H = \frac{1}{2} \int (|\dot{\alpha}_1|^2 + \mathbf{k}^2 |\alpha_1|^2 + |\dot{\alpha}_2|^2 + \mathbf{k}^2 |\alpha_2|^2) \, d^3\mathbf{k}. \tag{7.20}$$

Comparing this with the energy of a harmonic oscillator,

$$H = \tfrac{1}{2} m \dot{x}^2 + \tfrac{1}{2} m \omega^2 x^2, \tag{7.21}$$

and bearing in mind (7.14), we see that the total energy of the electromagnetic field is the sum of four harmonic-oscillator terms (one each for the real and imaginary parts of α_1 and α_2), with mass $m = 1$ and angular frequency $\omega = |\mathbf{k}|$, for each pair of vectors $(\mathbf{k}, -\mathbf{k})$. Each oscillator corresponds to a photon state, with two polarisation states for each direction of propagation. The energy of each photon state is the quantum of the corresponding oscillator, namely (reinstating $\hbar \neq 1$ for the moment) $\hbar\omega = h\nu$ where $\nu = 2\pi\omega$ is the frequency. Thus Planck's equation $E = h\nu$ also emerges from the quantum mechanics of the electromagnetic field.

The above argument refers only to energy differences; the increase in energy due to adding one photon is equal to the quantum of the corresponding oscillator, but the total energy of the collection of oscillators is not the same as that of the collection of photons because of the zero-point energy of the oscillators (the odd half in the formula $(n + \tfrac{1}{2})\hbar\omega$ for the oscillator energy eigenvalues). With an infinite number of oscillators, this gives rise to an infinite zero-point energy, which, however, is constant and has no physical significance. It can be eliminated by taking the Hamiltonian for each oscillator to be not $\tfrac{1}{2}(a^\dagger a + a a^\dagger)\omega$ but $a^\dagger a \omega$. This is just another way of resolving the ambiguity which attends any attempt to find a quantum-mechanical version of a classical expression, because of the necessity to specify the order of the factors in quantum mechanics. This choice of order is called **normal ordering**: all creation operators are put to the left of annihilation operators.

In order to disentangle the two directions of propagation $\mathbf{k}$ and $-\mathbf{k}$ it is more convenient to work in the Heisenberg picture rather than the Schrödinger picture. In the Heisenberg picture the operator A representing a given dynamical quantity is time-dependent, satisfying the differential equation

$$i \, dA/dt = [A, H] \tag{7.22}$$

(see (3.170)). Thus for the lowering operator a of a harmonic oscillator, for which

$$[H, a] = -\omega a \tag{7.23}$$

where ω is the frequency of the oscillator (see (4.129)), we have

$$da/dt = -i\omega a \tag{7.24}$$

and so

$$a(t) = e^{-i\omega t} a(0). \tag{7.25}$$

For the raising operator $a^\dagger$ we have

$$a^\dagger(t) = e^{i\omega t} a^\dagger(0). \qquad (7.26)$$

The real and imaginary parts of the Fourier components $\alpha_1(\mathbf{k})$ and $\alpha_2(\mathbf{k})$ are dynamical variables analogous to the position variable x of the harmonic oscillator. In the quantum theory they can therefore be expressed in terms of raising and lowering operators. Let $\mathbf{a_R}^\dagger$ and $\mathbf{a_I}^\dagger$ be the raising operators at $t = 0$ corresponding to the real and imaginary parts of $\boldsymbol{\alpha}(\mathbf{k})$; then from (4.126) we have

$$\boldsymbol{\alpha}(\mathbf{k}) = \frac{\mathbf{a_R}^\dagger - \mathbf{a_R}}{2i\sqrt{|\mathbf{k}|}} + i\left(\frac{\mathbf{a_I}^\dagger - \mathbf{a_I}}{2i\sqrt{|\mathbf{k}|}}\right). \qquad (7.27)$$

Write

$$\mathbf{a}^\dagger(\mathbf{k}) = \frac{\mathbf{a_R}^\dagger + i\mathbf{a_I}^\dagger}{\sqrt{2i}}, \quad \mathbf{a}^\dagger(-\mathbf{k}) = \frac{\mathbf{a_R}^\dagger - i\mathbf{a_I}^\dagger}{\sqrt{2i}}; \qquad (7.28)$$

then

$$\boldsymbol{\alpha}(\mathbf{k}, 0) = \frac{\mathbf{a}^\dagger(\mathbf{k}) + \mathbf{a}(-\mathbf{k})}{\sqrt{(2|\mathbf{k}|)}}, \qquad (7.29)$$

so that (7.25)–(7.26) give (in the Heisenberg picture)

$$\boldsymbol{\alpha}(\mathbf{k}, t) = \frac{1}{\sqrt{(2|\mathbf{k}|)}} \{e^{i|\mathbf{k}|t}\mathbf{a}^\dagger(\mathbf{k}) + e^{-i|\mathbf{k}|t}\mathbf{a}(-\mathbf{k})\}. \qquad (7.30)$$

Note that

$$\boldsymbol{\alpha}(\mathbf{k}, t)^\dagger = \boldsymbol{\alpha}(-\mathbf{k}, t), \qquad (7.31)$$

which is the quantum version of (7.14).

The Fourier analysis of the vector potential $\mathbf{A}$ can now be written as

$$\mathbf{A}(\mathbf{r}, t) = \frac{1}{(2\pi)^{\frac{3}{2}}} \int \{\mathbf{a}^\dagger(\mathbf{k})e^{iEt - i\mathbf{k}\cdot\mathbf{r}} + \mathbf{a}(-\mathbf{k})e^{-iEt - i\mathbf{k}\cdot\mathbf{r}}\} \frac{d^3\mathbf{k}}{\sqrt{(2|\mathbf{k}|)}} \qquad (7.32)$$

where $E = |\mathbf{k}|$ is the energy of a photon with momentum $\mathbf{k}$. This expression for a field in terms of creation and annihilation operators for particles is characteristic of quantum field theory.

Relativistic formulation   We will now express the decomposition (7.32) in a form which is manifestly invariant under Lorentz transformations, using 4-vectors. Let x (with components x^μ) denote the space-time position 4-vector $x = (t, \mathbf{r})$; then the exponentials in (7.32) can be written as $e^{-ik \cdot x}$ where $k = (k_0, \mathbf{k})$ is a 4-vector with $k_0 = \pm|\mathbf{k}_0|$, and $k \cdot x$ denotes the Lorentz-invariant inner product of 4-vectors. Then k satisfies

$$k^2 = k_0^2 - |\mathbf{k}|^2 = 0; \qquad (7.33)$$

if $k_0 > 0$ then k can be regarded as the 4-momentum of a massless particle (the photon).

From (7.33) and the property of the δ-function (see problem 2.14) we have

$$\delta(k^2) = (2|\mathbf{k}|)^{-1}\{\delta(k_0 - |\mathbf{k}|) + \delta(k_0 + |\mathbf{k}|)\}; \tag{7.34}$$

hence (7.32) can be written

$$\mathbf{A}(x) = \frac{1}{(2\pi)^{\frac{3}{2}}} \int \mathbf{b}(k)\delta(k^2)e^{ik \cdot x}d^4k \tag{7.35}$$

where

$$\mathbf{b}(k) = \begin{cases} (2|\mathbf{k}|)^{\frac{1}{2}}\mathbf{a}^\dagger(\mathbf{k}) \text{ if } k_0 = |\mathbf{k}| \\ (2|\mathbf{k}|)^{\frac{1}{2}}\mathbf{a}(-\mathbf{k}) \text{ if } k_0 = -|\mathbf{k}| \\ 0 \text{ if } k^2 \neq 0. \end{cases} \tag{7.36}$$

Thus

$$\mathbf{c}(k) = \mathbf{b}(k)\delta(k^2) \tag{7.37}$$

is the *four-dimensional* Fourier transform of $\mathbf{A}(x)$.

The commutation relations of the components of $\mathbf{a}(\mathbf{k})$ and $\mathbf{a}^\dagger(\mathbf{k})$ are those of annihilation and creation operators:

$$[a_i(\mathbf{k}), a_j(\mathbf{k}')^\dagger] = \delta_{ij}\delta(\mathbf{k} - \mathbf{k}'). \tag{7.38}$$

In terms of the operators $\mathbf{c}(k)$, depending on the 4-vector k, these can be written as

$$[c_i(k), c_j(k')] = \delta_{ij}\delta(k^2)\delta(k + k')\varepsilon(k_0) \tag{7.39}$$

where $\varepsilon(k_0) = \pm 1$ is the sign of k_0, which is invariant under Lorentz transformations.

Eqs. (7.35) and (7.39) show that the division of the 4-vectors x and k into space and time components which is apparent in (7.32) is not essential. There remains the fact that $\mathbf{A}(x)$ (and therefore $\mathbf{a}(\mathbf{k})$ and $\mathbf{b}(\mathbf{k})$) is a 3-vector, not a 4-vector; this appears to make the theory non-relativistic. The reason for this is the special choice (7.15) of potentials, which can only be valid in one frame of reference. In a general frame both ϕ and $\mathbf{A}$ will be non-zero; they make up a 4-vector $A^\mu = (\phi, \mathbf{A})$. The condition (7.7) is invariant, as can be seen by writing it as

$$\partial_\mu A^\mu = 0 \tag{7.40}$$

where $\partial_\mu = \partial/\partial x^\mu = (\partial/\partial t, \nabla)$. This means that for the theory to be valid in a general frame of reference the 3-vector $\mathbf{c}(k)$, satisfying $\mathbf{k} \cdot \mathbf{c} = 0$, must be replaced by a 4-vector $c_\mu(k)$ satisfying $k^\mu c_\mu = 0$. Then c_μ has three independent components, even though it only describes two independent physical states (the polarisation states of the photon). The redundant component can be eliminated by imposing an extra requirement like $c_0 = 0$, corresponding to (7.15), but there is no canonical way of doing this; in particular it cannot be done without mentioning a frame of reference. Imposing such an extra requirement is called making a choice of **gauge**. The freedom to make this choice is a highly significant feature of electromagnetic theory, and we will return to it in §7.3.

The condition $k^\mu c_\mu = 0$, although Lorentz-invariant, is of the same kind as the condition $c_0 = 0$ and can also be relaxed; this must be done if the commutation relations (7.39) are to be put in covariant form, namely

$$[c_\mu(k), c_\nu(k')] = -g_{\mu\nu}\delta(k^2)\delta(k+k')\varepsilon(k_0). \tag{7.41}$$

For each k $c_\mu(k)$ now has four independent components, but only two of these describe physically independent states.

To summarise:

●7.1 The electromagnetic field is described in the Heisenberg picture by a 4-vector quantum field $A_\mu(x)$ (an operator-valued function of space-time), whose Fourier components $c_\mu(k)$, defined by

$$A_\mu(x) = \int c_\mu(k)e^{ik\cdot x}\, d^4x, \tag{7.42}$$

consist of creation and annihilation operators for photons. If $k_0 > 0$, $c_\mu(k)$ is a creation operator for a photon with 4-momentum k; if $k_0 < 0$, $c_\mu(k)$ is an annihilation operator for a photon with 4-momentum $-k$. ∎

From (7.41) we can obtain commutation relations for the fields $A_\mu(x)$, namely

$$[A_\mu(x), A_\nu(x')] = g_{\mu\nu}\Delta(x - x') \tag{7.43}$$

where

$$\Delta(x) = \frac{1}{(2\pi)^3}\int e^{ik\cdot x}\delta(k^2)\varepsilon(k_0)\, d^4k. \tag{7.44}$$

It follows that the fields have the following property:

●7.2 Locality of the electromagnetic field

$$[A_\mu(x), A_\nu(x')] = 0 \quad \text{if } (x - x')^2 < 0. \tag{7.45}$$

Proof. From (7.43) we have

$$\Delta(x) = -\Delta(-x), \tag{7.46}$$

while the covariant form of the integral (7.44) shows that

$$\Delta(\Lambda x) = \Delta(x) \tag{7.47}$$

where Λ is any Lorentz transformation. Now if $x^2 < 0$ there is a Lorentz transformation Λ taking x to $-x$; hence $\Delta(x) = 0$ if $x^2 < 0$. ∎

According to ●5.4, this means that a measurement of A_μ at an event x cannot affect the result of a measurement of A_μ at x' if $(x - x')^2 < 0$, i.e. if a signal would have to travel faster than light to get from x to x'. Thus the theory obeys the requirements of causality as imposed by special relativity.

This theory has been developed in the Heisenberg picture. The Schrödinger picture is not so well adapted to relativistic description, since its time-dependent states require a separation between space and time and therefore a particular frame of reference. The fact that the Heisenberg picture, applied to the quantum mechanics of fields, gives a satisfactory relativistic theory suggests that field theory is the natural framework for a relativistic quantum theory. We will now see how field theory emerges even if one starts with the quantum mechanics of particles.

(b) Second quantisation Consider the system of an indefinite number of particles, each of which is a simple particle moving in space. If the particles are bosons, the state space is

$$\mathcal{U} = \mathcal{V} \oplus \mathcal{W} \oplus \vee^2 \mathcal{W} \oplus \cdots \tag{7.48}$$

where $\mathcal{V}$ is the one-dimensional vacuum space and $\mathcal{W}$ is the space of wave functions (see (4.153)). This space has annihilation and creation operators a_ψ and $a_\psi{}^\dagger$ for each wave function $\psi \in \mathcal{W}$. Let

$$|\psi_1 \cdots \psi_n\rangle = S(|\psi_1\rangle \cdots |\psi_n\rangle) \tag{7.49}$$

be a typical n-particle state, S being the symmetrisation operator of (2.142); then the action of the annihilation operator a_ψ is given by

$$a_\psi |\psi_1 \cdots \psi_n\rangle = n^{-\frac{1}{2}} \{\langle \psi | \psi_1 \rangle |\psi_2 \cdots \psi_n\rangle + \langle \psi | \psi_2 \rangle |\psi_1 \psi_3 \cdots \psi_n\rangle + \cdots\} \tag{7.50}$$

by (4.157). This makes sense not only for wave functions $\psi \in \mathcal{W}$, but for any bra $\langle \psi |$; in particular we could take $\langle \psi | = \langle \delta_r |$. This gives an annihilation operator which we will denote by $\phi(\mathbf{r})$, for each point of space:

$$\phi(\mathbf{r})|\psi_1 \cdots \psi_n\rangle = n^{-\frac{1}{2}} \{\psi_1(\mathbf{r})|\psi_2 \cdots \psi_n\rangle + \psi_2(\mathbf{r})|\psi_1 \psi_3 \cdots \psi_n\rangle + \cdots\}. \tag{7.51}$$

Then ϕ is an operator-valued function of position, i.e. a quantum field.

In quantum mechanics the numbers of classical mechanics, like the coordinates of a particle, become operators. Now we are seeing the numbers of one-particle quantum mechanics, namely the values of the wave function, themselves become operators $\phi(\mathbf{r})$. For this reason the construction of the many-particle theory is called **second quantisation**.

In §2.5 we found that one consequence of a state space being infinite-dimensional was that not every bra corresponded to a state vector: the δ-function bra $\langle \delta_r |$ was the first example. A similar phenomenon is that not every operator on an infinite-dimensional space has a hermitian conjugate. The operator $\phi(\mathbf{r})$ is an example of this, for if $\phi(\mathbf{r})^\dagger$ existed its action on the vacuum state $|0\rangle$ would be defined by

$$\langle \Psi | \phi(\mathbf{r})^\dagger | 0 \rangle = \overline{\langle 0 | \phi(\mathbf{r}) | \Psi \rangle} \quad \text{for every state } |\Psi\rangle. \tag{7.52}$$

This requires $\phi(\mathbf{r})^\dagger |0\rangle$ to be a single-particle state whose inner product with any other single-particle state is given by

$$\langle \psi | \phi(\mathbf{r})^\dagger | 0 \rangle = \overline{\psi(\mathbf{r})}; \tag{7.53}$$

in other words, $\phi(\mathbf{r})^\dagger |0\rangle$ is the non-existent ket corresponding to $\langle \delta_r |$. As explained in §2.5, it is convenient to pretend that this ket exists and to write equations involving the δ-'function' $\delta_r(\mathbf{r}') = \delta(\mathbf{r}' - \mathbf{r})$ as shorthand for equations involving integrals. In the same way, it is convenient to pretend that the operator $\phi(\mathbf{r})^\dagger$ exists, though equations involving it only really make sense when multiplied by a smooth function and integrated. The operator

$$\phi^\dagger[f] = \int f(\mathbf{r})\phi(\mathbf{r})^\dagger \, d^3\mathbf{r} \tag{7.54}$$

is a genuine creation operator for the state described by the wave function f; the quantum field $\phi(\mathbf{r})^\dagger$ can be thought of as the creation operator for the idealised state of a particle precisely localised at $\mathbf{r}$.

A similar construction can be performed with the eigenbras of momentum $\langle \varepsilon_\mathbf{k} |$ replacing $\langle \delta_\mathbf{r} |$. We define an annihilation operator $a(\mathbf{k})$ by

$$a(\mathbf{k})|\psi_1 \cdots \psi_n\rangle = n^{-\frac{1}{2}} \sum \langle \varepsilon_\mathbf{k} | \psi_i \rangle |\psi_1 \cdots \psi_{i-1}\psi_{i+1} \cdots \psi_n\rangle, \tag{7.55}$$

and regard its fictitious hermitian conjugate $a^\dagger(\mathbf{k})$ as a creation operator for a particle in the fictitious momentum eigenstate with wave function $(2\pi)^{-\frac{3}{2}}e^{i\mathbf{k}\cdot\mathbf{r}}$. The operators $a(\mathbf{k})$ and $\phi(\mathbf{r})$ are related as follows. From (7.55) we have

$$a(\mathbf{k})|\psi_1 \cdots \psi_n\rangle = \frac{1}{(2\pi)^{\frac{3}{2}}} \int d^3\mathbf{r} \, e^{-i\mathbf{k}\cdot\mathbf{r}} \sum_i \psi_i(\mathbf{r}) |\psi_1 \cdots \psi_{i-1}\psi_{i+1} \cdots \psi_n\rangle$$

$$= \frac{1}{(2\pi)^{\frac{3}{2}}} \int d^3\mathbf{r} \, e^{-i\mathbf{k}\cdot\mathbf{r}} \phi(\mathbf{r}) |\psi_1 \cdots \psi_n\rangle. \tag{7.56}$$

Hence

$$a(\mathbf{k}) = \frac{1}{(2\pi)^{\frac{3}{2}}} \int \phi(\mathbf{r}) e^{-i\mathbf{k}\cdot\mathbf{r}} \, d^3\mathbf{r}; \tag{7.57}$$

$a(\mathbf{k})$ is the *Fourier transform* of $\phi(\mathbf{r})$. By the Fourier inversion theorem we can write

$$\phi(\mathbf{r}) = \frac{1}{(2\pi)^{\frac{3}{2}}} \int a(\mathbf{k}) e^{i\mathbf{k}\cdot\mathbf{r}} \, d^3\mathbf{k} \tag{7.58}$$

(see also problem 7.2).

Now let us consider these operators in the Heisenberg picture. The operators $\phi(\mathbf{r})$ must be replaced by time-dependent operators $\phi(\mathbf{r}, t)$ satisfying the differential equation

$$i\frac{\partial \phi}{\partial t} = [\phi, H]. \tag{7.59}$$

Suppose the Hamiltonian is that of a non-relativistic theory describing particles moving in a potential V, so that its restriction to the n-particle subspace is

$$H_n = \frac{\mathbf{p}_1^2}{2m} + \cdots + \frac{\mathbf{p}_n^2}{2m} + V(\mathbf{r}_1) + \cdots + V(\mathbf{r}_n). \tag{7.60}$$

For a one-particle wave function we write

$$\tilde{\psi} = H\psi = -\frac{1}{2m}\nabla^2\psi + V\psi. \tag{7.61}$$

Then

$$[\phi(\mathbf{r}), H]|\psi_1 \cdots \psi_n\rangle = \sum_i \phi(\mathbf{r})|\psi_1 \cdots \tilde{\psi}_i \cdots \psi_n\rangle$$

$$-\sum_i H\psi_i(\mathbf{r})|\psi_1 \cdots \psi_{i-1}\psi_{i+1}\cdots \psi_n\rangle$$

$$=\sum_i \tilde{\psi}_i(\mathbf{r})|\psi_1 \cdots \psi_{i-1}\psi_{i+1}\cdots \psi_n\rangle$$

$$=\left[-\frac{1}{2m}\nabla^2 + V(\mathbf{r})\right]\phi(\mathbf{r})|\psi_1 \cdots \psi_n\rangle. \qquad (7.62)$$

Thus

$$i\frac{\partial\phi}{\partial t} = -\frac{1}{2m}\nabla^2\phi + V\phi; \qquad (7.63)$$

in the second-quantised theory the field operators $\phi(\mathbf{r}, t)$ satisfy the Schrödinger equation, just like the wave function in the single-particle theory.

In the case of free particles ($V=0$) we can explicitly determine the time-dependence of the operators $a(\mathbf{k})$ in the Heisenberg picture. Applying the wave equation (7.63) to the Fourier integral (7.58) gives

$$\int i\frac{\partial}{\partial t} a(\mathbf{k}, t)e^{i\mathbf{k}\cdot\mathbf{r}} d^3\mathbf{k} = \int \frac{\mathbf{k}^2}{2m} a(\mathbf{k}, t)e^{i\mathbf{k}\cdot\mathbf{r}} d^3\mathbf{k}, \qquad (7.64)$$

so that

$$a(\mathbf{k}, t) = a(\mathbf{k})e^{-iEt} \qquad (7.65)$$

where

$$E = \frac{\mathbf{k}^2}{2m}. \qquad (7.66)$$

Thus for free non-relativistic particles

$$\phi(\mathbf{r}, t) = \frac{1}{(2\pi)^{\frac{3}{2}}}\int a(\mathbf{k})e^{-iEt+i\mathbf{k}\cdot\mathbf{r}}d^3\mathbf{k} \qquad (7.67)$$

where $E = E(\mathbf{k})$ is the kinetic energy of a particle with momentum $\mathbf{k}$. In general, an expression like this for $\phi(\mathbf{r}, t)$ in terms of time-independent annihilation operators requires knowledge of the solutions of the Schrödinger equation (7.63).

Let us consider how problems in quantum dynamics can be formulated in terms of the field ϕ. In the Schrödinger picture these problems are of the form 'At $t=0$ a single particle has wave function ψ_0; what is its wave function at time t?' The answer can be expressed as a matrix element:

$$\psi(\mathbf{r}, t) = \langle \delta_{\mathbf{r}}|e^{-iHt}|\psi_0\rangle. \qquad (7.68)$$

This can be written in terms of the Schrödinger field operators $\phi^S(\mathbf{r}) = \phi(\mathbf{r}, 0)$:

$$\psi(\mathbf{r}, t) = \langle 0|\phi^S(\mathbf{r})e^{-iHt}\int \psi_0(\mathbf{r}')^\dagger d^3\mathbf{r}'|0\rangle \qquad (7.69)$$

where $|0\rangle$ is the vacuum state. This has zero energy and is therefore the same in both Schrödinger and Heisenberg pictures:

$$e^{-iHt}|0\rangle = |0\rangle. \tag{7.70}$$

Hence in terms of the Heisenberg operators

$$\phi(\mathbf{r}, t) = e^{-iHt}\phi^S(\mathbf{r})e^{iHt}, \tag{7.71}$$

the wave function at time t is

$$\psi(\mathbf{r}, t) = \int d^3\mathbf{r}'\psi_0(\mathbf{r}')\langle 0|\phi(\mathbf{r}, t)\phi(\mathbf{r}', 0)^\dagger|0\rangle. \tag{7.72}$$

This shows that the vacuum expectation value $\langle 0|\phi(\mathbf{r}, t)\phi^\dagger(\mathbf{r}', 0)|0\rangle$ is the **Green's function** for the Schrödinger equation (7.61): it is the kernel of the integral operator which converts the initial conditions $\psi_0(\mathbf{r})$ into the solution of the differential equation at time t.

The Klein–Gordon equation   The non-relativistic Schrödinger equation for a free particle is formed from the non-relativistic relation between energy and momentum, eq. (7.66), by making the substitutions

$$E \to i\,\partial/\partial t, \quad \mathbf{k} \to -i\mathbf{\nabla}. \tag{7.73}$$

These can be written as

$$k_\mu \to i\,\partial_\mu \tag{7.74}$$

where $k^\mu = (E, \mathbf{k})$ is the (contravariant) energy-momentum 4-vector and $\partial_\mu = \partial/\partial x^\mu$ is the (covariant) space-time derivative. Thus the obvious way to make the theory relativistic is to keep these substitutions but to replace (7.66) by the relativistic relation

$$E = +\sqrt{(\mathbf{k}^2 + m^2)} \tag{7.75}$$

where m is the rest-mass of the particle. Because of the square root, however, this does not lead to a differential equation; it is therefore necessary to square the relation. The substitutions (7.73) then give the differential equation

$$\frac{\partial^2\phi}{\partial t^2} - \nabla^2\phi = -m^2\phi, \tag{7.76}$$

which is called the **Klein–Gordon equation**. The differential operator on the left-hand side is invariant under Lorentz transformations and is usually denoted by $\square^2$:

$$\square^2 = \frac{\partial^2}{\partial t^2} - \nabla^2 = \partial^\mu\,\partial_\mu. \tag{7.77}$$

The Klein–Gordon equation is not considered suitable to be a wave equation for a one-particle theory, as the Schrödinger equation is, because it has negative-energy solutions (a result of the squaring of (7.75)). The general solution of the Klein–Gordon equation is

$$\phi(\mathbf{r}, t) = \frac{1}{(2\pi)^{\frac{3}{2}}}\int \{a(\mathbf{k})e^{-iEt} + a'(\mathbf{k})e^{iEt}\}e^{i\mathbf{k}\cdot\mathbf{r}}d^3\mathbf{k} \tag{7.78}$$

where $a(\mathbf{k})$ and $a'(\mathbf{k})$ are arbitrary and E is given by (7.75). Regarded as wave functions in the Schrödinger picture, the terms $a'(\mathbf{k})e^{i\mathbf{k}\cdot\mathbf{r}}e^{iEt}$ have negative energy. But if ϕ is not a wave function but a quantum field, the form (7.78) can be acceptable. It differs from the form (7.67) of a non-relativistic quantum field, but the possibility of a reasonable physical interpretation for the extra ('negative energy') terms can be seen by comparing with the expression (7.32) for the electromagnetic field, in which such terms appear associated with the photon creation operator $a^{\dagger}(-\mathbf{k})$.

Thus in (7.78) $a'(\mathbf{k})$ should be interpreted as a creation operator for a state with momentum $-\mathbf{k}$:

$$a'(\mathbf{k}) = b^{\dagger}(-\mathbf{k}). \tag{7.79}$$

In the case of the electromagnetic field this is actually equal to $a^{\dagger}(-\mathbf{k})$, but that is because the field $\mathbf{A}$ is hermitian (being related to the observables $\mathbf{E}$ and $\mathbf{B}$). In general there is no reason why a quantum field should be hermitian; this means that the particle created by $b^{\dagger}$ may not be the same as the particle destroyed by a. These two particles are **antiparticles** of each other.

By changing variables from $\mathbf{k}$ to $-\mathbf{k}$ in the second term, we can write (7.78) as

$$\phi(\mathbf{r}, t) = \frac{1}{(2\pi)^{\frac{3}{2}}} \int a(\mathbf{k})e^{-i(Et-\mathbf{k}\cdot\mathbf{r})}\, d^3\mathbf{k} + \frac{1}{(2\pi)^{\frac{3}{2}}} \int b^{\dagger}(\mathbf{k})e^{i(Et-\mathbf{k}\cdot\mathbf{r})}\, d^3\mathbf{k}$$

$$= \phi_{+}(\mathbf{r}, t) + \phi_{-}(\mathbf{r}, t)^{\dagger} \tag{7.80}$$

where ϕ_{+} is an annihilation operator for the particle associated with the field (cf. (7.67)), and $\phi_{-}{}^{\dagger}$ is a creation operator for its antiparticle. This is the justification for the introduction of 'reduced quantum fields' in §4.6.

7.2. The Dirac equation The Dirac equation is a partial differential equation which is relativistic in the sense that it incorporates the relation (7.75) between energy and momentum, yet it is first-order in the time derivative. The equation is

$$i\gamma^{\mu}\, \partial_{\mu}\psi = m\psi \tag{7.81}$$

in which the meaning of the symbols is as follows:

(i) $\partial_{\mu} = \partial/\partial x^{\mu}$ where $x^{\mu} = (t, \mathbf{r})$ is the space-time position 4-vector.
(ii) γ^{μ} is a set of four 4×4 matrices which satisfy

$$\gamma^{\mu}\gamma^{\nu} + \gamma^{\nu}\gamma^{\mu} = 2g^{\mu\nu} \tag{7.82}$$

where $g^{\mu\nu} = \operatorname{diag}(1, -1, -1, -1)$ is the metric tensor of special relativity, and the 4×4 unit matrix is understood on the right-hand side. We will always take these matrices to be

$$\gamma^{0} = \begin{bmatrix} 1 & 0 \\ 0 & -1 \end{bmatrix}, \quad \gamma^{i} = \begin{bmatrix} 0 & \sigma_i \\ -\sigma_i & 0 \end{bmatrix}, \tag{7.83}$$

in which the entries denote 2×2 blocks and σ_i are the 2×2 Pauli matrices. Note that the γ^μ are fixed matrices, and do not change with the frame of reference.

(iii) $\psi(x)$ is a four-component wave function. It is not a 4-vector; it is to be regarded as a 4×1 column vector suitable for multiplication by a γ-matrix, and the number of its components is a different four from the dimension of space-time†. It is called a **Dirac spinor**. The symbol ψ is usual for Dirac spinors, although they are a different sort of object from the one-component wave function appearing in the Schrödinger equation. For the rest of this book ψ will always denote a Dirac spinor.

(iv) m is the rest-mass of the particle whose state is described by ψ.

Now we will derive some of the properties of the Dirac equation and the matrices γ^μ.

●**7.3** The Dirac equation implies the Klein–Gordon equation:

$$i\gamma^\mu\,\partial_\mu\psi = m\psi \;\Rightarrow\; \Box^2\psi = -m^2\psi. \tag{7.84}$$

Proof. If ψ satisfies the Dirac equation then

$$(\gamma^\mu\,\partial_\mu)(\gamma^\nu\,\partial_\nu)\psi = -im(\gamma^\mu\,\partial_\mu)\psi = -m^2\psi. \tag{7.85}$$

But (7.82) gives

$$(\gamma^\mu\,\partial_\mu)(\gamma^\nu\,\partial_\nu) = \tfrac{1}{2}(\gamma^\mu\gamma^\nu + \gamma^\nu\gamma^\mu)\partial_\mu\,\partial_\nu = g^{\mu\nu}\,\partial_\mu\,\partial_\nu = \Box^2.$$

Hence (7.85) shows that every component of ψ satisfies the Klein–Gordon equation. ∎

The components of the Dirac spinor ψ must depend on the frame of reference, for otherwise the Dirac equation would not hold in all frames of reference. If the frame of reference is changed by means of a Lorentz transformation Λ, so that the space-time coordinates change by

$$x^\mu \to x'^\mu = \Lambda^\mu{}_\nu x^\nu \tag{7.86}$$

(see Appendix I), then the corresponding change in ψ is given by

●**7.4** For each Lorentz transformation Λ there is a 4×4 matrix $S(\Lambda)$ satisfying

$$S(\Lambda)^{-1}\gamma^\mu S(\Lambda) = \Lambda^\mu{}_\nu\gamma^\nu. \tag{7.87}$$

Let ψ' be the spinor wave function defined by

$$\psi'(x') = S(\Lambda)\psi(x) \tag{7.88}$$

where x' is given by (7.86); then

$$i\gamma^\mu\,\partial_\mu'\psi' = m\psi' \tag{7.89}$$

where $\partial_\mu' = \partial/\partial x'^\mu$.

---

† If space-time had dimension $2n$, ψ would have 2^n components.

Proof. The condition on Λ to be a Lorentz transformation is

$$\Lambda^\mu{}_\nu \Lambda_{\mu\rho} = g_{\nu\rho}. \tag{7.90}$$

If $\Lambda(\lambda)$ is a sequence of Lorentz transformations labelled by a real parameter λ, with $\Lambda^\mu{}_\nu(0) = \delta^\mu{}_\nu$ and with generator

$$\omega^\mu{}_\nu = \left.\frac{d\Lambda^\mu{}_\nu}{d\lambda}\right|_{\lambda=0}, \tag{7.91}$$

then (7.90) gives

$$\omega_{\mu\nu} + \omega_{\nu\mu} = 0. \tag{7.92}$$

Conversely, if $\omega_{\mu\nu}$ satisfies this condition and Ω is the 4×4 (vector) matrix with entries $\omega^\mu{}_\nu$, then

$$\Lambda^\mu{}_\nu(\lambda) = (e^{\lambda\Omega})^\mu{}_\nu \tag{7.93}$$

is a sequence of Lorentz transformations, for

$$\frac{d}{d\lambda}(\Lambda^\mu{}_\nu \Lambda_{\mu\rho}) = \omega^\mu{}_\sigma \Lambda^\sigma{}_\nu \Lambda_{\mu\rho} + \Lambda^\mu{}_\nu \omega_{\mu\sigma} \Lambda^\sigma{}_\rho = (\omega_{\mu\sigma} + \omega_{\sigma\mu})\Lambda^\sigma{}_\nu \Lambda^\mu{}_\rho = 0.$$

Then the solution of the differential equation

$$\frac{dx^\mu}{d\lambda} = \omega^\mu{}_\nu x^\nu \tag{7.94}$$

is

$$x^\mu(\lambda) = \Lambda^\mu{}_\nu(\lambda)x^\nu(0). \tag{7.95}$$

In particular, if $\omega_{0i} = 0$ and $\omega_{ij} = \varepsilon_{ijk}n_k$ for some 3-vector $\mathbf{n}$, (7.94) becomes

$$\frac{dt}{d\lambda} = 0, \quad \frac{d\mathbf{r}}{d\lambda} = \mathbf{n} \times \mathbf{r}. \tag{7.96}$$

Comparing with (3.129) shows that in this case $e^{\lambda\Omega}$ is a rotation about the axis $\mathbf{n}$ through angle λ. On the other hand, if $\omega_{ij} = 0$ and $\omega_{0i} = n_i$, (7.94) becomes

$$\frac{dt}{d\lambda} = \mathbf{n} \cdot \mathbf{r}, \quad \frac{d}{d\lambda}(\mathbf{n} \cdot \mathbf{r}) = t; \quad \frac{d}{d\lambda}(\mathbf{m} \cdot \mathbf{r}) = 0 \quad \text{if } \mathbf{m} \cdot \mathbf{n} = 0. \tag{7.97}$$

The solution of this is

$$t = t_0 \cosh \lambda + \mathbf{n} \cdot \mathbf{r}_0 \sinh \lambda,$$
$$\mathbf{n} \cdot \mathbf{r} = t_0 \sinh \lambda + \mathbf{n} \cdot \mathbf{r}_0 \cosh \lambda,$$
$$\mathbf{m} \cdot \mathbf{r} = \mathbf{m} \cdot \mathbf{r}_0 \quad \text{if } \mathbf{m} \cdot \mathbf{n} = 0, \tag{7.98}$$

which is a boost in the direction $\mathbf{n}$ with velocity $\tanh \lambda$. Thus both rotations and boosts are Lorentz transformations of the form $e^{\lambda\Omega}$.

Now define the spinor matrix corresponding to $\Lambda = e^{\lambda\Omega}$ to be

$$S(\Lambda) = \exp\left[\tfrac{1}{2}\lambda\omega_{\nu\rho}\sigma^{\nu\rho}\right] \tag{7.99}$$

where

$$\sigma^{\nu\rho} = \tfrac{1}{4}[\gamma^\nu, \gamma^\rho].$$

Then

$$d/d\lambda[S(\Lambda)^{-1}\gamma^\mu S(\Lambda)] = -\tfrac{1}{2}S(\Lambda)^{-1}\omega_{\nu\rho}[\sigma^{\nu\rho}, \gamma^\mu]S(\Lambda). \tag{7.100}$$

Now

$$[\sigma^{\nu\rho}, \gamma^\mu] = \tfrac{1}{4}[[\gamma^\nu, \gamma^\rho], \gamma^\mu]$$
$$= \tfrac{1}{4}\{\gamma^\nu, \{\gamma^\rho, \gamma^\mu\}\} - \tfrac{1}{4}\{\gamma^\rho, \{\gamma^\nu, \gamma^\mu\}\}$$
$$= g^{\mu\rho}\gamma^\nu - g^{\mu\nu}\gamma^\rho$$

by (7.82). Hence (7.100) becomes

$$d/d\lambda[S(\Lambda)^{-1}\gamma^\mu S(\Lambda)] = \omega^\mu{}_\nu S(\Lambda)^{-1}\gamma^\nu S(\Lambda). \tag{7.102}$$

Comparing with (7.93)–(7.94), we see that

$$S(\Lambda)^{-1}\gamma^\mu S(\Lambda) = \Lambda^\mu{}_\nu\gamma^\nu,$$

as stated.

If $x'^\mu = \Lambda^\mu{}_\nu x^\nu$ then $\partial_\mu' = \Lambda_\mu{}^\nu\partial_\nu$ and so

$$\gamma^\mu\,\partial_\mu'\psi' = \gamma^\mu\Lambda_\mu{}^\nu\,\partial_\nu[S(\Lambda)\psi].$$

But

$$\Lambda_\mu{}^\nu\gamma^\mu S(\Lambda) = \Lambda_\mu{}^\nu\Lambda^\mu{}_\rho S(\Lambda)\gamma^\rho = S(\Lambda)\gamma^\nu$$

by (7.90); hence

$$i\gamma^\mu\,\partial_\mu'\psi' = iS(\Lambda)\gamma^\nu\,\partial_\nu\psi = S(\Lambda)m\psi = m\psi'. \quad\blacksquare$$

We can now use Postulate VII to determine the angular momentum of the system whose state vectors are the spinor wave functions $\psi(x)$. In a single-particle theory the state space will consist of all suitably well-behaved spinor functions of spatial position $\mathbf{r}$ (as with the Schrödinger equation, the fourth space-time coordinate t describes the change of the state vector in this state space). On this state space the unitary operator representing a rotation R is given by

$$[U(R)\psi](\mathbf{r}) = S(R)\psi(R^{-1}\mathbf{r}) \tag{7.103}$$

where $S(R)$ is the 4×4 (spinor) matrix assigned to R as a Lorentz transformation by ●7.4. Suppose R is a rotation about an axis $\mathbf{n}$ through angle λ; then the angular momentum component $\mathbf{n} \cdot \mathbf{J}$ is given by

$$[\mathbf{n} \cdot \mathbf{J}\psi](\mathbf{r}) = [(i\,d/d\lambda)S(R)\psi](\mathbf{r}) + (i\,d/d\lambda)\psi(R^{-1}\mathbf{r}). \tag{7.104}$$

The second term is calculated as in §3.3, and gives the usual orbital angular momentum term $-i\mathbf{n} \cdot (\mathbf{r} \times \nabla\psi)$. For the first term we use (7.99) with $\omega_{0i} = 0$, $\omega_{ij} = \varepsilon_{ijk}n_k$ to obtain $i\omega_{ij}\sigma_{ij}\psi(\mathbf{r})$. Hence

$$\mathbf{J} = \mathbf{s} + \mathbf{r} \times \mathbf{p} \tag{7.105}$$

where

$$s_k = \tfrac{1}{2}i\varepsilon_{ijk}\sigma_{ij} = \tfrac{1}{8}i\varepsilon_{ijk}[\gamma_i, \gamma_j]. \tag{7.106}$$

From the form (7.83) of the γ-matrices and the multiplication rules (4.39) for the Pauli σ-matrices we obtain

$$s_k = \frac{1}{2}\begin{bmatrix} \sigma_k & 0 \\ 0 & \sigma_k \end{bmatrix}. \tag{7.107}$$

This shows that the state space is of the form $\mathscr{S}_1 \oplus \mathscr{S}_2$ where both $\mathscr{S}_1$ and $\mathscr{S}_2$ are isomorphic to the state space of a single spin-$\tfrac{1}{2}$ particle: $\mathscr{S}_1$ consists of spinors in which the bottom two components vanish, while $\mathscr{S}_1$ consists of spinors whose top two components vanish.

In order to see the significance of these two subspaces, consider the eigenstates of momentum with eigenvalue **0**. These consist of spinor wave functions $\psi(x)$ satisfying $\nabla\psi = \mathbf{0}$, so ψ is a function of t only; the Dirac equation becomes

$$i\gamma^0 \, d\psi/dt = m\psi. \tag{7.108}$$

With γ^0 given by (7.83), $\mathscr{S}_1$ is an eigenspace of γ^0 with eigenvalue 1; a spinor ψ belonging to $\mathscr{S}_1$ and satisfying (7.108) therefore has energy m (as is to be expected for a particle of rest-mass m when it is at rest). If ψ belongs to $\mathscr{S}_2$, however, it is an eigenvector of γ^0 with eigenvalue -1, and so (7.108) implies that it has energy $-m$. Thus the Dirac equation, like the Klein–Gordon equation, has negative-energy solutions.

Parity From the spinor matrices $S(\Lambda)$ of ●7.4 we can construct such a matrix for any Lorentz transformation which can be written as a product of rotations and boosts. These matrices constitute a projective representation of a subgroup of the Lorentz group, whose elements are known as **proper** Lorentz transformations. Space inversion P, which takes $(t, \mathbf{r})$ to $(t, -\mathbf{r})$, is also a Lorentz transformation according to (7.90), but it does not belong to this subgroup. It can be represented by a spinor matrix by taking $S(P) = \gamma^0$, for then the basic relation (7.82) gives

$$S(P)\gamma^0 S(P)^{-1} = \gamma^0, \quad S(P)\gamma^i S(P)^{-1} = -\gamma^i, \tag{7.109}$$

in analogy with (7.87).

Spinor bilinears A wave function satisfying the non-relativistic Schrödinger equation gives rise to a probability density and a probability current (see (3.43)). Similar quantities can be constructed from Dirac spinors as follows:

●7.5 There is a hermitian spinor matrix ζ which satisfies $\zeta^2 = 1$ and

$$\zeta\gamma^\mu\zeta = (\gamma^\mu)^\dagger. \tag{7.110}$$

Let $\bar{\psi} = \psi^\dagger\zeta$; then

$\bar{\psi}\psi$ is a Lorentz scalar;

$\bar{\psi}\gamma^\mu\psi$ and $\bar{\psi}\,\partial_\mu\psi$ are 4-vectors.

If ψ satisfies the Dirac equation, then

$$\partial_\mu(\bar{\psi}\gamma^\mu\psi) = 0. \tag{7.111}$$

[The meaning of these statements is as follows: Let ψ' be the spinor obtained from ψ by a Lorentz transformation Λ according to (7.88). The statements that $\bar{\psi}\psi$ is a scalar and $\bar{\psi}\gamma^\mu\psi$ is a 4-vector mean that

$$\bar{\psi}'\psi' = \bar{\psi}\psi \quad \text{and} \quad \bar{\psi}'\gamma^\mu\psi' = \Lambda^\mu{}_\nu\bar{\psi}\gamma^\nu\psi \tag{7.112}$$

for all Lorentz transformations Λ.]

Proof. Although the existence of ζ can be shown to follow from the γ-matrix relations (7.82), we will be content to exhibit ζ for our particular choice (7.83). Since in this choice γ^0 is hermitian and the γ^i are antihermitian, (7.110) is satisfied by taking $\zeta = \gamma^0$.

Now we have (using $\zeta^2 = 1$)

$$\zeta \sigma^{\mu\nu} \zeta = \tfrac{1}{4} \zeta [\gamma^\mu, \gamma^\nu] \zeta = \tfrac{1}{4} [\zeta \gamma^\mu \zeta, \zeta \gamma^\nu \zeta] = \tfrac{1}{4} [\gamma^{\mu\dagger}, \gamma^{\nu\dagger}] \tag{7.113}$$
$$= -\tfrac{1}{4} [\gamma^\mu, \gamma^\nu]^\dagger = -\sigma^{\mu\nu\dagger}$$

so that if $\Lambda = e^{i\Omega}$ is a boost or a rotation,

$$\zeta S(\Lambda) \zeta = \zeta \exp(\omega_{\mu\nu} \sigma^{\mu\nu}) \zeta = \exp(\omega_{\mu\nu} \zeta \sigma^{\mu\nu} \zeta)$$
$$= \exp[-(\omega_{\mu\nu} \sigma^{\mu\nu})^\dagger] = [S(\Lambda)^{-1}]^\dagger. \tag{7.114}$$

If Λ is space inversion, both sides of (7.114) are equal to γ^0. The equation can now be extended to any product of boosts, rotations, and space inversions. Thus if Λ is any such product we have

$$\psi' = S(\Lambda)\psi \tag{7.115}$$

and

$$\bar{\psi}' = \psi^\dagger S(\Lambda)^\dagger \zeta = \psi^\dagger \zeta S(\Lambda)^{-1} = \bar{\psi} S(\Lambda)^{-1}. \tag{7.116}$$

Hence

$$\bar{\psi}' \psi' = \bar{\psi} \psi \tag{7.117}$$

and

$$\bar{\psi}' \gamma^\mu \psi' = \bar{\psi} S(\Lambda)^{-1} \gamma^\mu S(\Lambda) \psi = \Lambda^\mu{}_\nu \bar{\psi} \gamma^\nu \psi \tag{7.118}$$

by (7.87). Also, since $\partial_\mu' = \Lambda_\mu{}^\nu \partial_\nu$,

$$\bar{\psi}' \partial_\mu' \psi' = \bar{\psi} S(\Lambda)^{-1} \Lambda_\mu{}^\nu \partial_\nu S(\Lambda) \psi = \Lambda_\mu{}^\nu \partial_\nu \psi. \tag{7.119}$$

(7.117)–(7.119) show that $\bar{\psi}\psi$, $\bar{\psi}\gamma^\mu\psi$ and $\bar{\psi}\partial_\mu\psi$ are respectively a scalar, a contravariant 4-vector and a covariant 4-vector.

Now suppose that ψ satisfies the Dirac equation:

$$i\gamma^\mu \partial_\mu \psi = m\psi. \tag{7.120}$$

Then

$$-i \partial_\mu \psi^\dagger \gamma^{\mu\dagger} \zeta = m\bar{\psi}. \tag{7.121}$$

Multiplying (7.120) on the left by $\bar{\psi}$ and (7.121) on the right by ψ and subtracting gives

$$\partial_\mu(\bar{\psi} \gamma^\mu \psi) = 0. \quad \blacksquare$$

Let $j^\mu = \bar{\psi}\gamma^\mu\psi$ and write $j^0 = \rho$; then (7.121) becomes

$$\partial\rho/\partial t + \nabla \cdot \mathbf{j} = 0, \tag{7.122}$$

which is the same as the continuity equation (3.42). Because of this equation j^μ is called† a **conserved current**; its time component ρ can be interpreted as the density of some substance whose rate of flow is described by the current 3-vector $\mathbf{j}$. In fact

$$\rho = \bar{\psi}\gamma^0\psi = \psi^\dagger(\gamma^0)^2\psi = \psi^\dagger\psi \tag{7.123}$$

† Illogically; it should be called a conserv*ing* current.

(since $\zeta = \gamma^0$), so ρ is positive definite. Originally ρ was interpreted as a probability density; but in the second-quantised theory which is made necessary by the negative-energy solutions of the Dirac equation, ρ has a different interpretation (see p. 311).

Another significant set of bilinear quantities can be constructed by means of the matrix

$$\gamma_5 = i\gamma^0\gamma^1\gamma^2\gamma^3. \tag{7.124}$$

Since the γ^μ all anticommute, this can also be written as

$$\gamma_5 = (4!)^{-1} i\varepsilon_{\mu\nu\rho\sigma}\gamma^\mu\gamma^\nu\gamma^\rho\gamma^\sigma \tag{7.125}$$

where $\varepsilon_{\mu\nu\rho\sigma}$ is the totally antisymmetric tensor with $\varepsilon_{0123} = 1$. From this the Lorentz transformation properties of γ_5 follow:

$$S(\Lambda)\gamma_5 S(\Lambda)^{-1} = (4!)^{-1}\varepsilon_{\mu\nu\rho\sigma}\Lambda^\mu{}_\alpha\Lambda^\nu{}_\beta\Lambda^\sigma{}_\kappa\Lambda^\sigma{}_\lambda\gamma^\alpha\gamma^\beta\gamma^\kappa\gamma^\lambda = (\det\Lambda)\gamma_5. \tag{7.126}$$

Thus γ_5 is invariant under proper Lorentz transformations, but changes sign under reflections. Such an object is called a **pseudoscalar**. Similarly, $\gamma^\mu\gamma_5$ is a **pseudovector**:

$$S(\Lambda)\gamma^\mu\gamma_5 S(\Lambda)^{-1} = (\det\Lambda)\gamma^\mu\gamma_5. \tag{7.127}$$

The important properties of γ_5 are the following:

●7.6   (i) $\gamma_5{}^2 = 1$;   (7.128)

    (ii) $\gamma_5\gamma^\mu = -\gamma^\mu\gamma_5$;   (7.129)

    (iii) $\gamma^\mu\gamma^\nu\gamma_5 = g^{\mu\nu}\gamma_5 - \frac{1}{2}i\varepsilon^{\mu\nu}{}_{\kappa\lambda}\gamma^\kappa\gamma^\lambda$;   (7.130)

    (iv) $\zeta\gamma_5\zeta = -\gamma_5{}^\dagger$;   (7.131)

    (v) $\bar\psi\gamma_5\psi$ is a pseudoscalar,

          $\bar\psi\gamma_5\gamma^\mu\psi$ and $\bar\psi\gamma_5\partial_\mu\psi$ are pseudovectors;

    (vi) if $\psi(x) = u(p)e^{-ip\cdot x}$ is a plane-wave solution of the Dirac equation, its helicity is given by

$$\frac{\mathbf{p}\cdot\mathbf{J}}{|\mathbf{p}|}\psi = \left(\frac{m\gamma^0 - p^0}{2|\mathbf{p}|}\right)\gamma_5\psi. \tag{7.132}$$

Proof. (i) and (ii) are easily calculated from (7.124), using the fact that the γ^μ anticommute. On the other hand, using the form (7.125) for γ_5 and repeatedly applying the anticommutation relation (7.82), we find

$$\gamma^\nu\gamma_5 = \gamma_5\gamma^\nu + \frac{1}{3}i\varepsilon^\nu{}_{\beta\kappa\lambda}\gamma^\beta\gamma^\kappa\gamma^\lambda \tag{7.133}$$

so that

$$\gamma^\nu\gamma_5 = \frac{1}{6}i\varepsilon^\nu{}_{\beta\kappa\lambda}\gamma^\beta\gamma^\kappa\gamma^\lambda. \tag{7.134}$$

Thus $\gamma^\nu\gamma_5$ is a product of the three matrices γ^μ with $\mu \neq \nu$. Hence γ^μ commutes with $\gamma^\nu\gamma_5$ if $\mu \neq \nu$. But a calculation like the above, starting from (7.134) and using (7.82), gives

$$\gamma^\mu\gamma^\nu\gamma_5 = -\gamma^\nu\gamma_5\gamma^\mu - i\varepsilon^{\mu\nu}{}_{\kappa\lambda}\gamma^\kappa\gamma^\lambda \tag{7.135}$$

so that

$$\gamma^\mu\gamma^\nu\gamma_5 = -\tfrac{1}{2}i\varepsilon^{\mu\nu}{}_{\kappa\lambda}\gamma^\kappa\gamma^\lambda \quad \text{if } \mu \neq \nu. \tag{7.136}$$

Also

$$\gamma^\mu\gamma^\nu\gamma_5 = g^{\mu\nu}\gamma_5 \qquad \text{if } \mu = \nu. \tag{7.137}$$

(7.136) and (7.137) can be put together as (7.131).

(iv) follows from (7.110), which gives

$$\zeta\gamma_5\zeta = i\gamma^{0\dagger}\gamma^{1\dagger}\gamma^{2\dagger}\gamma^{3\dagger} = (-i\gamma^3\gamma^2\gamma^1\gamma^0)^\dagger = -\gamma_5{}^\dagger$$

since the γ^μ anticommute and reversal of order is an even permutation of four objects.

(v) is established in the same way as ●7.5; because of (7.126), we find that if $\psi' = S(\Lambda)\psi$, then

$$\bar\psi'\gamma_5\psi' = (\det \Lambda)\bar\psi\gamma_5\psi, \tag{7.138}$$

$$\bar\psi'\gamma_5\gamma^\mu\psi' = (\det \Lambda)\Lambda^\mu{}_\nu\bar\psi\gamma_5\gamma^\nu\psi, \tag{7.139}$$

and

$$\bar\psi'\gamma_5\,\partial_\mu\psi' = (\det \Lambda)\Lambda_\mu{}^\nu\bar\psi\gamma_5\,\partial_\mu\psi. \tag{7.140}$$

Finally, if $\psi(x) = u(p)e^{-ip\cdot x}$ is a solution of the Dirac equation, so that

$$i\,\partial_\mu\psi = p_\mu\psi \quad \text{and} \quad p_\mu\gamma^\mu\psi = m\psi, \tag{7.141}$$

then

$$\mathbf{p}\cdot\mathbf{J}\psi = \mathbf{p}\cdot(\mathbf{s}+\mathbf{r}\times\mathbf{p})\psi = \tfrac{1}{4}ip_i\varepsilon_{ijk}\gamma^j\gamma^k\psi \quad \text{by (7.106)}.$$

But (ii) gives

$$i\varepsilon_{ijk}\gamma^j\gamma^k = -i\varepsilon^{0i}{}_{jk}\gamma^j\gamma^k = 2\gamma^0\gamma^i\gamma_5 = 2\gamma_5\gamma^0\gamma^i,$$

so

$$\mathbf{p}\cdot\mathbf{J}\psi = \tfrac{1}{2}\gamma_5\gamma^0(p_i\gamma^i)\psi = \tfrac{1}{2}\gamma_5\gamma^0(m - p_0\gamma^0)\psi = \tfrac{1}{2}\gamma_5(m\gamma^0 - p_0)\psi. \quad \blacksquare$$

Note that if $m = 0$ and ψ is a positive-energy solution of the Dirac equation, so that $p_0 = |\mathbf{p}|$, the helicity operator is just $-\tfrac{1}{2}\gamma_5$. Now the property of being an eigenspinor of γ_5 is invariant under proper Lorentz transformations, for from (7.127) we have

$$\gamma_5\psi = \varepsilon\psi \;\Rightarrow\; \gamma_5 S(\Lambda)\psi = S(\Lambda)\gamma_5\psi = \varepsilon S(\Lambda)\psi. \tag{7.142}$$

Thus for a massless particle it is a Lorentz-invariant (but not parity-invariant) statement to say that it has negative helicity (in order to change to a frame of reference in which a particle has opposite helicity, it would be necessary to overtake the particle and look at it from the other side, as in Fig. 6.22, and this is not possible if the particle is travelling at the speed of light). This explains how it is possible for all neutrinos to be left-handed: a neutrino is described by a Dirac spinor satisfying the pair of equations

$$\gamma^\mu\,\partial_\mu\psi = 0, \quad \gamma_5\psi = \psi. \tag{7.143}$$

For a massive particle the Dirac equation is not compatible with the equation $\gamma_5\psi = \psi$. This can be seen by writing

$$\psi = \psi_L + \psi_R$$

where

$$\psi_L = \tfrac{1}{2}(1+\gamma_5)\psi, \quad \psi_R = \tfrac{1}{2}(1-\gamma_5)\psi; \tag{7.144}$$

then the Dirac equation becomes

$$i\gamma^\mu \, \partial_\mu \psi_L = m\psi_R, \quad i\gamma^\mu \, \partial_\mu \psi_R = m\psi_L. \tag{7.145}$$

The spinors ψ_L and ψ_R are the left-handed and right-handed state vectors introduced in §6.6. They are not eigenstates of helicity, but they are eigenstates of an operator $-\tfrac{1}{2}\gamma_5$ which tends to the helicity as the velocity of the particle tends to c (see ●7.6(vi); as $v \to c$, $p_0/|\mathbf{p}| \to 1$ and $m/|\mathbf{p}| \to 0$).

Note that from ●7.6(iv) we have

$$\tilde{\psi}_L = \psi_L{}^\dagger \zeta = \tfrac{1}{2}\psi^\dagger(1+\gamma_5)^\dagger \zeta = \tfrac{1}{2}\tilde{\psi}(1-\gamma_5),$$

and similarly

$$\tilde{\psi}_R = \tfrac{1}{2}\tilde{\psi}(1+\gamma_5). \tag{7.146}$$

Since $\gamma_5{}^2 = 1$, it follows that

$$\tilde{\psi}_L\psi_L = \tilde{\psi}_R\psi_R = 0. \tag{7.147}$$

Second quantisation As with the Klein–Gordon equation, the existence of negative-energy solutions of the Dirac equation raises a problem which can be resolved if the field is an operator in the Heisenberg picture. Following the lead given by (7.32) and (7.80), we write

$$\psi(x) = \int \{\alpha(\mathbf{p})e^{-i(p^0 t - \mathbf{p}\cdot\mathbf{r})} + \beta^\dagger(\mathbf{p})e^{i(p^0 t - \mathbf{p}\cdot\mathbf{r})}\} \, d^3\mathbf{p} \tag{7.148}$$

where $p^0 = +\sqrt{(\mathbf{p}^2 + m^2)}$, and $\alpha(\mathbf{p})$ and $\beta^\dagger(\mathbf{p})$ are Dirac spinors whose components are annihilation operators and creation operators respectively. To specify α and $\beta^\dagger$ more fully, for each 3-momentum $\mathbf{p}$ we define a basis $\{u_\pm(\mathbf{p}), v_\pm(\mathbf{p})\}$ for the space of Dirac spinors as follows:

$$\begin{aligned} s_3' u_\pm(\mathbf{p}) &= \pm u_\pm(\mathbf{p}), & p_\mu\gamma^\mu u_\pm(\mathbf{p}) &= m u_\pm(\mathbf{p}) \\ s_3' v_\pm(\mathbf{p}) &= \pm v_\pm(\mathbf{p}), & -p_\mu\gamma^\mu v_\pm(\mathbf{p}) &= m v_\pm(\mathbf{p}) \end{aligned}, \tag{7.149}$$

where $p^\mu = (p^0, \mathbf{p})$ and s_3' is the third component of spin in the rest frame of p^μ; we choose a Lorentz transformation Λ such that $\Lambda^\mu{}_\nu p^\nu = (m, \mathbf{0})$, and define $s_3' = S(\Lambda^{-1})s_3 S(\Lambda)$ where s_3 is the spin matrix of (7.107). In other words, $u_\pm(\mathbf{p})$ are positive-energy spinors and $v_\pm(\mathbf{p})$ are negative-energy spinors (for explicit formulae, see problem 7.7). Then we can write

$$\begin{aligned} \alpha(\mathbf{p}) &= a_-(\mathbf{p})u_+(\mathbf{p}) + a_+(\mathbf{p})u_-(\mathbf{p}) \\ \beta^\dagger(\mathbf{p}) &= b_+{}^\dagger(\mathbf{p})v_+(\mathbf{p}) + b_-{}^\dagger(\mathbf{p})v_-(\mathbf{p}) \end{aligned} \tag{7.150}$$

If ψ is the electron field, $a_\pm(\mathbf{p})$ are annihilation operators for electrons and $b_\pm{}^\dagger(\mathbf{p})$ are creation operators for positrons. The labelling correctly indicates the spin properties of the states that these operators create and annihilate. This can be seen by starting with the transformation of the field $\psi(x)$ under rotations, namely

$$U(R)\psi(x)U(R)^{-1} = S(R^{-1})\psi(Rx) \tag{7.151}$$

where $U(R)$ is the unitary operator on state space representing the rotation R, and $S(R)$ is the spinor matrix of ●7.3 (with $\Lambda = R$). This leads to a spin observable Σ which satisfies

$$[\Sigma, \psi(x)] = s\psi(x) \tag{7.152}$$

where s is the spinor matrix of (7.106). This must also hold with $\alpha(\mathbf{p})$ and $\beta^\dagger(\mathbf{p})$ in place of $\psi(x)$. Since $u_\pm$ and $v_\pm$ are eigenspinors of s_3, this leads to

$$[\Sigma_3, a_\mp(\mathbf{p})] = \pm a_\mp(\mathbf{p}), \tag{7.153}$$

$$[\Sigma_3, b_\pm{}^\dagger(\mathbf{p})] = \pm b_\pm{}^\dagger(\mathbf{p}). \tag{7.154}$$

The second of these equations shows that $b_\pm{}^\dagger$ create states with spin up and down ($s_z = \pm\tfrac{1}{2}$) respectively. The hermitian conjugate of the first equation,

$$[\Sigma_3, a_\mp{}^\dagger(\mathbf{p})] = \mp a_\mp{}^\dagger(\mathbf{p}),$$

shows that $a_\pm$ are also correctly labelled.

This means that in the left-handed and right-handed fields ψ_L and ψ_R defined by (7.144), particles and antiparticles have opposite helicity; ψ_L is the sum of an annihilation operator for a left-handed electron and a creation operator for a right-handed positron.

Since the Dirac equation describes particles with spin $\tfrac{1}{2}$, they are always fermions and so their creation and annihilation operators obey anticommutation relations. As described in §4.6, this can be extended to antiparticles. The result can be summarised, in a similar fashion to (7.39), as follows: Let

$$\eta(p) = \begin{cases} 2p^0\alpha(\mathbf{p}) & \text{if } p^0 > 0, \\ -2p^0\beta^\dagger(-\mathbf{p}) & \text{if } p^0 < 0; \end{cases} \tag{7.155}$$

and let $\chi(p) = \eta(p)\,\delta(p^2 - m^2)$. Then $\chi(p)$ is the four-dimensional Fourier transform of $\psi(x)$, and satisfies the anticommutation relations

$$\left. \begin{aligned} \{\chi_r(p), \chi_s(p')\} &= 0 = \{\chi_r(p)^\dagger, \chi_s(p')^\dagger\}, \\ \{\chi_r(p), \chi_s(p')^\dagger\} &= \delta_{rs}\,\delta(p^2 - m^2)\,\delta(p - p'). \end{aligned} \right\} \tag{7.156}$$

These have the consequence that

$$\left. \begin{aligned} \{\psi(x), \psi(x')\} &= 0 \quad \text{for all } x, x'; \\ \{\psi(x), \bar{\psi}(x')\} &= 0 \quad \text{if } (x - x')^2 < 0. \end{aligned} \right\} \tag{7.157}$$

Like the corresponding relations for the electromagnetic field, these anticommutation relations define the property of **locality** for a fermion field.

Finally, let us consider the conserved current $\bar{\psi}\gamma^\mu\psi$. The quantity which is conserved by this current is the space integral of its 0th component; at $t = 0$ this is, by (7.148) and Plancherel's theorem (problem 2.13),

$$\int \psi(\mathbf{r})^\dagger\psi(\mathbf{r})\,d^3\mathbf{r} = \int \{\alpha(\mathbf{p})^\dagger + \beta(-\mathbf{p})\}\{\alpha(\mathbf{p}) + \beta(-\mathbf{p})^\dagger\}\,d^3\mathbf{p}. \tag{7.158}$$

Now $v_\pm(-\mathbf{p})$ satisfy

$$p'_\mu \gamma^\mu v = -mv \quad \text{where } p'^\mu = (p^0, -\mathbf{p}).$$

From the properties of γ^0 we have

$$(p'_\mu \gamma^\mu)^\dagger = p'_\mu \gamma^0 \gamma^{\mu\dagger} \gamma^0 = p_\mu \gamma^\mu$$

so that

$$v^\dagger p_\mu \gamma^\mu = -mv^\dagger.$$

But $u_\pm(\mathbf{p})$ satisfy

$$p_\mu \gamma^\mu u = mu,$$

from which it follows that

$$v_\pm(-\mathbf{p})^\dagger u_\pm(\mathbf{p}) = 0.$$

Hence (7.154) becomes

$$\int \psi(\mathbf{r})^\dagger \psi(\mathbf{r}) \, d^3\mathbf{r} = \int \sum_{r=\pm} \{a_r(\mathbf{p})^\dagger a_r(\mathbf{p}) + b_r(\mathbf{p}) b_r(\mathbf{p})^\dagger\} \, d^3\mathbf{p}.$$

Since the anticommutator of b_r and $b_r^\dagger$ is a c-number, this is

$$\int \sum_r \{a_r(\mathbf{p})^\dagger a_r(\mathbf{p}) - b_r(\mathbf{p})^\dagger b_r(\mathbf{p})\} \, d^3\mathbf{p}, \tag{7.159}$$

apart from an (infinite!) c-number term which is similar to the infinite zero-point energy of the electromagnetic field, and is ignored with a similar justification. If $a_\pm{}^\dagger$ create electrons and $b_\pm{}^\dagger$ create positrons, this integral is the total number of electrons minus the total number of positrons. Thus $\bar{\psi}\gamma^0\psi$ is the density of a quantity which has opposite values for electrons and positrons. Multiplying by the electric charge e of an electron, we are led to the identification of $e\bar{\psi}\gamma^\mu\psi$ as the *electric current density 4-vector*.

7.3. Field dynamics

We have now constructed three types of quantum field: the scalar $\phi(x)$, the spinor $\psi(x)$, and the 4-vector $A_\mu(x)$. Each of these is an operator function of position which, being in the Heisenberg picture, depends on time as well as space. We know how they are made of certain time-independent creation and annihilation operators: each component θ of any of these fields is of the form

$$\theta(x) = \int \{a(\mathbf{p})e^{-ip\cdot x} + b(\mathbf{p})^\dagger e^{ip\cdot x}\} \, d^3\mathbf{p} \tag{7.160}$$

where $a(\mathbf{p})$ is an annihilation operator for a particle with 4-momentum p^μ (p^0 being given as a function of $\mathbf{p}$), and $b(\mathbf{p})^\dagger$ is the creation operator for its antiparticle. We can (although we have not done so) work out the commutation relations between field operators from the known commutation relations of the creation operators a, $a^\dagger$, b and $b^\dagger$, and, knowing that the Hamiltonian is of harmonic-oscillator type in terms of the creation and annihilation operators, we can express it as an integral of products of fields (see (7.17) and problem 7.8). Being Heisenberg operators, the fields satisfy

$$\theta(\mathbf{r}, t) = e^{-iHt}\theta(\mathbf{r}, 0)e^{iHt}. \tag{7.161}$$

Using (7.160), any state of a number of particles can be expressed in terms of products of field operators acting on the vacuum. As in (7.72), the amplitude for one such state at time t to become (in the Schrödinger picture) a different state at time t can be expressed in terms of vacuum expectation values of products of (Heisenberg-picture) operators. Thus to answer dynamical questions about the system it is not necessary to know the Hamiltonian explicitly; it is sufficient to know the time dependence of the field operators. This time dependence is implied by the differential equations satisfied by the fields, which are the same as the equations satisfied by the c-number fields of the classical or first-quantised theory (e.g. Maxwell's equations or the Schrödinger equation). Thus the quantum dynamics of a system described by quantum fields is determined by the field equations.

In classical field theory, as in classical mechanics, the field equations can usually be obtained from a principle of least action. The action for a system with a finite number of degrees of freedom is an integral $\int L \, dt$, where the Lagrangian L is a function of the coordinates q_i and their rates of change $\dot{q}_i$. In a field theory the 'coordinates' are the values of the field at each point of space, say $\theta(\mathbf{r})$; the Lagrangian, which is to be a function of these infinitely many variables, can be taken to be an integral

$$L = \int \mathcal{L}[\theta(\mathbf{r}), \, \boldsymbol{\nabla}\theta(\mathbf{r}), \, \dot{\theta}(\mathbf{r})] \, d^3\mathbf{r} \qquad (7.162)$$

where $\mathcal{L}$, the **Lagrangian density**, is a function of the values of the field and its derivatives at one point of space; and then the action is

$$S = \int L \, dt = \int \mathcal{L}(\theta, \, \partial_\mu \theta) \, d^4x. \qquad (7.163)$$

In this equation x, as usual, stands for the space-time position $(t, \mathbf{r})$. The principle of least action is the requirement that this fourfold integral should be stationary under arbitrary variations of the field which vanish at the boundary of the region of integration; this leads to the **Euler–Lagrange equations**

$$\frac{\partial \mathcal{L}}{\partial \theta} - \partial_\mu \left(\frac{\partial \mathcal{L}}{\partial(\partial_\mu \theta)} \right) = 0 \qquad (7.164)$$

(see Goldstein 1980 or problem 7.9). This is sometimes written as

$$\frac{\delta S}{\delta \theta(x)} = 0; \qquad (7.165)$$

the expression on the left is called the **functional derivative** of S with respect to $\theta(x)$. Its value at x, roughly speaking, gives the change in the integral S if the value of θ is changed at the one point x (by means of a δ-function).

If there are several fields in the theory $\mathcal{L}$ will be a function of all of them and there will be an equation like (7.164) for each field.

Eqs. (7.163)–(7.164) show that the Lagrangian formalism fits in with special relativity much better than the Hamiltonian formalism. The fundamental quantity S is an integral over space and time and does not require any separation between them; it is the same in all frames of reference, and the integrand $\mathscr{L}$ is a relativistic scalar. The equation of motion likewise puts space and time on an equal footing and is manifestly invariant under Lorentz transformations. On the other hand, Hamilton's equations in classical mechanics and field theory, and Schrödinger's equation in quantum mechanics, all specify time derivatives and assign a special role to the time coordinate. The Hamiltonian itself, being the total energy, is not a scalar but the time component of a 4-vector, and its form depends on the frame of reference. (In field theory it is the space integral of a Hamiltonian density which is even further from being a scalar, being the $(0,0)$ component of a tensor $T_{\mu\nu}$.)

In the context of field theory Feynman's formulation of quantum mechanics becomes particularly natural. This formulation is most appropriate for quantum systems which have a classical counterpart; it assigns amplitudes to paths which are defined in terms of classical coordinates. It is difficult to apply it to quantum observables like spin and isospin. But Feynman's formulation can be used in field theory by taking the first-quantised theory as the classical counterpart. For example, the spin of electrons and positrons is described by means of a Dirac spinor ψ. In quantum field theory $\psi(x)$ is an operator. Nevertheless, we can develop a theory of the Dirac equation as an equation for a c-number field ψ, construct an associated action S, and then apply Feynman's postulate to obtain an amplitude for a field configuration at one time to evolve to another configuration at a later time. ●3.14 can be generalised to show that this is equivalent to the development based on Postulate VI, in which the Hamiltonian is fundamental.

For all of these reasons the Lagrangian density $\mathscr{L}$ is regarded as the quantity of fundamental significance in quantum field theory. It is commonly referred to simply as the 'Lagrangian'.

The field equations we have considered so far can be obtained from Lagrangians as follows:

For the *Klein–Gordon equation*, take

$$\mathscr{L} = \tfrac{1}{2}(\partial_\mu \phi^\dagger \cdot \partial^\mu \phi - m^2 \phi^\dagger \phi). \tag{7.166}$$

Then

$$\frac{\partial \mathscr{L}}{\partial \phi} - \partial_\mu \left(\frac{\partial \mathscr{L}}{\partial(\partial_\mu \phi)} \right) = -m^2 \phi - \partial_\mu(\partial^\mu \phi),$$

so the Euler–Lagrange equation is the same as the Klein–Gordon equation.

For the *Dirac equation*, take

$$\mathscr{L} = \tfrac{1}{2}i(\bar{\psi}\gamma^\mu \partial_\mu \psi - \partial_\mu \bar{\psi} \cdot \gamma^\mu \psi) - m\bar{\psi}\psi, \tag{7.167}$$

which is often written as

$$\mathscr{L} = i\bar{\psi}\gamma^\mu \overleftrightarrow{\partial_\mu} \psi - m\bar{\psi}\psi.$$

The components of ψ are complex quantities whose real and imaginary parts can vary independently, so for each component there are two Euler–Lagrange equations which are the real and imaginary parts of a single complex equation. Now any function $f(x, y)$ of two real variables can be written in terms of the complex variable $z = x + iy$ as a function $g(z, \bar{z})$, and the real partial derivatives of f can be combined to form the complex derivative

$$\frac{\partial g}{\partial \bar{z}} = \frac{\partial f}{\partial x} + i \frac{\partial f}{\partial y} \tag{7.168}$$

which is calculated by treating z and $\bar{z}$ formally as independent quantities and differentiating partially with respect to $\bar{z}$. (The other partial derivative $\partial g / \partial z$ is just the complex conjugate of $\partial g / \partial \bar{z}$, if f is real.) Thus the Euler–Lagrange equations for the Lagrangian (7.167) can be written as

$$0 = \frac{\partial \mathscr{L}}{\partial \overline{\psi_r}} - \partial_\mu \left(\frac{\partial L}{\partial (\overline{\partial_\mu \psi_r})} \right) = \tfrac{1}{2} i [\zeta \gamma^\mu \, \partial_\mu \psi]_r - m[\zeta \psi]_r + \tfrac{1}{2} i \, \partial_\mu [\zeta \gamma^\mu \psi]_r$$

$$= [\zeta (i \gamma^\mu \, \partial_\mu \psi - m \psi)]_r,$$

which is the same as the Dirac equation.

For *Maxwell's equations*, take

$$\mathscr{L} = -\tfrac{1}{4} (\partial_\mu A_\nu - \partial_\nu A_\mu)(\partial^\mu A^\nu - \partial^\nu A^\mu). \tag{7.169}$$

Then

$$\frac{\partial \mathscr{L}}{\partial A_\nu} - \partial_\mu \left(\frac{\partial \mathscr{L}}{\partial (\partial_\mu A_\nu)} \right) = \partial_\mu (\partial^\mu A^\nu - \partial^\nu A^\mu), \tag{7.170}$$

so the Euler–Lagrange equations reproduce Maxwell's equations in empty space,

$$\partial_\mu F^{\mu\nu} = 0, \tag{7.171}$$

where $F_{\mu\nu} = \partial_\mu A_\nu - \partial_\nu A_\mu$ is the electromagnetic field tensor ($F_{0i} = E_i$ and $F_{ij} = \varepsilon_{ijk} B_k$ where **E** and **B** are the electric and magnetic fields). This is not the same as the equation on which we based our discussion of the electromagnetic field, namely $\square^2 A_\mu = 0$, unless one adds the condition $\partial_\mu A^\mu = 0$. It is permissible to add this condition, because of the arbitrariness of the potentials A_μ, but it is not essential; for the purposes of developing the theory it is more convenient to leave A_μ unrestricted but to bear in mind that it is the field tensor $F_{\mu\nu}$ that is physically meaningful and not the potentials A_μ. As we shall see in the next section, the fact that the theory can be cast in this form has great physical significance.

Note that the masslessness of the photon is expressed by the Lagrangian's being constructed entirely out of derivatives of the fields; there is no term like the ϕ^2 in (7.166) which gives the mass in the Klein–Gordon equation.

Invariances and conserved quantities

The discussion of invariances and conserved quantities in §3.2 was based on the Hamiltonian as the fundamental dynamical quantity. In order to apply these ideas to quantum field theory we will need to return to their source in Lagrangian classical mechanics.

In the classical mechanics of a system with a finite number of degrees of freedom, with coordinates $q_1, \ldots, q_n$ and a Lagrangian $L(q_1, \ldots, q_n, \dot{q}_1, \ldots, \dot{q}_n)$, **Noether's theorem** states that if $q_i \to q_i'(q_1, \ldots, q_n, \alpha)$ (with $q_i' = q_i$ if $\alpha = 0$) is a set of transformations which leave the Lagrangian invariant, so that $L(q', \dot{q}') = L(q, \dot{q})$ for all α, then

$$\sum_{i=1}^{n} \frac{\partial L}{\partial \dot{q}_i} \frac{\partial q_i'}{\partial \alpha}\bigg|_{\alpha=0} \tag{7.172}$$

is a constant of the motion. We will now prove two extensions of this theorem to Lagrangian field theory.

●**7.7a** **Noether's theorem I.** Let $\mathscr{L}(\theta_1, \ldots, \theta_n, \partial_\mu \theta_1, \ldots, \partial_\mu \theta_n)$ be a Lagrangian density which is invariant under the transformations

$$\theta_i(x) \to \theta_i'(\theta_1(x), \ldots, \theta_n(x), \alpha) \tag{7.173}$$

where α is a real parameter and $\theta_i' = \theta_i$ when $\alpha = 0$. Then the field equations (7.164) obtained from $\mathscr{L}$ imply that $\partial_\mu j^\mu = 0$ where

$$j^\mu(x) = \sum_{i=1}^{n} \frac{\partial \mathscr{L}}{\partial(\partial_\mu \theta_i)} \frac{\partial \theta_i'}{\partial \alpha}\bigg|_{\alpha=0}. \tag{7.174}$$

Proof. Writing $\theta = (\theta_1, \ldots, \theta_n)$, we have

$$\mathscr{L}(\theta'(\theta, \alpha), \partial_\mu \theta'(\theta, \alpha)) = \mathscr{L}(\theta, \partial_\mu \theta) \tag{7.175}$$

for all α. Differentiating with respect to α,

$$\sum_i \frac{\partial \mathscr{L}}{\partial \theta_i'} \frac{\partial \theta_i'}{\partial \alpha} + \frac{\partial \mathscr{L}}{\partial(\partial_\mu \theta_i')} \frac{\partial(\partial_\mu \theta_i')}{\partial \alpha} = 0. \tag{7.176}$$

When $\alpha = 0$ the field equations give

$$\frac{\partial \mathscr{L}}{\partial \theta_i'} = \frac{\partial \mathscr{L}}{\partial \theta_i} = \partial_\mu \left(\frac{\partial \mathscr{L}}{\partial(\partial_\mu \theta_i)} \right).$$

Also

$$\frac{\partial}{\partial \alpha}(\partial_\mu \theta_i') = \frac{\partial}{\partial \alpha}\left(\frac{\partial \theta_i'}{\partial \theta_j} \partial_\mu \theta_j \right) = \frac{\partial^2 \theta_i'}{\partial \alpha \, \partial \theta_j} \partial_\mu \theta_j = \partial_\mu \left(\frac{\partial \theta_i'}{\partial \alpha} \right).$$

Hence when $\alpha = 0$ (7.176) becomes

$$\partial_\mu \left(\frac{\partial \mathscr{L}}{\partial(\partial_\mu \theta_i)} \frac{\partial \theta_i'}{\partial \alpha} \right) = 0,$$

i.e. $\partial_\mu j^\mu = 0$. ∎

The equation $\partial_\mu j^\mu = 0$ is a continuity equation (see (7.122) and the sentence after it); if j^μ vanishes at spatial infinity, it implies that

$$Q = \int j^0(\mathbf{r}, t) \, d^3\mathbf{r} \tag{7.177}$$

is constant in time. This conserved quantity is called the **charge** associated with the invariance $\theta_i \to \theta_i'$ and the 'conserved' current j^μ.

As an example of Noether's theorem, consider the transformation of the Dirac field consisting of multiplication by a phase factor:

$$\psi'(\psi, \alpha) = e^{-ie\alpha} \psi \tag{7.178}$$

where the e in the exponent is the charge on the electron. Clearly the Lagrangian (7.167) is invariant under these transformations. As explained after (7.167), we can take ψ_r and its complex conjugate $\bar{\psi}_r$ as independent variables; to simplify the algebra still further, we can change variables from $\bar{\psi}_r$ (the components of $\psi^\dagger$) to $\bar{\psi}_s \zeta_{rs}$ (the components of $\bar{\psi}$). Then

$$\left.\frac{\partial \psi'}{\partial \alpha}\right|_{\alpha=0} = -ie\psi, \quad \left.\frac{\partial \bar{\psi}}{\partial \alpha}\right|_{\alpha=0} = ie\bar{\psi} \tag{7.179}$$

so that (7.174), with $\mathscr{L}$ taken from (7.167), gives

$$\begin{aligned}
j^\mu &= \frac{\partial \mathscr{L}}{\partial(\partial_\mu \psi_r)}(-ie\psi_r) + ie\bar{\psi}_r \frac{\partial \mathscr{L}}{\partial(\partial_\mu \bar{\psi}_r)} \\
&= \tfrac{1}{2} i \bar{\psi} \gamma^\mu (-ie\psi) + i\bar{\psi}(-\tfrac{1}{2} ie\gamma^\mu \psi) \\
&= e\bar{\psi} \gamma^\mu \psi,
\end{aligned}$$

which is the electric current 4-vector. Thus the conserved quantity associated with invariance under the phase transformations (7.178) is the total electric charge.

A phase transformation like (7.178) can be defined for any additive quantum number A; the field of a particle for which $A = a$ (with an antiparticle for which $A = -a$) transforms by

$$\psi' \to \psi'(\alpha) = e^{-ia\alpha}\psi. \tag{7.180}$$

We will call this an **A phase transformation**; thus we have baryon number phase transformations, lepton number phase transformations, hypercharge phase transformations, and so on.

This derivation of the existence of a conserved quantity associated with an invariance appears quite different from what was proved for non-relativistic quantum mechanics, based on a Hamiltonian, in §3.2. However, it can be shown that in fact there is the same relation between the invariance and the conserved quantity: one can construct a unitary operator $U(\alpha)$ such that

$$U(\alpha)\theta_i U(\alpha)^{-1} = \theta_i'(\theta(x), \alpha) \tag{7.181}$$

and then

$$Q = i\left.\frac{dU}{d\alpha}\right|_{\alpha=0}, \tag{7.182}$$

i.e. Q is the hermitian generator of the transformations $\theta \to \theta'$. It follows that if we consider transformations depending on several parameters and forming a Lie group G, the various charges associated with them satisfy the commutation relations of the Lie algebra of G.

The transformations (7.173) involve no change of space-time points and therefore cannot describe 'external' symmetries like invariance under rotations or translations. Indeed the Lagrangian density $\mathscr{L}(\theta, \partial_\mu \theta)$ is not invariant under translations $\theta(x) \to \theta(x+a)$. By analogy with systems with a finite number of degrees of freedom, however, we would expect a form of Noether's theorem to hold for such transformations, since the total

Lagrangian $L = \int \mathscr{L}\, d^3x$ is invariant if $\mathscr{L}$ does not depend explicitly on x. We will prove the theorem for translations; for rotations and Lorentz transformations see problem 7.11.

●**7.7b Noether's theorem for translations.** Let $\mathscr{L}(\theta_1, \ldots, \theta_n, \partial_\mu\theta_1, \ldots, \partial_\mu\theta_n)$ be a Lagrangian density which does not depend explicitly on x. Then the field equations (7.164) imply that

$$\partial_\mu T^{\mu\nu} = 0 \tag{7.183}$$

where

$$T^{\mu\nu} = \sum_{i=1}^{n} \frac{\partial\mathscr{L}}{\partial(\partial_\mu\theta_i)}\, \partial^\nu\theta_i - \mathscr{L}g^{\mu\nu}. \tag{7.184}$$

Proof. If $\mathscr{L}$ does not depend explicitly on x,

$$\frac{\partial\mathscr{L}}{\partial x^\nu} = \frac{\partial\mathscr{L}}{\partial\theta_i}\,\partial_\nu\theta_i + \frac{\partial\mathscr{L}}{\partial(\partial_\mu\theta_i)}\,\partial_\mu\partial_\nu\theta_i = \partial_\mu\left(\frac{\partial\mathscr{L}}{\partial(\partial_\mu\theta_i)}\right)\partial_\nu\theta_i + \frac{\partial\mathscr{L}}{\partial(\partial_\mu\theta_i)}\,\partial_\mu\partial_\nu\theta_i$$

by the field equations

$$= \partial_\mu\left(\frac{\partial\mathscr{L}}{\partial(\partial_\mu\theta_i)}\,\partial_\nu\theta_i\right).$$

It follows that $\partial_\mu T^{\mu\nu} = 0$. ■

For a fixed ν we can regard the four quantities $T^{\mu\nu}$ ($\mu = 0, 1, 2, 3$) as the components of a conserved current 4-vector. The associated charge

$$P^\nu = \int T^{0\nu}\, d^3\mathbf{r} \tag{7.185}$$

is then a conserved quantity. Since it is associated with invariance under translations in space and time, we can identify it as a component of the total *energy-momentum* 4-vector. In particular, (7.184)–(7.185) enable us to construct the Hamiltonian $H = P^0$ from the Lagrangian.

The discrete operations of space reflection, time reversal and charge conjugation are related in quantum field theory by the **CPT theorem**, which states that any Lorentz-invariant Lagrangian field theory with fields which satisfy local commutation or anticommutation relations must be invariant under the combined operation CPT (for a proof see Itzykson & Zuber 1980).

Quantum electrodynamics   Each of the three Lagrangians described on p. 313–14 involves a single kind of field, and yields a field equation appropriate to free particles. We will now see how these Lagrangians can be modified to yield field equations describing particles which exert forces on each other; that is to say, we will introduce *interactions* between the fields.

The paradigm theory of interacting fields is the theory of electrons and positrons together with the electromagnetic field. The electrons and positrons provide a charge and current density j^μ which must be included in Maxwell's equations, whose full form is

$$\partial_\mu F^{\mu\nu} = j^\nu. \tag{7.186}$$

In §7.2 we found that the current density 4-vector for electrons and positrons is $j^\mu = e\bar{\psi}\gamma^\mu\psi$; thus the field equation for the electromagnetic field becomes

$$\partial_\mu F^{\mu\nu} = e\bar{\psi}\gamma^\mu\psi. \tag{7.187}$$

The right-hand side can be regarded as a force term in the equation of motion of the photons.

On the other hand, of course, the electrons and positrons experience a force exerted by the electromagnetic field. In order to see how this can be incorporated in the field equation for the Dirac field ψ, let us first go back to the classical equation of motion of a charged particle in an electric field $\mathbf{E}$ and a magnetic field $\mathbf{B}$, which is

$$m\frac{d^2\mathbf{r}}{dt^2} = e\mathbf{E} + e\frac{d\mathbf{r}}{dt} \times \mathbf{B}. \tag{7.188}$$

If $\mathbf{E}$ and $\mathbf{B}$ are given in terms of potentials $(\phi, \mathbf{A})$ by (7.5)–(7.6), this equation of motion is equivalent to Hamilton's equations (3.5) with the Hamiltonian

$$H(\mathbf{r}, \mathbf{p}) = (2m)^{-1}[\mathbf{p} - e\mathbf{A}(\mathbf{r}, t)]^2 + e\phi(\mathbf{r}, t) \tag{7.189}$$

(see problem 7.12). With the usual substitutions $H \to i\partial/\partial t$ and $\mathbf{p} \to -i\nabla$, this gives a non-relativistic Schrödinger equation

$$i\frac{\partial\psi}{\partial t} = \frac{1}{2m}(-i\nabla - e\mathbf{A})^2\psi + e\phi\psi \tag{7.190}$$

which is obtained from the free-particle Schrödinger equation by making the substitutions

$$i\partial/\partial t \to i\partial/\partial t - e\phi, \quad -i\nabla \to -i\nabla - e\mathbf{A}. \tag{7.191}$$

This suggests that the corresponding relativistic equation for a simple particle in an electromagnetic field should be obtained from the Klein–Gordon equation by making the same substitutions, which can be written in 4-vector form as

$$i\partial_\mu \to i\partial_\mu - eA_\mu. \tag{7.192}$$

(Note that $A^\mu = (\phi, \mathbf{A})$, so $A_\mu = (\phi, -\mathbf{A})$.) Finally, we obtain the relativistic wave equation for a spin-$\frac{1}{2}$ particle in an electromagnetic field by making these substitutions in the Dirac equation:

$$\gamma^\mu(i\,\partial_\mu - eA_\mu)\psi = m\psi. \tag{7.193}$$

In the first-quantised theory of a single electron, this equation is to be solved for $\psi(x)$ to find the wave function of an electron in a specified field $A_\mu(x)$. On applying second quantisation, as we saw in connection with the Schrödinger equation (7.63), $\psi(x)$ becomes an operator but continues to satisfy the same equation. If we want a completely quantum-mechanical description, the electromagnetic field $A_\mu(x)$ must also become an operator.

Thus (7.187) and (7.193) are coupled equations for the quantum fields $A_\mu(x)$

and $\psi(x)$. They can be obtained as Euler–Lagrange equations from the Lagrangian

$$\mathscr{L} = -\tfrac{1}{4} F_{\mu\nu} F^{\mu\nu} + i\bar{\psi}\gamma^{\mu} \overset{\leftrightarrow}{\partial}_{\mu} \psi + m\bar{\psi}\psi + e\bar{\psi}\gamma^{\mu} A_{\mu}\psi, \qquad (7.194)$$

which is the sum of the Lagrangians (7.167) and (7.169) for free photons and free electrons, with $i\partial_{\mu}$ in the electron Lagrangian replaced by $i\partial_{\mu} - eA_{\mu}$, as in the Dirac equation.

The Lagrangian (7.194) gives rise, according to Noether's theorem ●7.7b, to an energy–momentum density

$$T^{\mu\nu} = \frac{\partial \mathscr{L}}{\partial(\partial_{\mu} A_{\rho})} \partial^{\nu} A_{\rho} + \frac{\partial \mathscr{L}}{\partial(\partial_{\mu}\psi)} \partial^{\nu}\psi + \partial^{\nu}\bar{\psi} \frac{\partial \mathscr{L}}{\partial(\partial_{\mu}\bar{\psi})} - \mathscr{L} g^{\mu\nu}$$

$$= -\tfrac{1}{2} F^{\mu\rho} F^{\nu}{}_{\rho} + i\bar{\psi}\gamma^{\mu} \overset{\leftrightarrow}{\partial^{\nu}}\psi - \mathscr{L} g^{\mu\nu}, \qquad (7.195)$$

which yields the Hamiltonian

$$H = \int T^{00}\, d^3\mathbf{r} = \int \{\tfrac{1}{2}(\mathbf{E}^2 + \mathbf{B}^2) + i\bar{\psi}(\gamma^0 \overset{\leftrightarrow}{\partial_0} + \gamma \cdot \overset{\leftrightarrow}{\mathbf{V}})\psi - m\bar{\psi}\psi\}$$

$$- e \int \bar{\psi}\gamma^{\mu}\psi A_{\mu}\, d^3\mathbf{r}. \qquad (7.196)$$

The first integral is the sum of the Hamiltonians given by the Lagrangians (7.167) and (7.169) which describe the free motion of electrons and photons respectively; the second integral contains the interaction between them. When the fields are expressed in terms of annihilation and creation operators by means of their Fourier transforms, the first integral corresponds to the free Hamiltonian (4.183) in the simplified theory of §4.6; the second integral contains annihilation and creation operators for electrons, positrons and photons in combinations like $a_e a_e^\dagger a_\gamma$ as in (4.184), which cause processes described by the Feynman diagrams of §4.6.

Renormalisation   It is beyond the scope of this book to begin to describe the calculations based on the Hamiltonian (7.196) (or equivalently the Lagrangian (7.194)). We will, however, give a brief qualitative mention to one of the most striking and notorious features of these calculations.

One of the Feynman diagrams required by the Hamiltonian (7.196) is shown in Fig. 7.1. This shows a process whose initial and final states both consist of a single electron and which, therefore, we might expect to be adequately described by the theory of an electron on its own (i.e. the Dirac equation). However, the transition amplitudes given by the perturbation theory calculation associated with this diagram do not agree with those given by the

Fig. 7.1.
The self-energy diagram.

Dirac equation derived from the free-electron part of the Lagrangian (7.194); instead, they relate to the Dirac equation with a different mass $m + \delta m$. It is this, not the parameter m in the Lagrangian, that is measured as the mass of the electron.

In the perturbation theory calculation based on the Lagrangian (7.194) δm is given by an integral which diverges. Nevertheless, it is possible to carry out consistent calculations concerning interacting electrons and photons by supposing that the original mass m is infinite in such a way that the observed mass $m_0 = m + \delta m$ is finite, for it is only m_0 that appears in the final results.

There are other divergent integrals occurring in calculations of electromagnetic processes, but these can be removed by redefining the charge e to $e_0 = Ke$. The constant K is infinite, so that the original parameter e must be regarded as infinitesimal, but as with the mass it is only the finite quantity e_0 that appears in the final answers. This procedure is called **renormalisation**. A similar procedure must also be applied to the normalisation constant in the wave functions of the particles involved; when this has been done the results of all calculations are finite, and agree with experiment so well (to one part in 10^{11}) as to make this one of the most accurate of all physical theories.

The fact that the infinities in the perturbation theory solution can be removed by means of a finite number of renormalisation constants is a special feature of quantum electrodynamics; it is said to be a **renormalisable** theory. Most Lagrangians lead to theories which do not have this property, and have an infinite number of essentially different divergent integrals in their solutions.

7.4. Gauge theories   The Lagrangian (7.194) of quantum electrodynamics is, like the free-electron Lagrangian, invariant under the phase transformations $\psi \to e^{-ie\alpha}\psi$. The conserving current given by Noether's theorem is the same as for the free case, namely the electric current 4-vector $j^\mu = \bar{\psi}\gamma^\mu\psi$.

The Lagrangian is also invariant under the wider class of *local* phase transformations

$$\psi(x) \to \psi'(x) = e^{-ie\alpha(x)}\psi(x), \tag{7.197}$$

in which the phase α varies from point to point of space-time, provided they are accompanied by a transformation of the electromagnetic potentials:

$$A_\mu(x) \to A_\mu'(x) = A_\mu(x) - \partial_\mu\alpha(x). \tag{7.198}$$

This transformation is known in classical electromagnetic theory as merely an alternative permissible choice of potentials, for both A_μ and A_μ' give the same electric and magnetic fields:

$$F_{\mu\nu} = \partial_\mu A_\nu - \partial_\nu A_\mu = \partial_\mu A_\nu' - \partial_\nu A_\mu'. \tag{7.199}$$

The choice of potential is called a **gauge**, and (7.198) is called a **gauge transformation**.

The free-electron Lagrangian (7.167) is not invariant under local phase transformations, but only under *global* phase transformations in which the phase α is constant. The passage from global to local invariance is accomplished by replacing ∂_μ by

$$D_\mu = \partial_\mu + ieA_\mu; \tag{7.200}$$

the electromagnetic Lagrangian is

$$\mathscr{L} = F^{\mu\nu}F_{\mu\nu} + i\tilde{\psi}\gamma^\mu \overset{\leftrightarrow}{D}_\mu \psi \tag{7.201}$$

and $D_\mu\psi$, unlike $\partial_\mu\psi$, transforms under the local phase transformations (7.197)–(7.198) in the same way as ψ:

$$D_\mu\psi \to (D_\mu\psi)' = \partial_\mu\psi' + ieA_\mu'\psi' = e^{-ie\alpha(x)}D_\mu\psi. \tag{7.202}$$

$D_\mu\psi$ is called the **covariant derivative** of ψ.

Thus the existence of the photon field A_μ, with its transformation law (7.198), can be regarded as a consequence of invariance under local phase transformations for charged particles. It is called the **gauge field** of these transformations.

A **Yang–Mills theory** results from applying similar considerations to a theory of free particles with some other symmetry replacing phase transformations. Let us consider, as Yang and Mills originally did, a theory with isospin symmetry. A pair of spin-$\frac{1}{2}$ particles forming a doublet with isospin $\frac{1}{2}$, like the proton and the neutron, can be described by a pair of Dirac spinor fields ψ_p, ψ_n which can be put together to form an isospinor field Ψ:

$$\Psi = \begin{pmatrix} \psi_p \\ \psi_n \end{pmatrix}, \quad \tilde{\Psi} = (\tilde{\psi}_p, \tilde{\psi}_n). \tag{7.203}$$

The free-particle Lagrangian is

$$\mathscr{L} = i\tilde{\Psi}\gamma^\mu \overset{\leftrightarrow}{\partial}_\mu \Psi, \tag{7.204}$$

which is nothing but the sum of two free-particle Lagrangians with the same mass. It is invariant under SU(2) (isospin) transformations

$$\Psi \to \Psi' = U\Psi = e^{ig\mathbf{a}\cdot\boldsymbol{\tau}}\Psi \tag{7.205}$$

where g is a coupling constant (like the electric charge e in (7.197)), and the 2×2 matrix $U = e^{ig\mathbf{a}\cdot\boldsymbol{\tau}}$ is an element of SU(2) which can be written in terms of the Pauli matrices (τ_1, τ_2, τ_3) by means of three real parameters $(a_1, a_2, a_3) = \mathbf{a}$. Each of these parameters can play the role of α in Noether's theorem; putting $\mathbf{a} = (\alpha, 0, 0)$ and applying Noether's theorem we obtain a conserving current j_1^μ. Similarly there are conserving currents j_2^μ and j_3^μ. These three currents form an isovector

$$\mathbf{j}^\mu = \tilde{\Psi}\boldsymbol{\tau}\gamma^\mu\Psi. \tag{7.206}$$

We can form local SU(2) transformations by letting the parameters a_i, and therefore the SU(2) element U, depend on the space-time point x:

$$\Psi \to \Psi' = U(x)\Psi = e^{ig\mathbf{a}(x)\cdot\boldsymbol{\tau}}\Psi. \tag{7.207}$$

The Lagrangian (7.204) is not invariant under such local transformations; it will become so if the derivative ∂_μ (acting on the isospinor Ψ) is replaced by the covariant derivative

$$D_\mu = \partial_\mu + ig\mathbf{A}_\mu \cdot \boldsymbol{\tau} = \partial_\mu + ig A_\mu \qquad (7.208)$$

where $\mathbf{A}_\mu$ is a set of three 4-vector fields (like the photon field) which form an isospin triplet and whose transformation under the local SU(2) transformations (7.207) is best expressed in terms of the matrix $A_\mu = \mathbf{A}_\mu \cdot \boldsymbol{\tau}$ as

$$A_\mu \to A_\mu' = U(x)A_\mu U(x)^{-1} + ig^{-1}(\partial_\mu U)U^{-1}. \qquad (7.209)$$

Then the covariant derivative transforms by

$$D_\mu\Psi \to (D_\mu\Psi)' = U(x)D_\mu\Psi, \qquad (7.210)$$

which guarantees the invariance of the Lagrangian term

$$i\bar{\Psi}\gamma^\mu \overleftrightarrow{D}_\mu \Psi + m\bar{\Psi}\Psi. \qquad (7.211)$$

The transformations (7.207) and (7.209) are called SU(2) gauge transformations. $\mathbf{A}_\mu$ is the SU(2) gauge field; the particles created by it are the SU(2) **gauge bosons**.

The Lagrangian must also contain a term describing the free motion of the $\mathbf{A}_\mu$ field, corresponding to the term $-\frac{1}{4}F^{\mu\nu}F_{\mu\nu}$ in the electrodynamic Lagrangian. In the electrodynamic case this term is gauge-invariant because the field tensor $F_{\mu\nu}$ is itself invariant under gauge transformations, as is shown by (7.199). In the case of SU(2) gauge theory the same definition of $F_{\mu\nu}$ would not give a gauge-invariant result, because of the homogeneous part of the transformation of A_μ (the first term in (7.209)), which rotates the isovector $\mathbf{A}_\mu$ in isospin space. An appropriate SU(2) version of $F_{\mu\nu}$ can be obtained by noting that in electromagnetic theory $F_{\mu\nu}$ is the commutator of covariant derivatives:

$$[D_\mu, D_\nu]\psi = ieF_{\mu\nu}\psi. \qquad (7.212)$$

Applying this to SU(2) Yang–Mills theory gives

$$[D_\mu, D_\nu]\Psi = igF_{\mu\nu}\Psi \qquad (7.213)$$

where $F_{\mu\nu}$ is the 2×2 hermitian matrix

$$F_{\mu\nu} = \partial_\mu A_\nu - \partial_\nu A_\mu + ig[A_\mu, A_\nu]. \qquad (7.214)$$

To see how this behaves under SU(2) transformations, note that if Φ is any isospinor which transforms like Ψ (i.e. according to (7.207)), then $D_\mu\Phi$ transforms in the same way (see (7.210)). In particular, we could take Φ to be $D_\nu\Psi$ to show that $D_\mu D_\nu\Psi$ transforms in this way. The same applies to $D_\nu D_\mu\Psi$; hence

$$([D_\mu, D_\nu]\Psi)' = U(x)[D_\mu, D_\nu]\Psi,$$

i.e.

$$F_{\mu\nu}'\Psi' = U(x)F_{\mu\nu}\Psi,$$

from which we get

$$F_{\mu\nu}' = U(x)F_{\mu\nu}U(x)^{-1}. \qquad (7.215)$$

It follows that the Lagrangian term

$$-\tfrac{1}{2}\operatorname{tr}\left(F^{\mu\nu}F_{\mu\nu}\right)=\mathscr{L}_A \tag{7.216}$$

is invariant under SU(2) gauge transformations. The full gauge-invariant Lagrangian is the sum of (7.211) and (7.216).

Since $F_{\mu\nu}$ is a hermitian 2×2 matrix, it can be written as $F_{\mu\nu}=\mathbf{F}_{\mu\nu}\cdot\boldsymbol{\tau}$ where the isovector $\mathbf{F}_{\mu\nu}$ is rotated in isospace by the transformation (7.215). The Lagrangian term (7.216) is then equal to $-\tfrac{1}{4}\mathbf{F}_{\mu\nu}\cdot\mathbf{F}^{\mu\nu}$, i.e. it is the sum of three terms like the free electromagnetic Lagrangian, one for each component of the SU(2) gauge field $\mathbf{A}_\mu$. However, because of the extra commutator term in the field tensor $F_{\mu\nu}$, the Lagrangian term $\mathscr{L}_A$ also contains products of three and four A-fields or their derivatives. These give rise to interactions between the particles which are described by the Feynman diagrams of Fig. 7.2.

In comparison with electromagnetic theory, we can say that the SU(2) gauge bosons are the quanta of a force which they also experience themselves since they have non-zero values of the charge (namely isospin) to which this force is coupled. This arises because of the non-zero commutator in the definition of $F_{\mu\nu}$, i.e. because the gauge group SU(2) is *non-abelian*.

Like the photon, the gauge bosons governed by the SU(2) gauge field A_μ, with the Lagrangian (7.216), are massless. A non-zero mass would require a term $m^2A_\mu A^\mu$ in the Lagrangian (like the term $m^2\phi^2$ in the Klein–Gordon Lagrangian (7.166)); but because of the inhomogeneity of the gauge transformation (7.209), such a term would not be invariant under gauge transformations. Gauge invariance requires massless gauge bosons.

A gauge theory can be constructed for any Lie group. The general construction is as follows. Let G be a Lie group of $m\times m$ matrices, and let the $m\times m$ matrices $T_1,\dots,T_l$ be a basis for the Lie algebra of G. If G is compact, these can be chosen so that

$$\operatorname{tr}\left(T_iT_j\right)=\lambda\delta_{ij} \tag{7.217}$$

for some constant λ. Let ρ be an n-dimensional representation of G (so that for each $Q\in G$, $\rho(Q)$ is an $n\times n$ matrix) with generators $X_i=\rho(T_i)$ $(i=1,\dots,l)$. The **gauge field** is a set of 4-vector fields $A_{i\mu}(x)$ $(i=1,\dots,l)$ from which we can form matrices whose entries are fields:

$$\left.\begin{aligned} A_\mu(x)&=\sum_i A_{i\mu}(x)T_i \\ \rho(A)_\mu(x)&=\sum_i A_{i\mu}(x)X_i \end{aligned}\right\}. \tag{7.218}$$

Fig. 7.2.
Feynman vertices in a gauge theory.

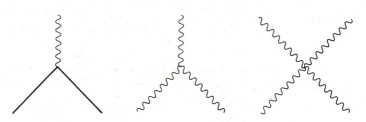

The **field strength** tensor is the matrix field

$$F_{\mu\nu}(x) = \partial_\mu A_\nu - \partial_\nu A_\mu + ig[A_\mu, A_\nu]. \tag{7.219}$$

The other fields in the theory may be either Dirac fields or Klein–Gordon fields. Let $\Psi(x)$ denote an n-component column vector whose entries are Dirac fields, $\Phi(x)$ one whose entries are Klein–Gordon fields. For both of these the covariant derivative is

$$D_\mu = \partial_\mu + ig\rho(A)_\mu. \tag{7.220}$$

A gauge transformation is defined in terms of a function $Q(x)$ from space-time to the group G:

$$\Psi \to \Psi' = \rho(Q(x))\Psi, \quad \Phi \to \Phi' = \rho(Q(x))\Phi, \tag{7.221}$$

$$A_\mu \to A_\mu{}' = Q(x)A^\mu Q(x)^{-1} + ig^{-1}(\partial_\mu Q)Q^{-1}. \tag{7.222}$$

The right-hand side of (7.222) belongs to the Lie algebra of G, so this defines a transformation of the fields $A_{i\mu}$. Its effect on the $n \times n$ matrix $\rho(A)_\mu$ is

$$\rho(A)_\mu \to \rho(Q(x))\rho(A)_\mu\rho(Q(x))^{-1} + ig^{-1}[\partial_\mu\rho(Q(x))]\rho(Q(x))^{-1}. \tag{7.223}$$

Now we have

●**7.8** The gauge transformations (7.221)–(7.222) make the field strength $F_{\mu\nu}$ and the covariant derivatives $D_\mu\Phi, D_\mu\Psi$ transform by

$$F_{\mu\nu} \to Q(x)F_{\mu\nu}Q(x)^{-1}, \tag{7.224}$$

$$D_\mu\Psi \to \rho(Q(x))D_\mu\Psi \tag{7.225}$$

(and similarly for $D_\mu\Phi$). Any function of $F_{\mu\nu}, \Phi, \Psi, D_\mu\Phi$ and $D_\mu\Psi$ which is invariant under constant transformations by elements of G is also invariant under gauge transformations; in particular,

$$\mathcal{L} = -\tfrac{1}{4}\lambda^{-1}\,\mathrm{tr}\,(F_{\mu\nu}F^{\mu\nu}) + i\bar{\Psi}\gamma^\mu\overleftrightarrow{D}_\mu\Psi + m\bar{\Psi}\Psi + \tfrac{1}{2}(D_\mu\Phi^\dagger D^\mu\Phi - m^2\Phi^\dagger\Phi) \tag{7.226}$$

is a gauge-invariant Lagrangian. If the Lagrangian is gauge-invariant, the gauge bosons are massless.

The proof is implicit in the preceding discussion, and follows the same lines as the special case of $G = \mathrm{SU}(2)$. ∎

The only known massless boson is the photon (though presumably the graviton also exists). Nevertheless, both the strong and the electroweak interactions are thought to be governed by gauge theories. The main reason for this is that gauge theories have been proved to be renormalisable (they are the only forms of quantum field theory which are known to have this property, and therefore to be consistent). The manner in which the lack of observed massless bosons is reconciled with the theory is different for the two interactions.

Quantum chromodynamics, the field theory of the strong force, is obtained by taking $G = \mathrm{SU}(3)$ (the colour group). There are then eight gauge bosons, corresponding to the eight generators of SU(3); these are the gluons of §6.6. The basis T_i can be taken to be the Gell-Mann matrices λ_i, which satisfy (7.217) with $\lambda = 3$; the gauge field A_μ then consists of 3×3 hermitian matrices. The other fields are the colour triplets Ψ_f of quark fields, one for each flavour $f = $ u, d, s, c, b, t. The Lagrangian is then

$$\mathscr{L} = -\tfrac{1}{12} \mathrm{tr}\,(F^{\mu\nu} F_{\mu\nu}) + \sum_f (i \tilde{\Psi}_f \gamma_\mu \overleftrightarrow{D}_\mu \Psi)_f + m_f \tilde{\Psi}_f \Psi_f). \tag{7.227}$$

This Lagrangian has the peculiar feature that all the particles in it, both quarks and gluons, have failed to manifest themselves as free particles. (As discussed in §6.5, quarks have been observed just as well as atomic nuclei have been observed, as bound particles.) This is thought to be a consequence of SU(3) gauge theory; the forces it represents increase with distance so as to prevent quarks and gluons escaping from combination with other quarks and gluons. Confirmation of this idea has been obtained from numerical calculations in which the space-time continuum is replaced by a discrete lattice of points.

The application of gauge theory to the electroweak force, with no corresponding massless boson, requires a further theoretical development which is described in the next section.

7.5. **Hidden symmetry**
Self-interacting fields

In the previous section we encountered Feynman diagrams in which three or more lines representing the same type of field meet at a single vertex. These arise from terms in the Lagrangian which are cubic or of higher order in the field and its derivatives. We will now examine this phenomenon more carefully, in the simple case of a single scalar field ϕ which, if free, would satisfy the Klein–Gordon equation. Suppose the self-interaction arises from terms in the Lagrangian containing only the field ϕ and not its derivatives; adding these to the Klein–Gordon Lagrangian (7.166) gives a full Lagrangian of the form

$$\mathscr{L} = \tfrac{1}{2}(\partial_\mu \phi)(\partial^\mu \phi) - V(\phi) \tag{7.228}$$

where V is a scalar function of a scalar variable. We will examine the quantum field theory arising from a Lagrangian of this type for a general function V; the only assumption we will make is that V is a twice differentiable function of ϕ.

The field equation (7.164) obtained from this Lagrangian is

$$\Box^2 \phi = -V'(\phi). \tag{7.229}$$

If V is twice differentiable, we can expand the right-hand side to get

$$\Box^2 \phi = a + b\phi + R(\phi) \tag{7.230}$$

where a and b are constants ($a = V'(0)$, $b = V''(0)$), and R is a function of ϕ satisfying $R(0) = R'(0) = 0$. On taking the Fourier transform of ϕ we expect to get an equation of motion for the Fourier components $\tilde{\phi}(\mathbf{k}, t)$ of the form

$$\frac{d^2}{dt^2} \tilde{\phi}(\mathbf{k}, t) = -(\mathbf{k}^2 + m^2)\tilde{\phi}(\mathbf{k}, t) + \tilde{R}(\tilde{\phi}) \tag{7.231}$$

where $\tilde{R}$ is of higher order than the linear term. This represents a harmonic oscillator (leading to the particle interpretation of the field) with an extra interaction described by $\tilde{R}(\bar{\phi})$. However, this does not happen unless the constants a and b in (7.230) satisfy

$$a = 0, \quad b \leqslant 0, \tag{7.232}$$

so that $V(\phi)$ has a minimum at $\phi = 0$.

In general, the appropriate variable is not ϕ but $\phi - \phi_0$ where ϕ_0 is a value of ϕ which makes V a minimum. This can also be understood by looking at the Hamiltonian which, according to (7.184)–(7.185), is

$$H = \int \{ \dot{\phi}^2 + (\nabla\phi)^2 + V(\phi) \} \, d^3\mathbf{r}. \tag{7.233}$$

In the classical theory of a c-number field $\phi(x)$ satisfying the field equation (7.229) this is the total energy in the field; it is a minimum if $\phi(x)$ has the constant value ϕ_0. Now the state of minimum energy is the vacuum; thus the field describing departures from the vacuum is $\phi(x) - \phi_0$. This is the field which, in the quantum theory, is an integral of annihilation and creation operators as in (7.80). It follows that $[\phi(x) - \phi_0]|0\rangle$ is orthogonal to $|0\rangle$ and so

$$\langle 0|\phi(x)|0\rangle = \phi_0. \tag{7.234}$$

This is the quantum counterpart of the classical statement that $\phi(x)$ takes the value ϕ_0 in the vacuum.

If the minimum value of $V(\phi)$ occurs for two different values of ϕ, say ϕ_1 and ϕ_2, then the vacuum state is no longer uniquely defined. In the classical theory the field configurations $\phi(x) = \phi_1$ and $\phi(x) = \phi_2$ are both possibilities for the vacuum, since they both have less energy than any other state of affairs. In the quantum theory $\phi(x) - \phi_1$ and $\phi(x) - \phi_2$ can both be written as integrals of annihilation and creation operators. The two sets of annihilation operators, say $a_1(\mathbf{k})$ and $a_2(\mathbf{k})$, annihilate different vacuum states $|0_1\rangle$ and $|0_2\rangle$. These can be shown to be orthogonal to each other, as are the finitely many-particle states $(a_1^\dagger)^n|0_1\rangle$ and $(a_2^\dagger)^n|0_2\rangle$ constructed on them. Thus the two minima of the potential $V(\phi)$ give rise to two orthogonal worlds which are unrelated to each other as far as perturbation theory goes.

Global symmetry:
Goldstone bosons
The situation considered in the last paragraph arises when there is a symmetry operation which preserves the potential $V(\phi)$ but not the position ϕ_0 of its minimum. In this case the symmetry operation must take ϕ_0 to another value of ϕ which minimises V. For example, consider

$$V(\phi) = \lambda(\phi^2 - a^2)^2 \tag{7.235}$$

(Fig. 7.3(a)). This is invariant under the reflection $\phi \rightarrow -\phi$, but its minima occur at the two values $\phi = \pm a$ which are not invariant but are taken to each other by this operation. To construct a quantum field theory we must work with one of the fields $\phi \pm a$, say $\theta(x) = \phi(x) - a$. The relevant state space is the space $\mathscr{S}_+$ of many-particle states constructed on the appropriate vacuum state $|0_+\rangle$, viz. that for which

$$\langle 0_+|\phi(x)|0_+\rangle = a, \quad \text{i.e.} \quad \langle 0_+|\theta(x)|0_+\rangle = 0. \tag{7.236}$$

The Lagrangian becomes

$$\mathscr{L} = \tfrac{1}{2}(\partial_\mu \theta)(\partial^\mu \theta) - \lambda(\theta^2 + 2a\theta)^2, \tag{7.237}$$

which is still invariant under the symmetry operation, now appearing as $\theta \to -\theta - 2a$. However, the corresponding operation on states would take the vacuum state $|0_+\rangle$ to the other vacuum state $|0_-\rangle$, which satisfies

$$\langle 0_-|\phi(x)|0_-\rangle = -a, \quad \text{i.e.} \quad \langle 0_-|\theta(x)|0_-\rangle = -2a. \tag{7.238}$$

For the state space $\mathscr{S}_+$ with vacuum state $|0_+\rangle$ there is no unitary operator representing the symmetry operation. Such an operation, which leaves the Lagrangian invariant but not the vacuum state, is called a **hidden symmetry** or a **spontaneously broken symmetry**. An operation which preserves both the Lagrangian and the vacuum state we will call an **overt** (or **unbroken**) symmetry.

If there is more than one field ϕ there may be a continuous set of minima of the potential $V(\phi)$, and correspondingly a continuous set of vacuum states (all orthogonal to each other). For example, suppose there are two fields forming a two-dimensional (column) vector $\Phi = (\phi_1, \phi_2)^{\mathrm{T}}$, and the potential is

$$V(\Phi) = \lambda(\Phi^{\mathrm{T}}\Phi - a^2)^2 = \lambda(\phi_1{}^2 + \phi_2{}^2 - a^2)^2. \tag{7.239}$$

This potential is shown in Fig. 7.3(b); its minima form the circle $\phi_1{}^2 + \phi_2{}^2 = a^2$ in the Φ plane. The potential is invariant under the rotations

$$(\phi_1, \phi_2) \to (\phi_1 \cos \alpha + \phi_2 \sin \alpha, -\phi_1 \sin \alpha + \phi_2 \cos \alpha), \tag{7.240}$$

but no individual minimum point Φ_0 is invariant. The particle interpretation of the quantum field theory must be based on a particular point Φ_0 on the circle by means of the field

$$\Theta(x) = \Phi(x) - \Phi_0, \tag{7.241}$$

in terms of which the Lagrangian becomes

$$\mathscr{L} = \tfrac{1}{2}(\partial_\mu \Theta)^{\mathrm{T}}(\partial^\mu \Theta) - \lambda(\Theta^{\mathrm{T}}\Theta + 2\Theta^{\mathrm{T}}\Phi_0)^2. \tag{7.242}$$

Let θ_1 and θ_2 be the components of Θ in the direction of Φ_0 and perpendicular

Fig. 7.3.
(a) One field; (b) two fields.

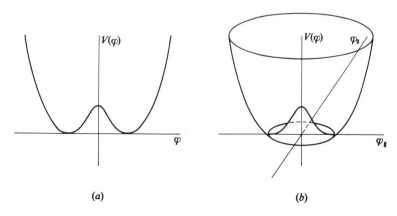

(a) (b)

to it, i.e. $\theta_1 = u^T\Theta$ and $\theta_2 = v^T\Theta$ where u and v are vectors satisfying

$$u^T u = v^T v = 1, \quad u^T v = 0, \quad u = \Phi_0/a. \tag{7.243}$$

Then the Lagrangian can be written in terms of θ_1 and θ_2 as

$$\mathscr{L} = \tfrac{1}{2}(\partial_\mu\theta_1)(\partial^\mu\theta_1) - 4\lambda a^2\theta_1{}^2 + \tfrac{1}{2}(\partial_\mu\theta_2)(\partial^\mu\theta_2) + f(\theta_1,\theta_2) \tag{7.244}$$

where $f(\theta_1, \theta_2)$ contains third-order and fourth-order terms and describes an interaction between the particles created by the fields θ_1 and θ_2. The other terms show that θ_1 creates a particle with mass $2a\sqrt{\lambda}$, while the particle created by θ_2 is massless.

Note that θ_2 is the component of Θ tangential to the circle of minima of V, i.e. in the direction in which the symmetry operation of rotation moves the point Φ. This appearance of a massless particle in connection with a continuous hidden symmetry is a general phenomenon, as is shown by

●**7.9   Goldstone's theorem.** Consider the Lagrangian

$$\mathscr{L} = \tfrac{1}{2}(\partial_\mu\Phi)^T(\partial^\mu\Phi) - V(\Phi) \tag{7.245}$$

where Φ is an n-component column vector of real fields subject to transformations $\Phi \to \rho(Q)\Phi$ by an orthogonal representation ρ of a Lie group G, and $V: \mathbb{R}^n \to \mathbb{R}$ is a function which is invariant under these transformations. Suppose V takes its minimum value on a k-dimensional manifold $M \subset \mathbb{R}^n$, any two points of which are connected by a transformation in G, and let $\Phi_0 \in M$. Then the field $\Theta(x) = \Phi(x) - \Phi_0$ creates n interacting particles of which k are massless.

The resulting theory has an overt symmetry group H which is a subgroup of G with dimension dim $G - m$.

Proof. Since V has a minimum at Φ_0, all its partial derivatives vanish there. Hence Taylor's theorem gives the expansion of V about Φ_0 as

$$V(\Phi) = V(\Phi_0) + \tfrac{1}{2}\Theta^T V''(\Phi_0)\Theta + W(\Theta) \tag{7.246}$$

where $V''(\Phi_0)$ is the matrix of second derivatives of V at Φ_0, and $W(\Theta)$ contains terms of third and higher orders. Since $V''(\Phi_0)$ is symmetric, it has n orthogonal eigenvectors u_i; hence

$$\mathscr{L} = \frac{1}{2}\sum_{i=1}^n \{(\partial_\mu\theta_i)(\partial^\mu\theta_i) - m_i{}^2\theta_i{}^2\} - W(\Theta) \tag{7.247}$$

where $\theta_i = \Theta^T u_i$ and $m_i{}^2$ are the eigenvalues of $V''(\Phi_0)$, which are non-negative since V has a minimum at Φ_0.

Since the minimum set M has dimension k, there are k curves $\Phi(s)$ through Φ_0, with independent tangent vectors at Φ_0, on which $V(\Phi(s))$ is constant. We can choose the parameter s so that $\Phi''(s) = 0$ at Φ_0. Then differentiating $V(\Phi(s))$ twice with respect to s gives

$$\Phi'(s)^T V''(\Phi_0)\Phi'(s) = 0. \tag{7.248}$$

Since $V''(\Phi_0)$ is positive semi-definite, each $\Phi'(s)$ must be an eigenvector of

$V''(\Phi_0)$ with eigenvalue 0. Thus k of the eigenvalues $m_i{}^2$ are 0, and so k of the particles created by the fields θ_i are massless.

Let H be the subgroup of G which keeps Φ_0 fixed:

$$H = \{Q \in G : \rho(Q)\Phi_0 = \Phi_0\}. \tag{7.249}$$

Then for each $Q \in H$ we can define a unitary operator $U(Q)$ to act on the many-particle space constructed on the vacuum $|0\rangle$ for which $\langle 0|\Phi(x)|0\rangle = \Phi_0$ and to satisfy

$$U(Q)|0\rangle = |0\rangle, \quad \text{all } Q \in H, \tag{7.250}$$

$$U(Q)\Theta(x)U(Q)^{-1} = \rho(Q)\Theta(x). \tag{7.251}$$

Since these operators leave the vacuum invariant, H is an overt symmetry of the theory.

To see that dim $H =$ dim $G -$ dim M, take a neighbourhood N of Φ_0 in M and for each $\Phi \in N$ choose an element $Q_\Phi \in G$ such that $\rho(Q_\Phi)\Phi_0 = \Phi$. Then if $Q \in G$ is close enough to the identity it can be written as $Q = Q_\Phi R$ with $\Phi \in N$, $R \in H$. Thus locally G is like $M \times H$, so dim $G =$ dim $M +$ dim H. ■

●7.9 refers only to real fields, but complex fields are included by treating the real and imaginary parts separately.

The k massless particles are called **Goldstone bosons**.

Hidden local symmetry:
the Higgs mechanism

In the last two sections we have seen two attractive theoretical ideas which are promising as formats for a theory of fundamental forces: gauge theories, which give a renormalisable quantum field theory, and hidden symmetry, which offers a way of having a symmetry in a theory while being realistic by not having exact symmetry in its observable consequences. Both ideas, however, suffer from the same drawback in comparison to reality: they both require the existence of massless bosons which have not been found. We shall now see that when these two ideas are put together the two kinds of massless boson magically disappear.

●**7.10** **The Higgs mechanism.** If the theory of ●7.9 is made invariant under local transformations $\Phi(x) \to \rho(Q(x))\Phi(x)$, then there is a gauge in which the fields of the Goldstone bosons vanish and an equal number of the gauge fields create massive particles.

Proof. We make the Lagrangian (7.245) gauge-invariant by introducing l gauge fields $A_{i\mu}$ $(i = 1, \dots, l)$, where $l =$ dim G, and the corresponding $n \times n$ field matrix $\rho(A)_\mu$ (see (7.218)), and by replacing the derivative $\partial_\mu\Phi$ by the covariant derivative (7.220). This gives

$$\mathscr{L} = -(4\lambda)^{-1} \operatorname{tr}(F_{\mu\nu}F^{\mu\nu}) + \tfrac{1}{2}(D_\mu\Phi)^\mathsf{T}(D^\mu\Phi) - V(\Phi). \tag{7.252}$$

As in ●7.9, let Φ_0 be a point at which V has a minimum, and let $\Theta = \Phi - \Phi_0$; then

$$D_\mu\Phi = D_\mu\Theta + ig\rho(A)_\mu\Phi_0, \tag{7.253}$$

so that in terms of Θ the Lagrangian becomes

$$\mathcal{L} = -(4\lambda)^{-1} \operatorname{tr} (F_{\mu\nu} F^{\mu\nu}) + \tfrac{1}{2}(D_{\mu}\Theta)^{\mathrm{T}}(D^{\mu}\Theta) - V(\Phi_0 + \Theta)$$
$$+ ig(D^{\mu}\Theta)^{\mathrm{T}}(\rho(A)_{\mu}\Phi_0) - g^2(\rho(A)^{\mu}\Phi_0)^{\mathrm{T}}(\rho(A)_{\mu}\Phi_0), \qquad (7.254)$$

using the fact that the generators X_i, and therefore the field matrix $\rho(A)_{\mu}$, are antisymmetric since the representation ρ is orthogonal.

In a gauge theory we can always adjust the values of the field $\Phi(x)$ by means of a gauge transformation which takes it to $\rho(Q(x))\Phi(x)$. If the group G is compact, we can choose $Q(x)^{-1}$ to be the maximum of the function $f_x : G \to \mathbb{R}$ where

$$f_x(Q) = \Theta(x)^{\mathrm{T}} \rho(Q)\Phi_0. \qquad (7.255)$$

Let $R(s)$ be a set of elements of G depending on a real parameter s, with $R(0)$ being the identity of G; then $Q(s) = Q(x)^{-1} R(s)$ passes through the maximum of f_x when $s = 0$, so that

$$\frac{d}{ds} f_x(Q(s)) = \frac{d}{ds} [\Theta(x)^{\mathrm{T}} \rho(Q(x)^{-1} R(s))] \Phi_0 = 0. \qquad (7.256)$$

But since ρ is an orthogonal representation, this is

$$[\rho(Q(x))\Theta(x)]^{\mathrm{T}} \frac{d}{ds} [\rho(R(s))\Phi_0]_{s=0} = 0. \qquad (7.257)$$

Now in the conditions of ●7.9, $\Phi(s) = \rho(R(s))\Phi_0$ is a curve of points on the minimum set M of $V(\Phi)$, and any such curve can be obtained in this way; hence $\Phi'(0)$ is a null vector of $V''(\Phi_0)$, and the gauge-transformed field $\Theta^*(x) = \rho(Q(x))\Theta(x)$ satisfies

$$(\Theta^*)^{\mathrm{T}} u_i = 0 \qquad (7.258)$$

where u_i are the k eigenvectors of $V''(\phi_0)$ with zero eigenvalue whose existence was demonstrated in ●7.9. But the combinations $\theta_i = \Theta^{\mathrm{T}} u_i$ were the fields of the Goldstone bosons; thus there is a gauge in which these fields vanish.

If $\Theta^*(x)$ satisfies (7.258) for all x, it is normal to the manifold M at the point Φ_0. Since M is invariant under the transformations $\rho(Q)$ and these are orthogonal, $\rho(Q)\Theta^*$ is also normal to M and therefore so is $X_i\Theta^*$ where X_i is a generator of the representation ρ, since a generator is obtained by differentiating $\rho(Q(s))$ for a set of group elements $Q(s)$. Also $X_i\Phi_0$ is a tangent vector to M, since $\rho(Q(s))\Phi_0$ is a curve in M. It follows that in the gauge in which the Goldstone fields vanish,

$$(D^{\mu}\Theta^*)^{\mathrm{T}} \rho(A)_{\mu}\Phi_0 = 0. \qquad (7.259)$$

Now let us examine the last term of the Lagrangian (7.254). We can find a basis $X_1, \ldots, X_l$ such that the last $l-k$ elements form a basis for the Lie algebra of the subgroup H; these satisfy $X_i\Phi_0 = 0$ since $\rho(Q)\Phi_0 = \Phi_0$ if $Q \in H$. But $X_i\Phi_0$ is non-zero for $i = 1, \ldots, k$; hence this term becomes

$$-\sum_{i=1}^{k} m_i^2 A_{i\mu} A_i^{\mu} \quad \text{where } m_i^2 = g^2(X_i\Phi_0)^{\mathrm{T}}(X_i\Phi_0), \qquad (7.260)$$

assuming that the X_i have been chosen so that $\Phi_0{}^T X_i{}^T X_j \Phi_0 = 0$ for $i \neq j$. This gives non-zero masses to the k gauge fields corresponding to the k Goldstone bosons. ∎

In a full treatment of gauge field theory it is shown that the time components of the 4-vector fields A_μ can be eliminated by the remaining freedom to make gauge transformations. This leaves each field with three components, which is the right number for a massive spin-1 particle. The full freedom of gauge transformation would make it possible to eliminate another component, leaving two components, the right number (helicity $= \pm 1$) for a massless particle. By insisting on a gauge which satisfies (7.258) we have renounced this possibility, and the fields retain three components. In the course of the gauge transformation which works the Higgs mechanism, as Coleman has put it, the gauge bosons eat the Goldstone bosons and become heavy.

7.6. Quantum flavourdynamics

We are now in a position to describe the basic structure of the Salam–Weinberg theory of the electroweak force. It is a gauge theory, with spontaneous symmetry breaking, based on the group

$$G_{ew} = SU(2) \times U(1) \tag{7.261}$$

in which the subgroup $SU(2)$ is the group of weak isospin transformations, and $U(1)$ (the multiplicative group of complex numbers with unit modulus) is the group of weak hypercharge transformations. The fields on which the gauge transformations act are the fermion fields of the leptons and quarks, as well as the Higgs field which must be described separately.

The classification of quark and lepton states by weak isospin was given in Table 6.5. The left-handed helicity states form a number of doublets, the right-handed states are singlets. Now if ψ is any Dirac field, the fields which destroy left-handed and right-handed states are, according to (7.144),

$$\psi_L = \tfrac{1}{2}(1 + \gamma_5)\psi \quad \text{and} \quad \psi_R = \tfrac{1}{2}(1 - \gamma_5)\psi.$$

For a weak isospin doublet like the quarks (u, d), therefore, we can form the fields

$$\Psi_L = \begin{bmatrix} \tfrac{1}{2}(1 + \gamma_5)\psi_u \\ \tfrac{1}{2}(1 + \gamma_5)\psi_d \end{bmatrix}, \quad \begin{aligned} \psi_{uR} &= \tfrac{1}{2}(1 - \gamma_5)\psi_u, \\ \psi_{dR} &= \tfrac{1}{2}(1 - \gamma_5)\psi_d, \end{aligned} \tag{7.262}$$

which are subject to the local weak isospin transformations

$$\Psi_L \to e^{-\frac{1}{2}ig\mathbf{a}(x)\cdot\boldsymbol{\tau}}\Psi_L, \quad \psi_{uR} \to \psi_{uR}, \quad \psi_{dR} \to \psi_{dR} \tag{7.263}$$

where g is a coupling constant.

The direct product structure of the group G_{ew} makes it possible for the two subgroups to have different coupling constants; thus weak hypercharge phase transformations can be defined as

$$\psi_{L/R} \to e^{-\frac{1}{2}ig'y\alpha(x)}\psi_{L/R} \tag{7.264}$$

for a left-handed or right-handed field $\psi_{L/R}$ with weak hypercharge y (which will have different values for ψ_L and ψ_R: see Table 6.5). Note that the product

$g'y$ (coupling constant × multiple of $\frac{1}{6}$) plays the same role as the electric charge in the electromagnetic phase transformation (7.197), the electric charge of a quark or lepton being the product of a coupling constant e and a multiple of $\frac{1}{3}$.

The group G_{ew} has four generators, so to form a gauge theory we need four gauge fields: the components of a weak isovector $\mathbf{W}_\mu$, corresponding to the generators of SU(2), and a weak isoscalar B_μ for the generator of U(1). The covariant derivative is then

$$D_\mu\psi_L = (\partial_\mu + \tfrac{1}{2}ig\boldsymbol{\tau}\cdot\mathbf{W}_\mu + \tfrac{1}{2}ig'yB_\mu)\Psi_L \quad \text{for a doublet } \Psi_L,$$

$$D_\mu\psi_R = (\partial_\mu + \tfrac{1}{2}ig'yB_\mu)\psi_R \qquad\qquad \text{for a singlet } \psi_R. \tag{7.265}$$

The fields $\mathbf{W}_\mu$ and B_μ can be put together in a 2×2 matrix

$$W_\mu = \mathbf{W}_\mu\cdot\boldsymbol{\tau} + B_\mu\mathbf{1}, \tag{7.266}$$

from which we can form the field matrix

$$F_{\mu\nu} = \partial_\mu W_\nu - \partial_\nu W_\mu + ig[W_\mu, W_\nu]. \tag{7.267}$$

A gauge-invariant Lagrangian for the gauge fields and fermions alone is

$$\mathscr{L}_0 = -\tfrac{1}{8}\,\mathrm{tr}\,(F_{\mu\nu}F^{\mu\nu}) + \sum i\bar{\Psi}_L\gamma^\mu\overrightarrow{D}_\mu\Psi_L + \sum i\bar{\psi}_R\gamma^\mu\overrightarrow{D}_\mu\psi_R \tag{7.268}$$

where the sums extend over all left-handed doublets Ψ_L and all right-handed singlets ψ_R. Since the symmetry is unbroken, the gauge fields are all massless. Moreover, invariance under transformations which act in different ways on the left-handed and right-handed components of a fermion field requires that the fermions should also be massless; for the Lagrangian term describing the mass of a fermion is

$$m\bar{\psi}\psi = m(\bar{\psi}_L + \bar{\psi}_R)(\psi_L + \psi_R) = m(\bar{\psi}_L\psi_R + \bar{\psi}_R\psi_L) \tag{7.269}$$

using (7.147), and since ψ_R is a scalar under weak isospin transformations while ψ_L is a member of a doublet, this is not invariant unless $m=0$.

The Higgs mechanism produces masses for the fermions as well as the gauge fields. There are four Higgs fields $\phi_1,\ldots,\phi_4$ forming the real and imaginary parts of a two-component complex vector

$$\Phi = \begin{bmatrix} \phi_1 + i\phi_2 \\ \phi_3 + i\phi_4 \end{bmatrix} \tag{7.270}$$

which responds to the SU(2) × U(1) transformations (7.263)–(7.264) by

$$\Phi \to e^{-\frac{1}{2}ig\mathbf{a}(x)\cdot\boldsymbol{\tau}}\Phi, \quad \Phi \to e^{-ig'\alpha(x)}\Phi. \tag{7.271}$$

These are orthogonal transformations of the real vector $(\phi_1, \phi_2, \phi_3, \phi_4)$ with the usual inner product, which can be written in terms of Φ as

$$\langle\Phi_1, \Phi_2\rangle = \mathrm{Re}\,(\Phi_1{}^\dagger\Phi_2). \tag{7.272}$$

The covariant derivative of Φ is

$$D_\mu\Phi = (\partial_\mu + ig\mathbf{W}_\mu\cdot\boldsymbol{\tau} + ig'B_\mu)\Phi. \tag{7.273}$$

The Higgs potential $V(\Phi)$ can only have quadratic and quartic terms in a renormalisable theory. It is taken to be a function of $\Phi^\dagger\Phi = \sum \phi_i{}^2$ which has a minimum at a non-zero value v^2 of its argument, so that the minimum manifold in Φ-space is the sphere $\Phi^\dagger\Phi = v^2$. We can choose the reference point Φ_0 on this manifold to be $(0, v)^\mathrm{T}$ and expand the fields as

$$\Phi = \Phi_0 + \Theta = \begin{bmatrix} \theta_1 + i\theta_2 \\ v + \chi + i\theta_4 \end{bmatrix} \tag{7.274}$$

– we use χ instead of θ_3 because this is the component which cannot be generated by the action of the group $\mathrm{SU}(2) \times \mathrm{U}(1)$, and which will remain in the theory as the field of a particle. The Higgs mechanism ●7.10 makes the other components disappear and gives masses to the corresponding components of the gauge fields; the mass term (the last term of (7.254)) is

$$-\mathrm{Re}\,[\Phi_{0\mu}{}^\dagger\Phi_0{}^\mu] \quad \text{where} \quad \Phi_{0\mu} = D_\mu\Phi_0 = (ig\mathbf{W}_\mu\cdot\boldsymbol{\tau} + ig'B_\mu)\Phi_0$$
$$= v^2 g^2(W_{1\mu}W_1{}^\mu + W_{2\mu}W_2{}^\mu) + v^2\gamma^2 Z_\mu Z^\mu \tag{7.275}$$

where $\gamma^2 = g^2 + g'^2$, $Z_\mu = W_{3\mu}\cos\theta_\mathrm{W} - B_\mu\sin\theta_\mathrm{W}$, and $\tan\theta_\mathrm{W} = g'/g$ as in (6.147)–(6.148). Thus the Higgs mechanism produces the massive bosons $\mathrm{W}^\pm$ and Z^0 with masses in the ratio

$$\frac{m_\mathrm{W}}{m_\mathrm{Z}} = \frac{g}{\gamma} = \cos\theta_\mathrm{W}. \tag{7.276}$$

The last Goldstone boson, which remains massless, is given by the gauge field $A_\mu = W_{3\mu}\sin\theta_\mathrm{W} + B_\mu\cos\theta_\mathrm{W}$. It is associated with the subgroup of $\mathrm{SU}(2) \times \mathrm{U}(1)$ keeping Φ_0 fixed, which consists of the transformations

$$\Phi \to \begin{bmatrix} e^{i\theta} & 0 \\ 0 & 1 \end{bmatrix}\Phi = \exp\left[\tfrac{1}{2}i\theta(1 + \tau_3)\right]\Phi \tag{7.277}$$

and is therefore a one-dimensional subgroup with generator $\tfrac{1}{2}(1 + \tau_3)$.

The covariant derivatives of the fermion fields are

$$D_\mu\Psi_\mathrm{L} = \{\partial_\mu + ig(W_1\tau_1 + W_2\tau_2) + \tfrac{1}{2}ieZ_\mu(\cot\theta_\mathrm{W}\tau_3 - y\tan\theta_\mathrm{W})$$
$$+ \tfrac{1}{2}ieA_\mu(y + \tau_3)\}\Psi_\mathrm{L} \tag{7.278}$$

$$D_\mu\psi_\mathrm{R} = \{\partial_\mu - \tfrac{1}{2}iye\tan\theta_\mathrm{W}\,Z_\mu + \tfrac{1}{2}iyeA_\mu\}\psi_\mathrm{R}$$

where $e = g\sin\theta_\mathrm{W}$. The formula for the doublet Ψ_L applies also to the Higgs fields Φ with $y = 1$; this shows the association of the massless gauge field A_μ with the generator $\tfrac{1}{2}(1 + \tau_3)$ of (7.277). Since the electric charge in units of e is $\tfrac{1}{2}y + i_3$, where i_3 is the eigenvalue of $\tfrac{1}{2}\tau_3$ for a member of a doublet and 0 for a singlet, (7.278) shows that the massless gauge field A_μ is the electromagnetic field.

Finally, the fermion masses arise from an interaction between the fermion fields and the Higgs field. Consider the leptons first. In each family there is a massive lepton l and a massless left-handed neutrino v, giving three fields ψ^l_L, ψ^l_R, ψ^v_L which are classified by weak isospin into a doublet Ψ_L with $y = -1$ and a singlet with $y = -2$ (see Table 6.5). From these, together with the Higgs doublet Φ with $y = 1$, we can form a Lagrangian term

$$\mathcal{L}_l = \alpha_l(\bar{\psi}_\mathrm{R}\Phi^\dagger\Psi_\mathrm{L} + \bar{\Psi}_\mathrm{L}\Phi\psi_\mathrm{R}) \tag{7.279}$$

which is invariant under $SU(2)_w \times U(1)_y$ transformations. Expanding Φ about the minimum point Φ_0 as in (7.274) and choosing the gauge in which $\theta_1 = \theta_2 = \theta_4 = 0$, we obtain (using (7.269))

$$\mathscr{L}_l = m_l \bar{\psi}_l \psi_l + \alpha_l \chi \bar{\psi}_l \psi_l \tag{7.280}$$

where

$$m_l = \alpha_l v.$$

Thus $\mathscr{L}$ gives a mass to the charged lepton while keeping the neutrino massless, and also describes an interaction between the charged lepton and the Higgs particle whose field is χ.

The quarks in each generation form a left-handed doublet Ψ_L with $y = \frac{1}{3}$ and two singlets ψ^+_R, ψ^-_R with $y = \frac{4}{3}$ and $y = -\frac{2}{3}$ respectively. The difference between the weak hypercharges of Ψ_L and ψ_R is the same as in the case of the leptons, and the same method can be used to obtain mass terms. This gives masses to the u, c and t quarks. A mass term for the d, s and b quarks can be constructed as

$$\mathscr{L}_q^+ = \alpha_q^+ \{ \bar{\psi}^+_R \det(\Phi, \Psi_L) + \det(\Phi^\dagger, \tilde{\Psi}_L) \psi^+_R \} \tag{7.281}$$

where $\det(\Phi, \Psi_L)$ is the determinant of the 2×2 matrix whose columns are Φ and Ψ_L, and $\det(\Phi^\dagger, \tilde{\Psi}_L)$ similarly has rows $\Phi^\dagger$ and $\tilde{\Psi}_L$.

Thus the full electroweak Lagrangian is

$$\mathscr{L}_{ew} = \mathscr{L}_0 + \tfrac{1}{2}(D_\mu \Phi^\dagger)(D^\mu \Phi) + V(\Phi) + \sum \mathscr{L}_l + \sum \mathscr{L}_q^- + \sum \mathscr{L}_q^+ \tag{7.282}$$

where the sums extend over all families. $\mathscr{L}_0$ is the gauge-invariant Lagrangian (7.268), containing the kinetic terms for the gauge bosons and the fermions and the gauge interactions between them. The second term contains the kinetic term for the Higgs field, the mass terms for the gauge fields, and the interaction between them. The third term $V(\Phi)$ contains the mass of the Higgs boson and its self-interaction. The last three terms contain the masses of the quarks and leptons and their interactions with the Higgs boson ($\mathscr{L}_q^-$ having the same structure as $\mathscr{L}_l$).

The undetermined parameters in the theory are the electromagnetic coupling constant e, the Weinberg angle θ_W, the Higgs vacuum expectation value v, and the masses of the Higgs boson and the fermions. All other parameters can be expressed in terms of these as follows:

W-fermion coupling constant:	$g = e \operatorname{cosec} \theta_W$
Z-fermion coupling constant:	$g' \sin \theta_W = e \tan \theta_W$
W mass:	$m_W = ev \operatorname{cosec} \theta_W$
Z mass:	$m_Z = 2ev \operatorname{cosec} 2\theta_W$
Higgs-fermion coupling constant:	$= v^{-1} \times$ mass of fermion.

Further reading   The reader who wants to study quantum field theory seriously will find an introduction in Mandl & Shaw 1984 or Ryder 1985. A different approach, based on Feynman's formulation, is adopted in Ramond 1981. Itzykson & Zuber 1980 is a comprehensive treatise. For gauge theories see Aitchison 1982, and for the applications to quantum chromo- and flavourdynamics Leader & Predazzi 1982. A complementary account, not using the full apparatus of quantum field theory, is given by Aitchison & Hey 1982. The lattice approach to calculations in quantum chromodynamics is treated in Creutz 1983; there are elementary accounts by Rebbi (1983) and Wallace (1983). Gauge theories have interesting mathematical aspects (the theory of fibre bundles) which have been described by Atiyah (1978); see also Bernstein & Phillips 1981. On the cosmological relevance of the Higgs field see Guth & Steinhardt 1984.

Problems on Chapter 7

1. Prove (7.19).

2. Prove (7.58) by starting with (7.51) and writing each ψ_i as a Fourier transform.

3. Show that the Lorentz group has generators $M_{\mu\nu} = -M_{\nu\mu}$ satisfying
   $$[M_{\mu\nu}, M_{\rho\sigma}] = g_{\mu\rho}M_{\nu\sigma} - g_{\nu\rho}M_{\mu\sigma} - g_{\mu\sigma}M_{\nu\rho} + g_{\nu\sigma}M_{\mu\rho}$$
   and that these commutation relations are satisfied by $\sigma_{\mu\nu} = \frac{1}{4}[\gamma_\mu, \gamma_\nu]$.

4. Show that the Poincaré group (consisting of the Lorentz group together with translations in space and time) has generators $M_{\mu\nu}, P_\mu$, where $M_{\mu\nu}$ are as in q.3 and $[M_{\mu\nu}, P_\rho] = g_{\mu\rho}P_\nu - g_{\nu\rho}P_\mu$, $[P_\mu, P_\nu] = 0$.

   The **Poincaré superalgebra** consists of the Lie algebra L_0 of the Poincaré group, as above, together with an eight-dimensional space L_1 with basis ψ_α, $\bar{\psi}^\alpha$ ($\alpha = 1, \ldots, 4$). Brackets are defined as follows:
   $$[M_{\mu\nu}, \psi_\alpha] = (\sigma_{\mu\nu})_\alpha{}^\beta \psi_\beta, \quad [M_{\mu\nu}, \bar{\psi}^\alpha] = -\bar{\psi}^\beta(\sigma_{\mu\nu})_\beta{}^\alpha, \quad [P_\mu, \psi_\alpha] = 0 = [P_\mu, \bar{\psi}^\alpha],$$
   $$\{\psi_\alpha, \psi_\beta\} = 0 = \{\bar{\psi}^\alpha, \bar{\psi}^\beta\}, \quad \{\psi_\alpha, \bar{\psi}^\beta\} = (\gamma^\mu)_\alpha{}^\beta P_\mu.$$
   Show that these satisfy the conditions (6.161) to be a Lie superalgebra.

5. Show that if $\psi(x)$ is a Dirac spinor, $\bar{\psi}\gamma^\mu \partial_\mu \psi$ is a Lorentz scalar.

6. For any 4-vector p^μ, let $\Pi(p)$ be the 4×4 (spinor) matrix $\Pi(p) = \frac{1}{2}(1 - m^{-1}\gamma^\mu p_\mu)$. Show that $\Pi(p)^2 = (p^2 - m^2)/(4m^2) + \Pi(p)$, $\Pi(p)\Pi(-p) = (p^2 - m^2)/m^2$, and $\Pi(p) + \Pi(-p) = 1$.

   Deduce that any spinor can be written as $u(p) + v(p)$ where $u(p)e^{-ip \cdot x}$ and $v(p)e^{ip \cdot x}$ are solutions of the Dirac equation.

   Show that $\Pi(p)$ commutes with $S(\Lambda^{-1})s_i S(\Lambda)$ where s_i are the spin matrices of (7.107) and Λ is a Lorentz transformation such that $\Lambda^\mu{}_\nu p^\nu = (m, \mathbf{0})$. Deduce the existence of the basis $u_\pm(\mathbf{p})$, $v_\pm(\mathbf{p})$ of (7.149).

7. Find the spinor matrix $S(\Lambda)$ representing the Lorentz transformation Λ consisting of a boost in the direction $\mathbf{n}$ with velocity $\tanh \lambda$. By applying such a transformation to the eigenspinors of s_3, find explicit formulae for the basic spinors $u_{\pm}(\mathbf{p})$, $v_{\pm}(\mathbf{p})$.

8. Show that if the creation operators for the Klein–Gordon field are taken to be $E^{\frac{1}{2}}a^{\dagger}(\mathbf{k})$ and $E^{\frac{1}{2}}b^{\dagger}(\mathbf{k})$, the harmonic oscillator Hamiltonian (after normal ordering) is

$$H = \frac{1}{2} \int (m^2 \phi^{\dagger}\phi + \dot{\phi}^{\dagger}\dot{\phi} + \nabla\phi^{\dagger}\cdot\nabla\phi)\, d^3\mathbf{r},$$

and show that this is the same as the Hamiltonian given by applying Noether's theorem to the Lagrangian (7.166).

9. Let $S[\theta]$ denote the integral (7.162), calculated with a function $\theta(\mathbf{r}, t)$, and suppose that to first order in ε, $S[\theta + \varepsilon\eta] = S[\theta]$ for all continuous functions $\eta(\mathbf{r}, t)$. Show that θ satisfies the Euler–Lagrange equations (7.164).

10. Prove Noether's theorem for a Lagrangian system with a finite number of degrees of freedom (see p. 315).

11. Let $\mathscr{L}(\phi, A_{\mu}, \partial_{\mu}\phi, \partial_{\mu}A_{\nu})$ be a Lagrangian density which depends on a scalar field ϕ, a 4-vector field A_{μ}, and their derivatives, but not on x explicitly, and which is Lorentz-invariant in the sense that $\mathscr{L}(\phi, A_{\mu}', \partial_{\mu}'\phi, \partial_{\mu}'A_{\nu}') = \mathscr{L}(\phi, A_{\mu}, \partial_{\mu}\phi, \partial_{\mu}A_{\nu})$ where $A_{\mu}' = \Lambda_{\mu}^{\nu}A_{\nu}$, $\partial_{\mu}'\phi = \Lambda_{\mu}^{\nu}\partial_{\nu}\phi$, $\partial_{\mu}'A_{\nu}' = \Lambda_{\mu}^{\rho}\Lambda_{\nu}^{\sigma}\partial_{\rho}A_{\sigma}$, for any Lorentz transformation Λ. By putting $\Lambda = e^{\lambda\Omega}$ and differentiating with respect to λ, show that the field equations imply that $\partial_{\mu}S^{\mu\nu\rho} = 0$ where

$$S^{\mu\nu\rho} = A^{\rho} \frac{\partial \mathscr{L}}{\partial(\partial_{\mu}A_{\nu})} - A^{\nu} \frac{\partial \mathscr{L}}{\partial(\partial_{\mu}A_{\rho})} + x^{\nu}T^{\mu\rho} - x^{\rho}T^{\mu\nu}.$$

12. Show that the equation of motion (7.188) for a charged particle in an electromagnetic field deriving from given potential functions $\phi(\mathbf{r}, t)$, $\mathbf{A}(\mathbf{r}, t)$ can be obtained as Hamilton's equations from the Hamiltonian (7.189).

13. Show that the effect of translation operators on a quantum field $\psi(x)$ is given by

$$U(T_{\mathbf{a}})\psi(\mathbf{r}, t)U(T_{\mathbf{a}})^{-1} = \psi(\mathbf{r} + \mathbf{a}, t)$$

and explain the difference between this and the action (3.75) on a wave function.

14. Write down the equation which replaces the Klein–Gordon equation for a charged relativistic particle in an electromagnetic field, and show how the Klein–Gordon Lagrangian (7.166) can be adapted to this situation. Show that this adapted Lagrangian is invariant under local electromagnetic phase transformations. Find the electric current 4-vector.

Appendix I

3-vector and 4-vector algebra

3-vectors The components of a vector $\mathbf{u}$, with respect to a particular set of axes, are denoted by u_i; the indefinite index i stands for 1, 2 or 3 (if the components of $\mathbf{u}$ are written as (u_1, u_2, u_3)) or for x, y or z (if the component of $\mathbf{u}$ are written as (u_x, u_y, u_z)). The **summation convention** is that if a term is written as a product of such components in which the same indefinite index occurs twice, it is to be understood as a sum over all values of the repeated index. Thus the scalar product between two vectors is

$$\mathbf{u} \cdot \mathbf{v} = u_i v_i \quad \text{which means} \quad \sum_{i=1}^{3} u_i v_i. \tag{I.1}$$

If two (or more) indices are repeated in a term, it is to be understood as a double (or multiple) sum.

Symbols with two or more indices denote the components of **tensors**. Normally these, like vectors, have different components with respect to different coordinate systems. Two special tensors whose components are the same for all coordinate systems are

$$\delta_{ij} = \begin{cases} 1 & \text{if } i = j, \\ 0 & \text{if } i \neq j. \end{cases} \tag{I.2}$$

and

$$\varepsilon_{ijk} = \begin{cases} 0 & \text{if any two of } i, j, k \text{ are equal,} \\ 1 & \text{if } (i, j, k) = (1, 2, 3), (2, 3, 1) \text{ or } (3, 1, 2), \\ -1 & \text{if } (i, j, k) = (2, 1, 3), (1, 3, 2) \text{ or } (3, 2, 1) \end{cases} \tag{I.3}$$

i.e. ε_{ijk} is the signature of the permutation that takes $(1, 2, 3)$ to (i, j, k). It is completely specified by the statements that it is totally antisymmetric in i, j, k, and that $\varepsilon_{123} = 1$. The summation convention applies to terms containing tensors; thus the equation

$$t_{ij} = \varepsilon_{ijk} u_k \tag{I.4}$$

stands for the set of six equations

$$\begin{array}{cccc} t_{12} = u_3, & t_{23} = u_1, & t_{31} = u_2 \\ t_{21} = -u_3, & t_{32} = -u_1, & t_{13} = -u_2 \end{array} \Bigg\}. \tag{I.5}$$

The tensors δ_{ij} and ε_{ijk} are related by the identity

$$\varepsilon_{ijk}\varepsilon_{klm} = \delta_{il}\delta_{jm} - \delta_{im}\delta_{jl}, \tag{I.6}$$

which is equivalent to the vector identity

$$\mathbf{a} \times (\mathbf{b} \times \mathbf{c}) = (\mathbf{a} \cdot \mathbf{c})\mathbf{b} - (\mathbf{a} \cdot \mathbf{b})\mathbf{c}. \tag{I.7}$$

4-vectors   (only required in Chapter 7)

A **contravariant 4-vector** x consists of a pair of physical quantities, a scalar s and a 3-vector **u** which can be assigned definite values relative to a frame of reference, in such a way that the values assigned in different frames of reference are related by the Lorentz transformation: if F and F' are two frames of reference with a common space-time origin, F' moving relative to F with velocity v in the direction of the unit vector **n**, and if x has values $(s, \mathbf{u})$ relative to F and $(s', \mathbf{u}')$ relative to F', then (taking $c = 1$)

$$s' = \gamma(s - v\mathbf{u} \cdot \mathbf{n}), \quad \mathbf{u}' \cdot \mathbf{n} = \gamma(\mathbf{u} \cdot \mathbf{n} - vs), \quad \mathbf{u}_\perp{}' = \mathbf{u}_\perp \tag{I.8}$$

where $\gamma = (1 - v^2)^{-\frac{1}{2}}$ and $\mathbf{u}_\perp$ and $\mathbf{u}_\perp{}'$ are the components of **u** and **u**' perpendicular to **n**. For example, the coordinates $(t, \mathbf{r})$ of an event constitute a contravariant 4-vector.

The four components of a contravariant 4-vector x are denoted by x^μ ($\mu = 0, 1, 2, 3$): $x^0 = s$, $x^i = u_i$. If y is another contravariant 4-vector, consisting of a scalar t and a 3-vector **v**, then the inner product

$$x \cdot y = st - \mathbf{u} \cdot \mathbf{v} \tag{I.9}$$

is a **Lorentz scalar**, i.e. it has the same value in every frame of reference. This can be written as

$$x \cdot y = g_{\mu\nu} x^\mu y^\nu \tag{I.10}$$

(using the summation convention), where $g_{\mu\nu} = 0$ if $\mu \neq \nu$, $g_{00} = 1$ and $g_{11} = g_{22} = g_{33} = -1$.

A **covariant 4-vector** consists of a scalar s and a 3-vector **w** such that $(s, -\mathbf{w})$ constitute a contravariant 4-vector. The four components of a covariant 4-vector are denoted by a subscript index, as x_μ. If x^μ are the components of a contravariant 4-vector, the components of the corresponding covariant 4-vector are

$$x_\mu = g_{\mu\nu} x^\nu. \tag{I.11}$$

For brevity, we say 'x^μ is a contravariant 4-vector' and 'x_μ is a covariant 4-vector'.

If x^μ is a contravariant 4-vector and y_μ is a covariant one, $x^\mu y_\mu$ is a Lorentz scalar. In general, to ensure that equations are valid in all inertial frames of reference, the summation convention should only be applied to pairs of indices in which one index is in the upper position and the other is in the lower position.

The inverse of (I.11) is

$$x^\mu = g^{\mu\nu} x_\nu \tag{I.12}$$

where $g^{\mu\nu}$ has the same numerical values as $g_{\mu\nu}$. (I.11)–(I.12) exemplify the process of raising and lowering indices which can be applied to a quantity with any number of indices, some in the upper position and some in the lower. The effect is that raising or lowering the index 0 leaves the component unchanged, while raising or lowering an index i ($= 1, 2$ or 3) changes the sign of the component. In particular, raising an index of $g_{\mu\nu}$ gives $g^\mu{}_\nu = \delta^\mu{}_\nu$ (the usual Kronecker δ, defined as in (I.2)).

The general **Lorentz transformation** is a linear transformation of 4-vectors specified by a 4×4 matrix $\Lambda^\mu{}_\nu$,

$$x^\mu \rightarrow x'^\mu = \Lambda^\mu{}_\nu x^\nu, \tag{I.13}$$

such that

$$x' \cdot y' = x \cdot y \quad \text{for all contravariant 4-vectors } x, y. \tag{I.14}$$

From (I.10) it follows that the condition on $\Lambda^\mu{}_\nu$ is

$$g_{\mu\nu} \Lambda^\mu{}_\rho \Lambda^\nu{}_\sigma = g_{\rho\sigma}, \tag{I.15}$$

i.e.

$$\Lambda^{\mu}{}_{\rho}\Lambda_{\mu\sigma} = g_{\rho\sigma}. \tag{I.16}$$

The symbol $\varepsilon_{\mu\nu\rho\sigma}$ is totally antisymmetric with $\varepsilon_{0123} = +1$; hence it vanishes if any two of μ, ν, ρ, σ are equal, and for $i, j, k = 1, 2$ or 3 we have

$$\varepsilon_{0ijk} = -\varepsilon_{i0jk} = \varepsilon_{ij0k} = -\varepsilon_{ijk0} = \varepsilon_{ijk}. \tag{I.17}$$

Appendix II

Particle properties

Name	Mass (MeV)	Lifetime (s)	Spin	Parity	Electric charge	Iso-spin	Other flavours	Main decays
Gauge bosons								
Photon γ	0	∞	1	$-$	0			
W^+	81 000	$>0.6 \times 10^{-24}$	1	$-$	$+1$			$\mathrm{e}^+ \nu_e,\ \mu^+ \nu_\mu$
W^-				$-$	-1			$\mathrm{e}^- \bar{\nu}_e,\ \mu^- \bar{\nu}_\mu$
Z^0	93 000	$>0.5 \times 10^{-24}$	1	$-$	0			$\mathrm{e}^+ \mathrm{e}^-,\ \mu^+ \mu^-$
Leptons								
Electron e^-	0.511	∞	$\frac{1}{2}$	$+$	-1			
Muon μ^-	106	2×10^{-6}	$\frac{1}{2}$	$+$	-1			$\mathrm{e}^- \bar{\nu}_e \nu_\mu$
τ	1784	3×10^{-13}	$\frac{1}{2}$	$+$	-1			$\mathrm{e}^- \bar{\nu}_e \nu_\tau, \mu^- \bar{\nu}_\mu \nu_\tau, \text{hadrons}$
Neutrinos ν_e								
ν_μ	0	∞	$\frac{1}{2}$					
ν_τ								
$+$ antiparticles								
Mesons								
π^0	135	8.3×10^{-17}	0	$-$	0	1		2γ
$\pi^\pm$	140	2.6×10^{-8}	0	$-$	± 1	1		$\mu^{\pm}(\bar{\nu}/\nu)$
$\mathrm{K}^\pm$	494	1.2×10^{-8}	0	$-$	± 1	$\frac{1}{2}$	Strangeness ± 1	$\mu^{\pm}(\bar{\nu}/\nu)_\mu, \pi^\pm \pi^0$
K_L^0	498	5.2×10^{-8}	0	$-$	0	$\frac{1}{2}$		$3\pi, \pi^\pm \mu^\mp (\bar{\nu}/\nu)_\mu, \pi^\pm \mathrm{e}^\mp (\bar{\nu}/\nu)$
K_S^0	498	8.9×10^{-11}	0	$-$	0	$\frac{1}{2}$		$\pi^+ \pi^-, \pi^0 \pi^0$
η	549	4.7×10^{-18}	0	$-$	0	0		$2\gamma, 3\pi$
$\mathrm{D}^\pm$	1869	9.2×10^{-13}	0	$-$	± 1	$\frac{1}{2}$	Charm ± 1	$\mathrm{e}^\pm (\bar{\nu}/\nu)_e + \text{hadrons}$
D^0	1865	4.4×10^{-13}	0	$-$	0	$\frac{1}{2}$	Charm ± 1	$\mathrm{K}^\pm + \text{pions}$
$\bar{\mathrm{D}}^0$								$\mathrm{K}^0 \bar{\mathrm{K}}^0 + \text{pions}$
$\mathrm{F}^\pm$	1971	1.9×10^{-13}	0	$-$	± 1	0	Strangeness ± 1 Charm ± 1	Pions
$\mathrm{B}^\pm$	5271	1.4×10^{-12}	0	$-$	± 1	$\frac{1}{2}$	Beauty ± 1	$\mathrm{D}^0 + \text{pions}$
B^0	5274	1.4×10^{-12}	0	$-$	0	$\frac{1}{2}$	Beauty ± 1	$\mathrm{e}^\pm (\bar{\nu}/\nu)_e + \text{hadrons}$
$\bar{\mathrm{B}}^0$								

Baryons

Proton p	938.3	$> 10^{32}$ yr	$\frac{1}{2}$	$+$	$+1$	$\frac{1}{2}$		
Neutron n	939.6	898	$\frac{1}{2}$	$+$	0	$\frac{1}{2}$		$pe^-\bar{v}_e$
Λ^0	1116	2.6×10^{-10}	$\frac{1}{2}$	$+$	0	0	Strangeness -1	$p\pi^-$, $n\pi^0$
Σ^+	1189	8.0×10^{-11}	$\frac{1}{2}$	$+$	$+1$	1	Strangeness -1	$p\pi^0$, $n\pi^+$
Σ^0	1192	5.8×10^{-20}	$\frac{1}{2}$	$+$	0	1	Strangeness -1	$\Lambda^0\gamma$
Σ^-	1197	1.5×10^{-10}	$\frac{1}{2}$	$+$	-1	1	Strangeness -1	$n\pi^-$
Ξ^0	1315	2.9×10^{-10}	$\frac{1}{2}$	$+$	0	$\frac{1}{2}$	Strangeness -2	$\Lambda^0\pi^0$
Ξ^-	1321	1.6×10^{-10}	$\frac{1}{2}$	$+$	-1	$\frac{1}{2}$	Strangeness -2	$\Lambda^0\pi^-$
Ω^-	1672	8.2×10^{-11}	$\frac{3}{2}$	$+$	-1	0	Strangeness -3	$\Lambda^0 K^-$, $\Xi^0\pi^-$, $\Xi^-\pi^0$
$\Lambda_c{}^+$	2282	2.3×10^{-13}	$\frac{1}{2}$	$+$	$+1$	0	Charm $+1$	Λ + pions, $pK\pi$
+ antiparticles								

Adapted from the *Review of Particle Properties* (Particle Data Group 1984)

Appendix III

Clebsch–Gordan coefficients

Each table displays the coefficients $\langle J\,M|j_1\,m_1, j_2\,m_2\rangle$ for particular values of j_1, j_2 and M.

$j_1 = \frac{1}{2}, \quad j_2 = \frac{1}{2}$

$M = 1$

$M = 0$

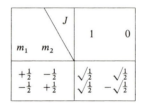

$M = 0$

$j_1 = 1, \quad j_2 = \frac{1}{2}$

$M = \frac{3}{2}$

$M = \frac{1}{2}$

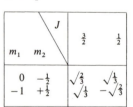

$M = -\frac{1}{2}$

$M = -\frac{3}{2}$

$j_1 = \frac{3}{2}, \quad j_2 = \frac{1}{2}$

M = 2

m_1	m_2	J 2
$\frac{3}{2}$	$\frac{1}{2}$	1

M = 1

m_1	m_2	J 2	1
$\frac{3}{2}$	$-\frac{1}{2}$	$\frac{1}{2}$	$\frac{\sqrt{3}}{2}$
$\frac{1}{2}$	$\frac{1}{2}$	$\frac{\sqrt{3}}{2}$	$-\frac{1}{2}$

M = 0

m_1	m_2	J 2	1
$\frac{1}{2}$	$-\frac{1}{2}$	$\sqrt{\frac{1}{2}}$	$\sqrt{\frac{1}{2}}$
$-\frac{1}{2}$	$\frac{1}{2}$	$\sqrt{\frac{1}{2}}$	$-\sqrt{\frac{1}{2}}$

M = −1

m_1	m_2	J 2	1
$-\frac{1}{2}$	$-\frac{1}{2}$	$\frac{\sqrt{3}}{2}$	$\frac{1}{2}$
$-\frac{3}{2}$	$\frac{1}{2}$	$\frac{1}{2}$	$-\frac{\sqrt{3}}{2}$

M = −2

m_1	m_2	J 2
$-\frac{3}{2}$	$-\frac{1}{2}$	1

$j_1 = 2, \quad j_2 = \frac{1}{2}$

M = $\frac{5}{2}$

m_1	m_2	J $\frac{5}{2}$
2	$\frac{1}{2}$	1

M = $\frac{3}{2}$

m_1	m_2	J $\frac{5}{2}$	$\frac{3}{2}$
2	$-\frac{1}{2}$	$\sqrt{\frac{1}{5}}$	$\sqrt{\frac{4}{5}}$
1	$\frac{1}{2}$	$\sqrt{\frac{4}{5}}$	$-\sqrt{\frac{1}{5}}$

M = $\frac{1}{2}$

m_1	m_2	J $\frac{5}{2}$	$\frac{3}{2}$
1	$-\frac{1}{2}$	$\sqrt{\frac{2}{5}}$	$\sqrt{\frac{3}{5}}$
0	$\frac{1}{2}$	$\sqrt{\frac{3}{5}}$	$-\sqrt{\frac{2}{5}}$

M = $-\frac{1}{2}$

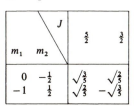

m_1	m_2	J $\frac{5}{2}$	$\frac{3}{2}$
0	$-\frac{1}{2}$	$\sqrt{\frac{3}{5}}$	$\sqrt{\frac{2}{5}}$
−1	$\frac{1}{2}$	$\sqrt{\frac{2}{5}}$	$-\sqrt{\frac{3}{5}}$

M = $-\frac{3}{2}$

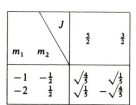

m_1	m_2	J $\frac{5}{2}$	$\frac{3}{2}$
−1	$-\frac{1}{2}$	$\sqrt{\frac{4}{5}}$	$\sqrt{\frac{1}{5}}$
−2	$\frac{1}{2}$	$\sqrt{\frac{1}{5}}$	$-\sqrt{\frac{4}{5}}$

M = $-\frac{5}{2}$

m_1	m_2	J $\frac{5}{2}$
−2	$-\frac{1}{2}$	1

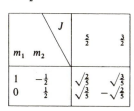

$j_1 = 1, \quad j_2 = 1$

$M = 2$

m_1	m_2	J / 2
1	1	1

$M = 1$

m_1	m_2	J	2	1
1	0		$\sqrt{\frac{1}{2}}$	$\sqrt{\frac{1}{2}}$
0	1		$\sqrt{\frac{1}{2}}$	$-\sqrt{\frac{1}{2}}$

$M = 0$

m_1	m_2	J	2	1	0
1	-1		$\sqrt{\frac{1}{6}}$	$\sqrt{\frac{1}{2}}$	$\sqrt{\frac{1}{3}}$
0	0		$\sqrt{\frac{2}{3}}$	0	$-\sqrt{\frac{1}{3}}$
-1	1		$\sqrt{\frac{1}{6}}$	$-\sqrt{\frac{1}{2}}$	$\sqrt{\frac{1}{3}}$

$M = -1$

m_1	m_2	J	2	1
0	-1		$\sqrt{\frac{1}{2}}$	$\sqrt{\frac{1}{2}}$
-1	0		$\sqrt{\frac{1}{2}}$	$-\sqrt{\frac{1}{2}}$

$M = -2$

m_1	m_2	J / 2
-1	-1	1

$j_1 = \frac{3}{2}, \quad j_2 = 1$

$M = \frac{5}{2}$

m_1	m_2	J / $\frac{5}{2}$
$\frac{3}{2}$	1	1

$M = \frac{3}{2}$

m_1	m_2	J	$\frac{5}{2}$	$\frac{3}{2}$
$\frac{3}{2}$	0		$\sqrt{\frac{2}{5}}$	$\sqrt{\frac{3}{5}}$
$\frac{1}{2}$	1		$\sqrt{\frac{3}{5}}$	$-\sqrt{\frac{2}{5}}$

$M = \frac{1}{2}$

m_1	m_2	J	$\frac{1}{2}$	$\frac{5}{2}$	$\frac{3}{2}$
$\frac{3}{2}$	-1		$\sqrt{\frac{1}{10}}$	$\sqrt{\frac{2}{5}}$	$\sqrt{\frac{1}{2}}$
$\frac{1}{2}$	0		$\sqrt{\frac{3}{5}}$	$\sqrt{\frac{1}{15}}$	$-\sqrt{\frac{1}{3}}$
$-\frac{1}{2}$	1		$\sqrt{\frac{3}{10}}$	$-\sqrt{\frac{8}{15}}$	$\sqrt{\frac{1}{6}}$

$M = -\frac{1}{2}$

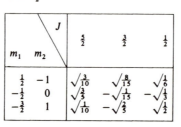

m_1	m_2	J	$\frac{5}{2}$	$\frac{3}{2}$	$\frac{1}{2}$
$\frac{1}{2}$	-1		$\sqrt{\frac{3}{10}}$	$\sqrt{\frac{8}{15}}$	$\sqrt{\frac{1}{6}}$
$-\frac{1}{2}$	0		$\sqrt{\frac{3}{5}}$	$-\sqrt{\frac{1}{15}}$	$-\sqrt{\frac{1}{3}}$
$-\frac{3}{2}$	1		$\sqrt{\frac{1}{10}}$	$-\sqrt{\frac{2}{5}}$	$\sqrt{\frac{1}{2}}$

$M = -\frac{3}{2}$

m_1	m_2	J	$\frac{5}{2}$	$\frac{3}{2}$
$-\frac{1}{2}$	-1		$\sqrt{\frac{3}{5}}$	$\sqrt{\frac{2}{5}}$
$-\frac{3}{2}$	0		$\sqrt{\frac{2}{5}}$	$-\sqrt{\frac{3}{5}}$

$M = -\frac{5}{2}$

m_1	m_2	J / $\frac{5}{2}$
$-\frac{3}{2}$	$-\frac{1}{2}$	1

$j_1 = 2, \quad j_2 = 1$

$M = 3$

m_1 m_2	J 3
2 1	1

$M = 2$

m_1 m_2	J 3	2
2 0	$\sqrt{\frac{1}{3}}$	$\sqrt{\frac{2}{3}}$
1 1	$\sqrt{\frac{2}{3}}$	$-\sqrt{\frac{1}{3}}$

$M = 1$

m_1 m_2	J 3	2	1
2 -1	$\sqrt{\frac{1}{15}}$	$\sqrt{\frac{1}{3}}$	$-\sqrt{\frac{3}{5}}$
1 0	$\sqrt{\frac{8}{15}}$	$\sqrt{\frac{1}{6}}$	$-\sqrt{\frac{3}{10}}$
0 1	$\sqrt{\frac{2}{5}}$	$-\sqrt{\frac{1}{2}}$	$\sqrt{\frac{1}{10}}$

$M = 0$

m_1 m_2	J 3	2	1
1 -1	$\sqrt{\frac{1}{5}}$	$\sqrt{\frac{1}{2}}$	$\sqrt{\frac{3}{10}}$
0 0	$\sqrt{\frac{3}{5}}$	0	$-\sqrt{\frac{2}{5}}$
-1 1	$\sqrt{\frac{1}{5}}$	$-\sqrt{\frac{1}{2}}$	$\sqrt{\frac{3}{10}}$

$M = -1$

m_1 m_2	J 3	2	1
0 -1	$\sqrt{\frac{2}{5}}$	$\sqrt{\frac{1}{2}}$	$\sqrt{\frac{1}{10}}$
-1 0	$\sqrt{\frac{8}{15}}$	$-\sqrt{\frac{1}{6}}$	$-\sqrt{\frac{3}{10}}$
-2 1	$\sqrt{\frac{1}{15}}$	$-\sqrt{\frac{1}{3}}$	$\sqrt{\frac{3}{5}}$

$M = -2$

m_1 m_2	J 3	2
-1 -1	$\sqrt{\frac{2}{3}}$	$\sqrt{\frac{1}{3}}$
-2 0	$\sqrt{\frac{1}{3}}$	$-\sqrt{\frac{2}{3}}$

$M = -3$

m_1 m_2	J 3
-2 -1	1

Taken from the *Review of Particle Properties* (Particle Data Group 1984)

Answers to and help with problems

Chapter 1

1. 6.03 kg^{-1}.
2. 9×10^{-31} kg.
3. 5.68×10^{-15} volts.
4. The total energy E and momentum $\mathbf{p}$ cannot satisfy $E = |\mathbf{p}|c$.

Chapter 2

1. Show that $|\int \bar{\phi}\psi \, dV|^2 \leqslant (\int |\phi|^2 \, dV)(\int |\psi|^2 \, dV)$.
2. Use ●2.1. Eigenstates of E with result α correspond to eigenvectors of P_α with eigenvalue 1.
4. Eigenvalues ± 25, eigenstates $\frac{3}{5}|a_0\rangle + \frac{4}{5}i|a_1\rangle$, $\frac{4}{5}|a_0\rangle - \frac{3}{5}i|a_1\rangle$. Probability 337/625.
5. Let $|a_0\rangle$, $|a_1\rangle$ be the eigenstates of A. If $\alpha|a_0\rangle + \beta|a_1\rangle$ is one eigenstate of B, the other must be $\beta|a_0\rangle - \alpha|a_1\rangle$, and the probability is always $|\alpha|^4 + |\beta|^4$.
6. Let $|\phi\rangle = A|\psi\rangle$ and use $\langle\phi|\phi\rangle \geqslant 0$.
8. Probabilistic argument: $\Delta A = 0 \Rightarrow A$ certainly has value $\langle A \rangle$. Algebraic argument: consider $|\phi\rangle = A|\psi\rangle - \langle a\rangle|\psi\rangle$.
9. $\cos^2\theta \sin^2\theta$.
11. Let $|\phi\rangle = A_1|\psi\rangle + zB_1|\psi\rangle$ where A_1 and B_1 are as in ●2.5 and z is arbitrary (or apply Cauchy–Schwarz to $A_1|\psi\rangle$ and $B_1|\psi\rangle$).
13. Use (2.111).
15. $\Delta x = a/\sqrt{2}$.
16. Use (2.80).
17. Show that $[D, x_i] = -i\hbar x_i$ and $[D, p_i] = i\hbar p_i$; then use (2.80).
18. Put $|\phi\rangle = |\phi_1\rangle + |\phi_2\rangle$ and $|\phi\rangle = |\phi_1\rangle + i|\phi_2\rangle$ in (2.123).
19. Fermion:

$$\binom{n}{r}.$$

Boson:

$$\binom{n+r}{n}$$

(the number of choices of r objects from n, with repetitions).

Chapter 3

1. $\cos^4 \beta T + \sin^4 \beta T$.

2. (a) $\frac{9}{16} \sin^2 \frac{1}{2}\omega t$, (b) $\frac{3}{16} \sin^2 \frac{1}{2}\omega t$, where $\omega = (m_1 - m_2)c^2/\hbar$.

3. For any x, $\cos^n (x/n) \to 1$ as $n \to \infty$, so $p_n \to p_0$.

5. Use ●2.5 and 3.1.

6. Use ●3.1 to find $d/dt(\Delta A)$.

7. $(\pi\hbar^2/2ma^2)(l^2 + m^2 + n^2)$, l, m, n integers > 0.

8. $\psi(x, t) = \pi^{-\frac{1}{4}}(a_0 + i\hbar t/ma_0)^{-\frac{1}{2}} \exp\{-x^2(a_0^2 + 2i\hbar t/m)^{-1}\}$.

9. No; if $k \neq k'$ there is a non-classical interference term.

10. $|A|^2 = |B|^2 + |C|^2$ shows conservation of probability.

13. Relative probability $= \frac{1}{2}(1 + k^2/K^2) + (4Ka)^{-1}(1 - k^2/K^2)\sin 2Ka$ where $k^2 = 2mE/\hbar^2$, $K^2 = 2m(E - V_0)/\hbar$. Limit as $\hbar \to 0$ is $\frac{1}{2}(1 + \lambda^2)$ where λ is the classical ratio.

14. $f(r) = C/r$ (C constant), $E = \hbar^2 k^2/2m$ [use $\nabla^2 f = f'' + 2f'/r$].
    $\mathbf{j} = (C^2\hbar k/mr^3)\mathbf{r}$. This is impossible in a stationary state for a region enclosing the origin, as the probability of finding the particle near the origin would be steadily decreasing.

15. $\langle x \rangle = \frac{1}{2}a$, $\Delta x^2 = (a^2/12)(1 - 6/n^2\pi^2)$. For a classical particle, equally likely to be anywhere, $\Delta x^2 = a^2/12$.

16. Use $(d/d\lambda)f(\lambda\mathbf{n}) = \mathbf{n} \cdot \nabla f$ with $f(\mathbf{a}) = U(T_\mathbf{a})$.

18. $U(T_\mathbf{a})VU(T_\mathbf{a})^{-1} = V'$ where $V'(\mathbf{r}) = V(\mathbf{r} - \mathbf{a})$.
    No; in classical mechanics the motion of a particle in a constant force is invariant under translations.

20. The plane of polarisation rotates with angular velocity $(E_+ - E_-)/2\hbar$.

24. Differentiate $U(\lambda)U(\lambda)^{-1} = 1$.

25. See the proof of ●3.6.

27. $\omega = e^{-iMv \cdot a}$; hermitian generators $-M\mathbf{R}$ (M = total mass, $\mathbf{R}$ = centre of mass).

28. Use the associative law.

29. For $X, Y \in L$ (generators of representation U) and $u, v \in V$, $[X, Y]$ is as in L, $[X, v] = Xv$ and $[u, v] = 0$.

32. $\psi_k = (2/a)^{\frac{1}{2}} \sin(k\pi x/a)$; $V = -eEx$;
    probability $= (2^{10}e^2E^2m^2a^4/\pi^2\hbar^4u^6)\sin^2(u\pi^2\hbar^2/4ma^2)$ where $u = k^2 - l^2$.

33. $P(t) = 4(\varepsilon/E)^4|\langle\psi_2|V|\psi_3\rangle\langle\psi_3|V|\psi_1\rangle|^2 \sin^4(\frac{1}{2}Et/\hbar)$.

34. $\varepsilon^2|\langle\psi_2|V_0|\psi_1\rangle|^2\left\{\left(\dfrac{\sin\frac{1}{2}\alpha_+ t}{\alpha_+}\right)^2 + \left(\dfrac{\sin\frac{1}{2}\alpha_- t}{\alpha_-}\right)^2 + 2\left(\dfrac{\sin\frac{1}{2}\alpha_- t}{\alpha_+}\right)\left(\dfrac{\sin\frac{1}{2}\alpha_- t}{\alpha_-}\right)\cos\omega t\right\}$

    where $\alpha_\pm = (E_2 - E_1)/\hbar \pm \omega$.

Chapter 4

1. $l = 2$, $m = 0$. $\psi_{22} = (J_+)^2\psi_{20} = -(x + iy)^2$.

3. R can be replaced by RS where S is a rotation about the z-axis; but $U(R)|jm\rangle$ and $d^j_{mn}(R)$ would only be multiplied by a phase factor.

4. $\cos^2 \frac{1}{2}\theta : \sin^2 \frac{1}{2}\theta$.

7. $2l + 1$ if $0 \leqslant l \leqslant j$; $3j - l + 1$ if $j \leqslant l \leqslant 3j$.

8. $\langle T_0 \rangle_{-2} : \langle T_0 \rangle_{-1} : \langle T_0 \rangle_0 : \langle T_0 \rangle_1 : \langle T_0 \rangle_2 = 2 : -1 : -2 : -1 : 2$
   $(\langle T_0 \rangle_m = \langle m|T_0|m\rangle \propto \langle 2\, m|2\, 0, 2\, m\rangle)$.

9. $\langle\frac{5}{2}|z|\frac{5}{2}\rangle : \langle\frac{3}{2}|z|\frac{3}{2}\rangle = -5:7$. $\langle x \rangle = \langle y \rangle = 0$.

12. Use q.1 to write z^2 in terms of r^2 and an operator of spin type 2.

13. $P_{-1} : P_0 : P_1 = 3:4:3$.

15. $l=0$, 1 or 2; probability $=\frac{1}{3}$.
22. Stationary states $|n\,l\,m\rangle$ (simultaneous eigenstates of H_0, $\mathbf{L}^2$ and L_z) with energy $E_n+\mu m\hbar$.
23. $\Delta l=0$, ± 1 by angular momentum conservation; Δl odd by parity conservation.
24. Use time-dependent perturbation theory with $V=am\omega^2 x$.
 $P_i=\lambda_i(m\omega a^2/\hbar)\sin^2\frac{1}{2}\omega t$ with $\lambda_0=2$, $\lambda_2=4$.
25. $\exp(-4|z_0|^2\sin^2\frac{1}{2}\omega t)$.
32. $-2\varepsilon^2\displaystyle\int_0^t \mathrm{d}t_2\int_0^{t_2}\mathrm{d}t_1\,e^{-2iE_{\gamma}t}\,e^{i(E_{\gamma}-2EX)t_2}\,e^{i(E_{\gamma}-EX)t_1}$.

Chapter 5

5. $\mathrm{tr}\,(A\rho)=\cos\theta$, $\mathrm{tr}\,(A\rho')=0$.
7. Use ●5.4.
9. Probability $=\cos^2\theta$. Inequality $\sin^2\phi\leqslant\cos^2\theta+\cos^2(\theta+\phi)$.
10. Subset $X\leftrightarrow$ proposition $x\in X$.
11. $(x_1,x_2)\vee(y_1,y_2)=(x_1\vee y_1,x_2\vee y_2)$; same for $\wedge$.
12. If z belongs to the centre of $\mathscr{L}$, $\mathscr{L}$ is the direct sum of $\{x:0\leqslant x\}$ and $\{x:x\leqslant z'\}$.

Chapter 6

1. $\pi^+\pi^-$ (the antisymmetric isospin state).
4. Neither of the $\mathrm{p\bar{p}}$ processes can occur if the p and $\bar{\mathrm{p}}$ have antiparallel spins; the $2\pi^0$ process cannot occur if the p and $\bar{\mathrm{p}}$ have even relative orbital angular momentum.
5. $\Gamma(\mathrm{p^*}\to\Delta^{++}\mathrm{p^-}):\Gamma(\mathrm{p^*}\to\Delta^+\pi^0):\Gamma(\mathrm{p^*}\to\Delta^0\pi^+)=3:2:1$.
6. $\Gamma(\Delta_2^{++}\to\Delta_1^+\pi^+):\Gamma(\Delta_2^+\to\Delta_1^+\pi^0):\Gamma(\Delta_2^+\to\Delta_1^0\pi^+)=6:1:8$.
7. $\sqrt{2}\langle\pi^0\mathrm{p}|e^{-iHt}|\pi^+\mathrm{n}\rangle+\langle\pi^+\mathrm{n}|e^{-iHt}|\pi^+\mathrm{n}\rangle=\langle\pi^+\mathrm{p}|e^{-iHt}|\pi^+\mathrm{p}\rangle$.
8. $CI_1C=-I_1$, $CI_2C=I_2$, $CI_3C=-I_3$.
11. $\tau(^{14}\mathrm{C}):\tau(^{14}\mathrm{N^*}):\tau(^{14}\mathrm{O})=e^2:g_\mathrm{w}^2:e^2$.
12. $\mathrm{X}^+\to\mathrm{n+p}$, $\mathrm{X}^{++}\to\mathrm{p+p}$.
13. There is no isospin analogue of orbital angular momentum.
14. $\sqrt{2}\langle\mathrm{p}\pi^0|H|\Sigma^+\rangle+\langle\mathrm{n}\pi^+|H|\Sigma^+\rangle=\langle\mathrm{n}\pi^-|H|\Sigma^-\rangle$.
15. $\tau(\Xi^-):\tau(\Xi^0)=1:2$.
17. For example, $\pi^-+\mathrm{p}\to\mathrm{n}+\mathrm{K}^++\mathrm{K}^-$.
20. $I_3|n l_0-r\mathbf{i}+s\mathbf{v}\rangle=\frac{1}{2}(n-2r-s)|n l_0-r\mathbf{i}+s\mathbf{v}\rangle$,
 $I_\pm|n l_0-r\mathbf{i}+s\mathbf{v}\rangle=\sqrt{(r(n-r-s\pm\frac{1}{2}))}|n l_0-(r\pm 1)\mathbf{i}+s\mathbf{v}\rangle$,
 with similar formulae for U_3, $U_\pm$, V_3 and $V_\pm$.
21. $Y=\frac{2}{3}(\mathbf{u}-\mathbf{v})\cdot\mathbf{H}$, $Q=\frac{2}{3}(\mathbf{i}-\mathbf{v})\cdot\mathbf{H}$.
22. (ii) $|\Sigma_\mathrm{U}^0\rangle=-\frac{1}{2}|\Sigma^0\rangle+\sqrt{\frac{3}{2}}|\Lambda^0\rangle$. The reason for the better fit of the Gell-Mann/Okubo formula to the meson octet with squared masses instead of masses is not fully understood.
23. u quark number $=I_3+\frac{1}{2}(3B+S-C-B'-T)$,
 d quark number $=-I_3+\frac{1}{2}(3B+S-C-B'-T)$.
29. (i), (ii), (iii) and (v): all $\cot^2\theta_\mathrm{C}$. (iv) $\tan^2\theta_\mathrm{C}$. (vi) and (vii): both $\cot^4\theta_\mathrm{C}:\cot^2\theta_\mathrm{C}:1$.
30. (i) 0.43. (ii) 321:13.4:15.6:1.

Chapter 7

6. $\Pi(\pm p)$ are projection operators onto eigenspaces of $p^\mu \gamma_\mu$.

7. $S(\Lambda) = \begin{pmatrix} \cosh\frac{1}{2}\lambda & (\sinh\frac{1}{2}\lambda)\mathbf{n}.\boldsymbol{\sigma} \\ (\sinh\frac{1}{2}\lambda)\mathbf{n}.\boldsymbol{\sigma} & \cosh\frac{1}{2}\lambda \end{pmatrix} = \begin{pmatrix} \frac{1}{2}\sqrt{(p_0+1)} & \dfrac{\sqrt{(p_0-1)}}{2|\mathbf{p}|}\mathbf{p}.\boldsymbol{\sigma} \\ \dfrac{\sqrt{(p_0-1)}}{2|\mathbf{p}|}\mathbf{p}.\boldsymbol{\sigma} & \frac{1}{2}\sqrt{(p_0+1)} \end{pmatrix}$

$u_\pm(p)$ and $v_\pm(p)$ are the columns of this matrix.

13. $D_\mu D^\mu \phi = -m^2 \phi$; $\mathcal{L} = \frac{1}{2}(D_\mu \phi)^\dagger (D^\mu \phi) - m^2 \phi^\dagger \phi$ where $D_\mu = \partial_\mu + ie A_\mu$. $j^\mu = e\phi^\dagger \partial_\mu \phi$.

Bibliography

Aharanov, Y. & Vardi, M. (1980). 'Meaning of an individual "Feynman path"'.
Physical Review, D **21**, 2235–40.

Aitchison, I. J. R. (1982). *An Informal Introduction to Gauge Field Theories*.
Cambridge University Press.

Aitchison, I. J. R. & Hey, A. J. G. (1982). *Gauge Theories in Particle Physics*.
Bristol: Hilger.

Andrade e Silva, J. & Lochak, G. (1969). *Quanta*. London: Weidenfeld &
Nicholson.

Aspect, A., Dalibard, J. & Roger, G. (1982). 'Experimental test of Bell's inequalities
using time-varying analyzers'. *Physical Review Letters*, **49**, 1804–7.

Atiyah, M. F. (1979). *Geometry of Yang–Mills Fields*. Pisa: Accademia Nazionale
dei Lincei Scuola Normale Superiore.

Ballentine, L. E. (1970). 'The statistical interpretation of quantum mechanics'.
Reviews of Modern Physics, **42**, 358–81.

Belinfante, F. J. (1973). *A Survey of Hidden-Variable Theories*. Oxford: Pergamon.

Bell, J. S. (1971). 'Introduction to the hidden-variable question'. In: *Foundations of
Quantum Mechanics*, ed. B. d'Espagnat, pp. 171–81. Reading, Mass.: Addison-
Wesley.

Bell, J. S. (1982). 'On the impossible pilot wave'. *Foundations of Physics*, **12**,
989–99.

Bell, J. S. (1984). '*B*eables for quantum field theory'. CERN preprint TH.4035/84.
Geneva: CERN.

Bernstein, H. J. & Phillips, A. V. (1981). 'Fibre bundles and quantum theory'.
Scientific American, July, pp. 95–109.

Berofsky, B. (1966) (ed.). *Free Will and Determinism*. New York: Harper & Row.

Birkhoff, G. & von Neumann, J. (1936). 'The logic of quantum mechanics'. *Annals
of Mathematics*, **37**, 823–43. Reprinted in von Neumann (1962): *Collected Works*,
Vol. IV. Oxford: Pergamon.

Bloom, E. D. & Feldman, G. J. (1982). 'Quarkonium'. *Scientific American*, May,
pp. 42–53.

Bogolubov, N. N. Logunov, A. A. & Todorov, I. T. (1975). *Introduction to
Axiomatic Quantum Field Theory*. Reading, Mass.: Benjamin.

Böhm, A. (1978). *The Rigged Hilbert Space and Quantum Mechanics*. Berlin:
Springer-Verlag.

Bohm, D. (1951). *Quantum Theory*. Englewood Cliffs: Prentice-Hall.

Bohr, N. (1958, reprinted 1963). 'Quantum physics and philosophy – causality and complementarity'. In: *Essays 1958/62 on Atomic Physics and Human Knowledge*. New York: Wiley.

Borges, J. L. (1941; English translation 1965). *The Garden of Forking Paths*. In *Fictions*. London: Calder.

Clauser, J. F. & Shimony, A. (1978). 'Bell's theorem: experimental tests and implications'. *Reports on Progress in Physics*, **41**, 1881–1927.

Cline, D. B., Rubbia, C. & van der Meer, S. (1982). 'The search for intermediate vector bosons'. *Scientific American*, March, pp. 38–49.

Cornwell, J. F. (1984). *Group Theory in Physics*. London: Academic Press.

Creutz, M. (1983). *Quarks, Gluons and Lattices*. Cambridge University Press.

d'Espagnat, B. (1976). *Conceptual Foundations of Quantum Mechanics* (2nd ed.). Reading, Mass.: Benjamin.

d'Espagnat, B. (1979). 'The quantum theory and reality'. *Scientific American*, May, pp. 158–81.

Daneri, A., Loinger, A. & Prosperi, G. M. (1962). 'Quantum theory of measurement and ergodicity conditions'. *Nuclear Physics*, **33**, 297–319. Reprinted in Wheeler & Zurek 1983.

Davies, P. C. W. (1979). *The Forces of Nature*. Cambridge University Press.

DeWitt, B. S. & Graham, N. (1973) (eds.). *The Many-Worlds Interpretation of Quantum Mechanics*. Princeton: University Press.

Dick, P. K. (1962, reprinted 1965). *The Man in the High Castle*. Harmondsworth: Penguin.

Dirac, P. A. M. (1930; 4th ed. 1958). *The Principles of Quantum Mechanics*. Oxford: Clarendon Press.

Dodd, J. E. (1984). *The Ideas of Particle Physics*. Cambridge University Press.

Einstein, A. & Infeld, L. (1938). *The Evolution of Physics*. Cambridge University Press.

Everett, H. III (1957). '"Relative state" formulation of quantum mechanics'. *Reviews of Modern Physics*, **29**, 454–62. Reprinted in Wheeler & Zurek 1983.

Feynman, R. P. & Hibbs, A. R. (1965). *Quantum Mechanics and Path Integrals*. New York: McGraw-Hill.

Feynman, R. P., Leighton, R. B. & Sands, M. (1965). *The Feynman Lectures on Physics*. Reading, Mass.: Addison-Wesley.

Fonda, L., Ghirardi, G. C. & Rimini, A. (1978). 'Decay theory of unstable quantum systems'. *Reports on Progress in Physics*, **41**, 587–631.

Freedman, D. Z. & van Nieuwenhuizen, P. (1978). 'Supergravity and the unification of the laws of nature'. *Scientific American*, February, pp. 126–38.

Gell-Mann, M. (1979). 'What are the building blocks of matter?' In: *The Nature of the Physical Universe*, ed. D. Huff & O. Prewett. New York: Wiley.

Georgi, H. (1981). 'A unified theory of elementary particles and forces'. *Scientific American*, April, pp. 40–55.

Ghirardi, G. C., Omero, C., Rimini, A. & Weber, T. (1978). 'The stochastic interpretation of quantum mechanics'. *Rivista del Nuovo Cimento*, Vol. 1, No. 3.

Gibson, W. M. & Pollard, B. R. (1976). *Symmetry Principles in Elementary Particle Physics*. Cambridge University Press.

Gilmore, R. (1974). *Lie Groups, Lie Algebras, and Some of Their Applications*. New York: Wiley.

Glashow, S. L. (1975). 'Quarks with color and flavor'. *Scientific American*, October, pp. 38–50.

Glashow, S. L. (1980). 'Towards a unified theory: threads in a tapestry'. *Reviews of Modern Physics*, **52**, 539–43.

Gnedenko, B. V. (1968). *The Theory of Probability*. New York: Chelsea.

Goldstein, H. (1980). *Classical Mechanics*, 2nd ed. Reading, Mass.: Addison-Wesley.

Gottfried, K. (1966). *Quantum Mechanics, Vol. I: Fundamentals*. New York: Benjamin.

Gottfried, K. & Weisskopf, V. F. (1984). *Concepts of Particle Physics*, Vol. I. Oxford: University Press.

Guth, A. H. & Steinhardt, P. J. (1984). 'The inflationary universe'. *Scientific American*, May, pp. 90–102.

Halzen, F. & Martin, A. D. (1984). *Quarks and Leptons*. New York: Wiley.

Harari, H. (1983). 'The structure of quarks and leptons'. *Scientific American*, April, pp. 48–60.

Heisenberg, W. (1930, reprinted 1949). *The Physical Principles of the Quantum Theory*. New York: Dover.

Heisenberg, W. (1959). *Physics and Philosophy*. London: Allen & Unwin.

Hughes, R. I. G. (1981). 'Quantum logic'. *Scientific American*, October, pp. 146–57.

Ishikawa, K. (1982). 'Glueballs'. *Scientific American*, November, pp. 122–35.

Itzykson, C. & Zuber, J. B. (1980). *Quantum Field Theory*. New York: McGraw-Hill.

Jacob, M. & Landshoff, P. V. (1980). 'The inner structure of the proton'. *Scientific American*, March, pp. 66–75.

Jammer, M. (1974). *The Philosophy of Quantum Mechanics*. New York: Wiley.

Jauch, J. M. (1968). *Foundations of Quantum Mechanics*. Reading, Mass.: Addison-Wesley.

Kaufmann, W. J. III (1980) (ed.). *Particles and Fields*. San Francisco: Freeman.

Leader, E. & Predazzi, E. (1982). *Gauge Theories and the 'New Physics'*. Cambridge University Press.

Lederman, L. M. (1978). 'The upsilon particle'. *Scientific American*, October, pp. 60–8.

Lighthill, M. J. (1958). *Fourier Analysis and Generalised Functions*. Cambridge University Press.

Lyons, L. (1983). 'An introduction to the possible substructure of quarks and leptons'. *Progress in Particle and Nuclear Physics*, **10**, 227–304.

Mandl, F. & Shaw, G. (1984). *Quantum Field Theory*. Chichester: Wiley.

Mistry, N. B., Poling, R. A. & Thorndike, E. H. (1983). 'Particles with naked beauty'. *Scientific American*, July, pp. 98–107.

Nambu, Y. (1976). 'The confinement of quarks'. *Scientific American*, November, pp. 48–60.

Nelson, E. (1985). *Quantum Fluctuations*. Princeton: University Press.

O'Brien, F. (1974). *The Third Policeman*. London: Pan.

Particle Data Group (1984). 'Review of particle properties'. *Reviews of Modern Physics*, **56**, no. 2, part 2.

Peres, A. (1980). 'Nonexponential decay'. *Annals of Physics*, **129**, 33–46.

Peres, A. (1984). 'What is a state vector?' *American Journal of Physics*, **52**, 644–50.

Perkins, D. H. (1982). *Introduction to High Energy Physics*. Reading, Mass.: Addison-Wesley.

Perl, M. L. & Kirk, W. T. (1978). 'Heavy leptons'. *Scientific American*, March, pp. 50–7.

Piron, C. (1976). *Foundations of Quantum Physics*. Reading, Mass.: Benjamin.

Polkinghorne, J. C. (1979). *The Particle Play*. London: Freeman.

Polkinghorne, J. C. (1984). *The Quantum World*. London: Longman.

Primas, H. (1981). *Chemistry, Quantum Mechanics and Reductionism*. Berlin: Springer-Verlag.

Putnam, H. (1965; reprinted 1975). 'A philosopher looks at quantum mechanics'. In: *Mathematics, Matter and Method*. Cambridge University Press.

Putnam, H. (1968; reprinted 1975). 'The logic of quantum mechanics'. *Ibid*.

Quigg, C. (1985). 'Elementary particles and forces'. *Scientific American*, April, pp. 64–75.

Ramond, P. (1981). *Field Theory: A Modern Primer*. Reading, Mass.: Benjamin/Cummings.

Rebbi, C. (1983). 'The lattice theory of quark confinement., *Scientific American*, February, pp. 36–47.

Reichenbach, H. (1944). *Philosophic Foundations of Quantum Mechanics*. Berkeley: University of California Press.

Ryder, L. (1985). *Quantum Field Theory*. Cambridge University Press.

Salam, A. (1980). 'Gauge unification of fundamental forces'. *Reviews of Modern Physics*, **52**, 525–38.

Schiff, L. I. (1968). *Quantum Mechanics*. Tokyo: McGraw-Hill Kogakusha.

Schwitters, R. F. (1977). 'Fundamental particles with charm', *Scientific American*, October, pp. 56–70.

Scully, M. O. & Sargent, M. (1972). 'The concept of the photon'. *Physics Today*, March, pp. 38–47.

Stapp, H. P. (1972). 'The Copenhagen interpretation'. *American Journal of Physics*, **40**, 1098–1116.

Sudbery, A. (1984). 'The observation of decay'. *Annals of Physics*, **157**, 512–36.

Sutton, C. (1985) (ed.). *Building the Universe*. Oxford: Basil Blackwell.

ter Haar, D. (1967) (ed.). *The Old Quantum Theory*. Oxford: Pergamon.

von Neumann, J. (1932; English translation 1955). *Mathematical Foundations of Quantum Mechanics*. Princeton: University Press.

Wallace, D. J. (1983; reprinted 1985). 'Computing the strong force'. In: Sutton 1985.

Wallace Garden, R. (1984). *Modern Logic and Quantum Mechanics*. Bristol: Hilger.

Weinberg, S. (1974). Unified theories of elementary-particle interaction'. *Scientific American*, July, pp. 50–59.

Weinberg, S. (1980). 'Conceptual foundations of the unified theory of weak and electromagnetic interactions'. *Reviews of Modern Physics*, **52**, 515–24.

Weinberg, S. (1981). 'The decay of the proton'. *Scientific American*, June, pp. 52–63.

Weinberg, S. (1983). *The Discovery of Subatomic Particles*. New York: Scientific American.

Wess, J. & Bagger, J. (1983). *Supersymmetry and Supergravity*. Princeton: University Press.

Weyl, H. (1928; English translation 1950). *The Theory of Groups and Quantum Mechanics*. New York: Dover.

Wheeler, J. A. & Zurek, W. H. (1983) (ed.). *Quantum Theory and Measurement*. Princeton: University Press.

Wigner, E. P. (1959). *Group Theory*. New York: Academic Press.

Wigner, E. P. (1983). 'Interpretation of quantum mechanics'. In: Wheeler & Zurek 1983.

Zukav, G. (1979). *The Dancing Wu Li Masters*. London: Hutchinson.

Index

Page numbers in bold type indicate definitions.
The index includes the authors of works cited in the text, but not of works cited under 'Further reading'.